Clostridia

Molecular Biology in the Post-genomic Era

Edited by

Holger Brüggemann

Max Planck Institute for Infection Biology
Department of Molecular Biology
Berlin
Germany

and

Gerhard Gottschalk

Göttingen Genomics Laboratory
Institute for Microbiology and Genetics
Goettingen
Germany

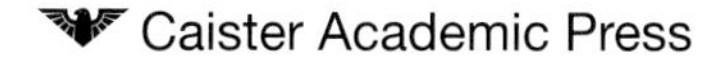

Caister Academic Press

Caister Academic Press
Norfolk, UK

www.caister.com

British Library Cataloguing-in-Publication Data
A catalogue record for this book is available from the British Library

ISBN: 978-1-904455-38-7

Printed and bound in Great Britain

Contents

Contributors

Ben Blount
Centre for Biomolecular Sciences
Institute of Infection Immunity and Inflammation
Energy Research Technologies Institute
School of Molecular Medical Sciences
University of Nottingham
Nottingham
UK

nixbb@nottingham.ac.uk

David A. Burns
Centre for Biomolecular Sciences
Institute of Infection Immunity and Inflammation
Energy Research Technologies Institute
School of Molecular Medical Sciences
University of Nottingham
Nottingham
UK

nixdb1@nottingham.ac.uk

Stephen T. Cartman
Centre for Biomolecular Sciences
Institute of Infection Immunity and Inflammation
Energy Research Technologies Institute
School of Molecular Medical Sciences
University of Nottingham
Nottingham
UK

stephen.cartman@nottingham.ac.uk

Justin A. Caserta
Department of Molecular Genetics and Biochemistry
University of Pittsburgh School of Medicine
Pittsburgh, PA
USA

jac86@pitt.edu

Clare M. Cooksley
Centre for Biomolecular Sciences
Institute of Infection Immunity and Inflammation
Energy Research Technologies Institute
School of Molecular Medical Sciences
University of Nottingham
Nottingham
UK

clare.cooksley@nottingham.ac.uk

Aurélie Couesnon
Unité des Bactéries anaérobies et Toxines
Institut Pasteur
Paris
France

couesnon@pasteur.fr

Lisa F. Dawson
Department of Infectious and Tropical Diseases
London School of Hygiene and Tropical Medicine
London
UK

Lisa.dawson@lshtm.ac.uk

Ludwig Dubois
University Maastricht
GROW Research School for Oncology and Developmental Biology
MAASTRO Lab
Maastricht
The Netherlands

ludwig.dubois@maastro.unimaas.nl

Peter Dürre
Institut für Mikrobiologie und Biotechnologie
Universität Ulm
Ulm
Germany

peter.duerre@uni-ulm.de

Jenny Emerson
Division of Cell and Molecular Biology
Imperial College London
London
UK

jenny.emerson02@imperial.ac.uk

Neil Fairweather
Division of Cell and Molecular Biology
Imperial College London
London
UK

n.fairweather@imperial.ac.uk

Maria Fredriksson-Ahomaa
Institute of Hygiene and Technology of Food of Animal Origin
Ludwig-Maximilians University Munich
Oberschleißheim
Germany

maria.fa@lmhyg.vetmed.uni-muenchen.de

John T. Heap
Centre for Biomolecular Sciences
Institute of Infection Immunity and Inflammation
Energy Research Technologies Institute
School of Molecular Medical Sciences
University of Nottingham
Nottingham
UK

john.heap@nottingham.ac.uk

Eileen M. Hotze
Department of Microbiology and Immunology
The University of Oklahoma Health Sciences Center
Oklahoma, OK
USA

eileen-hotze@ouhsc.edu

Hannu Korkeala
Department of Food and Environmental Hygiene
Faculty of Veterinary Medicine
University of Helsinki
Helsinki
Finland

hannu.korkeala@helsinki.fi

Philippe Lambin
University Maastricht
GROW Research School for Oncology and Developmental Biology
MAASTRO Lab
Maastricht
The Netherlands

philippe.lambin@maastro.unimaas.nl

Miia Lindström
Department of Food and Environmental Hygiene
Faculty of Veterinary Medicine
University of Helsinki
Helsinki
Finland

miia.lindstrom@helsinki.fi

Bruce A. McClane
Department of Molecular Genetics and Biochemistry
University of Pittsburgh School of Medicine
Pittsburgh, PA
USA

bamcc@pitt.edu

Paola Mastrantonio
Department of Infectious, Parasitic and Immune-mediated Diseases
Istituto Superiore di Sanità
Rome
Italy

paola.mastrantonio@iss.it

Christelle Mazuet
Unité des Bactéries anaérobies et Toxines
Institut Pasteur
Paris
France

cmazuet@pasteur.fr

Asferd Mengesha
University Maastricht
GROW Research School for Oncology and Developmental Biology
MAASTRO Lab
Maastricht
The Netherlands

asferd.mengesha@maastro.unimaas.nl

Nigel P. Minton
Centre for Biomolecular Sciences
Institute of Infection Immunity and Inflammation
Energy Research Technologies Institute
School of Molecular Medical Sciences
University of Nottingham
Nottingham
UK

nigel.minton@nottingham.ac.uk

Kim Paesmans
University Maastricht
GROW Research School for Oncology and Developmental Biology
MAASTRO Lab
Maastricht
The Netherlands

kim.paesmans@maastro.unimaas.nl

Oliver J. Pennington
Centre for Biomolecular Sciences
Institute of Infection Immunity and Inflammation
Energy Research Technologies Institute
School of Molecular Medical Sciences
University of Nottingham
Nottingham
UK

oliver.pennington@nottingham.ac.uk

Yannick Pereira
Unité des Bactéries anaérobies et Toxines
Institut Pasteur
Paris
France

yannickpereira@free.fr

Michel R. Popoff
Unité des Bactéries anaérobies et Toxines
Institut Pasteur
Paris
France

mpopoff@pasteur.fr

Stéphanie Raffestin
Unité des Bactéries anaérobies et Toxines
Institut Pasteur
Paris
France

stephanie.raffestin@pasteur.fr

Susan L. Robertson
Department of Molecular Genetics and Biochemistry
University of Pittsburgh School of Medicine
Pittsburgh, PA
USA

sur18@pitt.edu

Maja Rupnik
Institute of Public Health Maribor and Faculty of Medicine
University of Maribor
Maribor
Slovenia

maja.rupnik@uni-mb.si

Sameera Sayeed
Department of Molecular Genetics and Biochemistry
University of Pittsburgh School of Medicine
Pittsburgh, PA
USA

sas69@pitt.edu

Jamie C. Scott
Centre for Biomolecular Sciences
Institute of Infection Immunity and Inflammation
Energy Research Technologies Institute
School of Molecular Medical Sciences
University of Nottingham
Nottingham
UK

paxjcs@nottingham.ac.uk

Patrizia Spigaglia
Department of Infectious, Parasitic and Immune-mediated Diseases
Istituto Superiore di Sanità
Rome
Italy

patrizia.spigaglia@iss.it

Richard A. Stabler
Department of Infectious and Tropical Diseases
London School of Hygiene and Tropical Medicine
London
UK

Richard.stabler@lshtm.ac.uk

Bradley G. Stiles
U.S. Army Medical Research Institute of Infectious Diseases
Integrated Toxicology Division
Fort Detrick
Frederick, MD
USA

bradley.stiles@us.army.mil

Jan Theys
University Maastricht
GROW Research School for Oncology and Developmental Biology
MAASTRO Lab
Maastricht
The Netherlands

jan.theys@maastro.unimaas.nl

Rodney K. Tweten
Department of Microbiology and Immunology
The University of Oklahoma Health Sciences Center
Oklahoma, OK
USA

Rod-Tweten@ouhsc.edu

Brad Wouters
University Maastricht
GROW Research School for Oncology and Developmental Biology
MAASTRO Lab
Maastricht
The Netherlands

brad.wouters@maastro.unimaas.nl

Brendan W. Wren
Department of Infectious and Tropical Diseases
London School of Hygiene and Tropical Medicine
London
UK

Brendan.wren@lshtm.ac.uk

Preface

In recent years, clostridia have regained attention for various reasons, not only from microbiologists but also from researchers in other areas, such as biotechnology, genomics, cell biology, toxicology, pharmacology and clinical research. Studies on clostridial toxins have been reinforced in the light of possible bioterrorist attacks. This led to a wealth of insights in structure–function relationships of toxins, such as the rho-dependent glucosyltransferases TcdA and B of *Clostridium difficile*, or the binary toxins of *C. perfringens* and *C. difficile*. Cell biologists have benefited from this research, and use clostridial toxins as a tool to investigate elementary cellular processes, such as synaptic vesicle trafficking and retrograde transport pathways in motor neurons (tetanus and botulinum toxin) or rho-dependent signaling pathways and their implication in cell morphology and endo- and exocytosis (TcdAB). Botulinum toxin (sold commercially under the brand name Botox) is used for cosmetic purposes to smoothen facial lines, but is also used to treat conditions such as migraine headaches and cervical dystonia. Biotechnologically exploited clostridia, such as the acetone and butanol-producing *C. acetobutylicum*, have experienced a renaissance due to efforts to develop alternatives to petrol to reduce our reliance on finite crude-oil resources and to lower CO_2 emissions. In the course of large metagenomic projects to sequence the entire human microbiota it became apparent that clostridial species are major commensals of the human intestinal tract. Their mutualistic traits and contribution to our well-being is still poorly understood. Another example of a new research direction is the use of *C. novyi* in cancer therapy. When injected intravenously, *C. novyi* spores produce substantial antitumor effects in experimental animals without excessive toxicity. The proliferation of *C. novyi* depends on anoxic conditions, since the vegetative form is extremely sensitive to oxygen. Thus, tumours with their large hypoxic regions provide an ideal growth environment for *C. novyi*. Other reasons why clostridial research has intensified are based on recent advances in developing new genetic tools to study this rather reluctant genus. A knock-out system has been developed, which has been applied successfully to some pathogenic clostridial species, including *C. botulinum* and *C. difficile*.

Gene knock-out techniques and approaches of study systems biology depend on the availability of whole genome sequences. Prior to the genome-sequencing era, genetic information about clostridial species, if available, was largely restricted to a few loci, such as those encoding major toxin genes of pathogenic clostridia, or those encoding important fermentative pathways and unique or unusual enzymatic activities. Since the year 2000, the genomes of ten clostridial species have been completely sequenced, starting with the genome of the solvent-producing nonpathogenic *C. acteobutylicum* in 2001. Subsequently, the genomes of major clostridial pathogens were sequenced, including *C. perfringens, C. tetani, C. novyi, C. botulinum*, and *C. difficile*. More recently, genome sequencing has been applied to deduce the 'pan-genome' of a given species: several different strains of *C. perfringens*, *C. botulinum* and *C. difficile* were completely sequenced, which makes it possible to distinguish between the core genome and strain-specific genomic variation within a given species. These massive sequencing efforts enabled large-scale post-genome transcriptomic and proteomic projects.

For this book, we were happy to gather contributions from leading scientists in the clostridia research field. The book contains a variety of topics ranging from clostridial 'core' research topics, for instance the biology and genetics of clostridial toxins, to new research directions, such as clostridia in cancer therapy, post-genomic approaches and systems biology.

We hope you will enjoy reading the book.

Holger Brüggemann and Gerhard Gottschalk
Berlin, Göttingen, 2008

Botulinum and Tetanus Neurotoxins: Molecular Biology, Toxin Gene Regulation and Mode of Action

1

Stéphanie Raffestin, Aurélie Couesnon, Yannick Pereira, Christelle Mazuet and Michel R. Popoff

Abstract

Botulinum neurotoxins (BoNT) and tetanus toxin (TeNT) are potent toxins which are responsible for severe diseases, botulism and tetanus, in men and animals. BoNTs induce a flaccid paralysis, whereas TeNT causes a spastic paralysis. Both toxins are zinc-dependent metalloproteases, which specifically cleave one of the three proteins (VAMP, SNAP25 and syntaxin) forming the SNARE complex within target neuronal cells which have a critical function in the release of neurotransmitter. BoNTs inhibit the release of acetylcholine at peripheral cholinergic nerve terminals, whereas TeNT blocks neurotransmitter release at central inhibitory interneurons. Only a single form of TeNT is known, but BoNTs are divided in seven toxinotypes and various subtypes, which differ in amino acid sequences and immunological properties. In contrast to TeNT, BoNTs are associated to non-toxic proteins (ANTPs) to form highly stable botulinum complexes. TeNT is produced by *Clostridium tetani*, and BoNTs by *Clostridium botulinum* and atypical strains of *Clostridium baratii* and *Clostridium butyricum*. The genes encoding the neurotoxin and ANTPs are clustered in a DNA segment, called botulinum locus, which is located on chromosome, plasmid or phage. Neurotoxin synthesis is a highly regulated process, which occurs in late exponential growth phase and beginning of stationary phase, and which is dependent of alternative sigma factors (BotR or TetR). BotR and TetR are related to other clostridial sigma factors, TcdR and UviA, which are involved in the control of *Clostridium difficile* toxins A and B, and *Clostridium perfringens* bacteriocin, respectively. BotR, TetR, TcdR and UviA form a new subgroup of RNA polymerase sigma factors.

Introduction

Among the large number of *Clostridium* species, which are bacteria from the environment, some produce potent neurotoxins and are responsible for neurological disorders in human and animals. Botulinum neurotoxins (BoNTs) inhibit the release of acetylcholine at peripheral cholinergic nerve terminals, causing the flaccid paralysis of botulism, whereas tetanus neurotoxin (TeNT) blocks neurotransmitter release at synapses at the central nervous system, causing the spastic paralysis of tetanus. In the last decades, numerous works have been performed to elucidate the structure and mechanism of action of the clostridial neurotoxins as well as their genetic organization and regulation.

Neurotoxigenic *Clostridia*

Clostridium tetani

The TeNT-producing *Clostridia* display homogeneous bacteriological characteristics and belong to a uniform group forming the *C. tetani* species. Some *C. tetani* strains are non-toxigenic and are indistinguishable from the toxigenic strains by their phenotypic characteristics and DNA/DNA homology except the production of toxin.

The G+C content of *C. tetani* is 25–26% (Cato *et al.*, 1986). This species has been classified in the *Clostridium* group II by using 23S rRNA homology (Johnson and Francis, 1975). The *Clostridium* genus encompasses more than 100 species which display a wide range of phenotypes and genotypes (Hippe *et al.*, 1992). Phylogenetic analysis using 16S rRNA comparison indicate that the *Clostridium* genus should be restricted to the homology group I, as defined by Johnson and Francis (Johnson and Francis, 1975; Lawson *et al.*, 1993). According to these data, *C. tetani* should be classified in a different genus. However, a study based on restriction maps of 16S rRNA showed that *C. tetani* belongs to the same cluster as *C. perfringens*, *C. sporogenes*, and *C. botulinum* C and G which are members of the homology group I of Johnson and Francis (Gurtler *et al.*, 1991).

The complete genome sequence of a toxigenic *C. tetani* strain was recently determined (Brüggemann *et al.*, 2003). It consists of a 2,799,250-bp chromosome containing 2372 putative genes, and of a 74,082-bp plasmid containing 61 genes including *tent* gene.

Clostridium botulinum

In contrast, *Clostridia* capable of producing BoNT display heterogeneous bacteriological characters and are divided into several species and groups. The taxonomic position of the *C. botulinum* species was originally based on only one phenotype, the production of a BoNT, and non-toxic variant strains, although genetically related to *C. botulinum*, were assigned to different species such as *C. sporogenes* and *C. subterminale*. It appeared soon that BoNT are seven different protein neurotoxins which are serologically distinct and are termed with letters from A to G. However, all of them cause the typical flaccid paralysis of botulism in experimental animals, similar to that observed in human patients suffering from botulism. Later on, physiological differences between *C. botulinum* strains were identified, but the production of the different BoNT types do not necessarily correlate with the *C. botulinum* strain phenotypes. The species was divided into four physiological groups (Table 1.1) (Smith, 1975).

- group I: *C. botulinum* A, and proteolytic strains of *C. botulinum* B and F.
- group II: *C. botulinum* E, and glucidolytic strains of *C. botulinum* B and F.
- group III: *C. botulinum* C and D.
- group IV: *C. botulinum* G. This group which also includes non-toxic strains previously identified as *C. subterminale* and *C. hastiforme*, is metabolically distinct from the other groups and has been assigned to a different species called *C. argentinense* (Suen *et al.*, 1988).

The taxonomic position of *C. botulinum* became more ambiguous since it has been found that BoNT can be produced by *Clostridium* strains clearly distinct from already defined *C. botulinum*, and biochemically and genetically related to different species such as *C. butyricum* and *C. baratii* (Hall *et al.*, 1985; McCroskey *et al.*, 1986; McCroskey *et al.*, 1991). In each group, all the strains regardless of toxin types, are closely related according to their phenotypic properties, DNA/DNA homology and 16S rRNA analysis. All the BoNT producer *Clostridium* have a G+C content between 26% and 28% (Cato *et al.*, 1986), and belong to the group I of 23S rRNA homology (Johnson and Francis, 1975) which contains the type species *C. butyricum*. 16S rRNA comparison corroborated that the group I strains are homogeneous and form a distinct phylogenetic branch. It has been proposed that the genus *Clostridium* should be retained only for this taxon (Hippe *et al.*, 1992; Lawson *et al.*, 1993). The atypical toxigenic *C. butyricum* and *C. baratii* strains are phenotypically and genetically related to the type strains of these species and not with the other BoNT producing *Clostridia* (Popoff, 1995).

The complete genome sequence of a proteolytic group I *C. botulinum* strain Hall A was recently determined (Sebaihia *et al.*, 2007). The genome consists of a chromosome (3,886,916 bp) and a plasmid (163,44 bp), which carry 3650 and 19 predicted genes, respectively.

Genetics of neurotoxins

Neurotoxin gene organization

The genes encoding the neurotoxins and associated non-toxic proteins (ANTP) have been cloned and sequenced in representative clostridial strains of each BoNT type. The neurotoxin and ANTP genes are clustered in close vicinity and constitute the botulinum or tetanus neurotoxin

Table 1.1 Groups of *botulinum* neurotoxin-producing Clostridia, toxin types and subtypes, and main properties.

Neurotoxin-producing *Clostridium*	Group I	Group II	Group III	Group IV, *C. argentinense*	*C. butyricum*	*C. baratii*
Toxin type	A, proteolytic B, F	E, non-proteolytic B, F	C, D	G	E	F
Subtype	A1, A2, A3, A4, B1, B2, B3, bivalent B (Ba, Bf, Ab), proteolytic F	E1, E2, E3, E6, non-proteolytic B, F	C, D, C/D	G	E4, E5	*baratii* F
Proteolysis	+	–	–	+	–	–
Lipase production	+	+	+	–	–	–
Main physiological properties	Highly thermoresistant spores	Growth at low temperature (3°C)	Optimal growth around 40°C			
Botulism	Human	Human	Animal		Human, animal?	Human, animal?
Related non-botulinum toxin-producing *Clostridium*	*C. sporogenes*		*C. novyi*, *C. hemolyticum*	*C. subterminale*, *C. proteolyticus*, *C. schimacherense*	*C. butyricum*	*C. baratii*

locus. The organization of the botulinum locus is conserved in the 3′ part but differs slightly in the 5′ part in the different types of BoNT-producing *Clostridia*. The *bont* genes are at the 3′ part of the locus and are preceded by the genes of the non-toxic–non-haemagglutinin (NTNH) components. *ntnh* and *bont* genes are transcribed in the same orientation (Fig. 1.1), whereas the haemagglutinin (HA) genes (*ha*33, *ha*17, and *ha*70), which are upstream to the *ntnh-bont* genes, are transcribed in the opposite orientation. The *ha* genes are missing in the non-haemagglutinating toxinotypes A2, E and F. The *ha* genes of *C. botulinum* G comprise only *ha*17 and *ha*70. In the toxinotype A2, E and F, a gene (*p*47) encoding a 47 kDa protein is immediately upstream of the *ntnh* gene, and both genes are transcribed in the same orientation. In addition, two genes (*orfX1* and *orfX2*), which are not related to *ha* genes, lie upstream of *p47* in the opposite orientation. An additional gene, *orfX3*, was characterized in *C. botulinum* A2 strain kyotoF, downstream the gene *orfX2* and in the same orientation (Dineen *et al*., 2004).

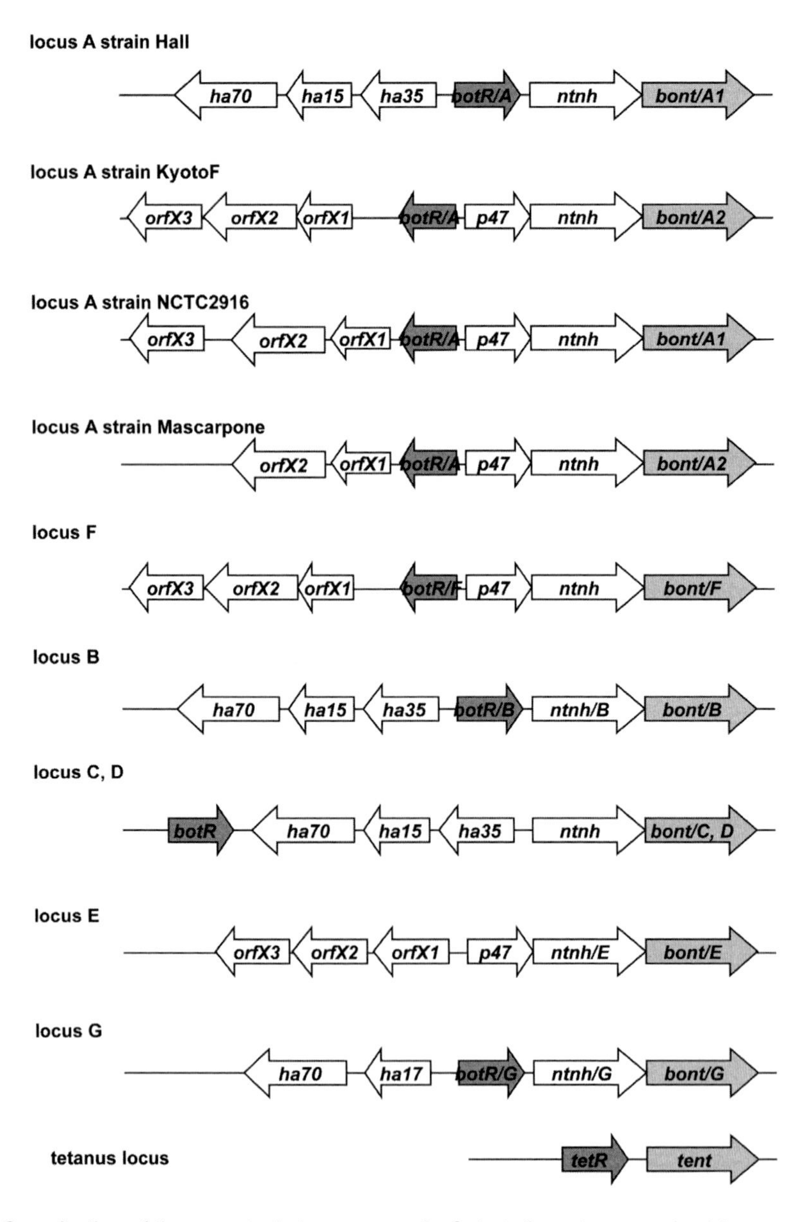

Figure 1.1 Organization of the neurotoxin locus genes in *C. botulinum* types and subtypes, and *C. tetani*.

A gene (*bot*R, previously called *orf*21 or *orf*22) encoding for a 21–22 kDa protein which presents features of a regulatory protein, is present in different positions in different strains of *C. botulinum* (Fig. 1.1). In *C. tetani*, a gene (*tet*R) equivalent to *bot*R is present upstream of the *tent* gene. No *antp*-like genes have been identified in *C. tetani* (Brüggemann *et al.*, 2003).

Usually, one clostridial strain produces only one type of neurotoxin and the botulinum locus is present in a single copy on the genome as suggested by Southern blotting of DNA fragments separated by pulse-field gel electrophoresis (Lin and Johnson, 1995; Hutson *et al.*, 1996). However, some rare strains synthesize two different BoNTs: BoNT/A-BoNT/B, BoNT/A-BoNT/F, and BoNT/B-BoNT/F producing strains have been isolated (Hutson *et al.*, 1996; Henderson *et al.*, 1997). The A-B strain contains two *bont* genes related to those of *C. botulinum* A2 and proteolytic *C. botulinum* B respectively (Fujinaga *et al.*, 1995). In such strains, the two neurotoxins are usually produced in different proportions. Thus, in Ba and Bf strains, BoNT/B is produced 10 times more than BoNT/A and BoNT/F (Henderson *et al.*, 1997). Some clostridial strains contain silent neurotoxin genes. Several *C. botulinum* A strains isolated from food-borne and infant botulism contain a silent *bont*/B gene. These strains are noted A(b). The characterization of strain NCTC2916 shows that it has two loci, A and B, which are 40 kbp distant within the chromosome. The botulinum B locus consists of *bont*/B, *ntnh*, *orf*21, *ha*33 and *ha*11 genes. The *bont*/A gene is identical to that of *C. botulinum* A1 strains, but the organization of the botulinum A locus is similar to that of *C. botulinum* A2 and F strains. The BoNT/B nucleotide sequence is related to that of *C. botulinum* B strains (97% identity), but it has a stop mutation in position 128 and two base deletions (positions 2839 and 2944) resulting in reading frameshifts and multiple stop codons (Hutson *et al.*, 1996; Jovita *et al.*, 1998; Dineen *et al.*, 2004). Silent *bont*/B has also been evidenced in non-toxigenic *C. subterminale* strains (Franciosa *et al.*, 1994; Jovita *et al.*, 1998). The strain *C. botulinum* 667, contains also two loci, A and B, 40 kbp distant within the chromosome. The genetic organization is the same than that in strain NCTC2916 and the *bont*/B gene is silenced by mutations and deletions (Hutson *et al.*, 1996).

Genetic diversity of *C. botulinum* strains and botulinum neurotoxin gene variation

Genetic analysis by 16S RNA gene sequence comparison or DNA/DNA homology have shown that *C. botulinum* strains form four distinct clusters which correspond to the physiological groups I to IV (Hutson *et al.*, 1994; Collins and East, 1997). Amplified fragment length polymorphism (AFLP) and pulsed-field gel electrophoresis (PFGE) analysis also confirm the classification of proteolytic types A, B and F strains in group I and the non-proteolytic types B, E and F strains in group II, but can differentiate individual strains into each group (Hielm *et al.*, 1998a,b; Hyytia *et al.*, 1999a,b; Keto-Timonen *et al.*, 2005; Nevas *et al.*, 2005; Hill *et al.*, 2007). These methods have been used in epidemiological studies and are useful tools to investigate relatedness between strains isolated from patients and food. For example, among proteolytic *C. botulinum* strains, PFGE analysis differentiate the toxinotypes A, B, and F at a 83–86% similarity level, and enable discrimination of most of individual strains. A greater diversity was observed between type A strains than in type B strains (Nevas *et al.*, 2005). These studies also indicate that each *C. botulinum* group is heterogeneous at the genome level.

Recently, the whole genome sequence is available from *C. botulinum* type A strain Hall (group I) (Sebaihia *et al.*, 2007). A high level of similarity was observed between strains from group I by using DNA hybridization with a DNA microarray including 94% of the coding sequences from strain Hall. Two type A strains share 95–96% of the strain Hall coding sequences, and seven other proteolytic strains have 87–91% common coding sequences. Interestingly, two *C. sporogenes* strains (physiologically related to *C. botulinum* group I but not toxigenic) are significantly similar to Hall strain and share 84–87% of the coding sequences (Sebaihia *et al.*, 2007).

Botulinum neurotoxin gene has been sequenced from a large number of strains and sequence comparison has permitted to identify sequence variations in each toxinotype. Thereby, botulinum toxinotypes are divided into subtypes, which are defined as toxin sequences differing by at least 2.6% identity at the amino acid level (Smith *et al.*, 2005). Botulinum neurotoxin genes from type A strains show 92 to 95% nucleotide identities corresponding to 84–90% amino acid identities and are divided into four subtypes termed A1, A2, A3, and A4 (Table 1.1). Type B genes differ from 2% to 4% at the nucleotide level and 4% to 6% at the amino acid level. They are classified in five subtypes, B1, B2, B3, bivalent B, and non-proteolytic B. Botulinum neurotoxin genes from non-proteolytic type B strains form only one subtype, whereas those from proteolytic strains show a greater variation leading to a four subtype division. Sequences of neurotoxin genes type B show an overall less variation than those of type A, but a more sequence variation is observed within members of each type B subtype compared to *bont*/A. BoNT/E sequences from *C. botulinum* type E (group II) fit in three subtypes (E1, E2, E3) sharing 99% nucleotide identity and 97–99% amino acid identity, and are more distantly related to BoNT/E sequences from *C. butyricum* strains which are into two subtypes (E4, E5) with 97–98% nucleotide and 95–96% amino acid identities between sequences from both *Clostridium* species. The low number of strains type C, D, F and G which were analysed does not permit to significantly evaluate the neurotoxin gene diversity. In group III, mosaic genes between BoNT genes types C and D can be distinguished from classical types C and D strains. BoNT/F sequences from *C. botulinum* type F form a different cluster of those from *C. baratii* (Hill *et al.*, 2007).

Signification of sequence diversity in each toxinotype is not yet well known, but could be important in diagnostic tests and development of therapeutic agents such as those based on immunotherapy. Thereby, BoNT/A1 and BoNT/A2, which differs by 10% at the amino acid sequence level, show large differences in monoclonal antibody-binding affinity. Among six monoclonal antibodies which bind to BoNT/A1 with high affinity, three show a marked decrease in binding affinity (500- to more than 1000-fold) to BoNT/A2. Only combinations of monoclonal antibodies which tightly bind to toxin subtype, potently neutralize the corresponding toxin *in vivo*. Association of the three monoclonal antibodies with high-affinity binding to subtypes A1 and A2, completely neutralizes A1 or A2 toxin, while replacement of two from three monoclonal antibodies by two having a low binding affinity to BoNT/A2 induces a decrease in BoNT/A2 neutralization (50-fold less) (Smith *et al.*, 2005). The impact of subtype variation in binding and neutralization potency of polyclonal antibodies remains to be determined. Thus, development of therapeutic polyclonal or monoclonal antibodies as well as vaccines based on single toxin subtype, need to be evaluated for their protection ability with the other related subtypes. Although two toxins show a low level of sequence difference, they can have marked difference in activity if amino acid variation occurs in strategic toxin sites. Subtypes A1, A2, A3 and A4 of BoNT/A have been analysed by sequence comparison, as well as molecular modelling and structure comparison with the crystal structure of subtypes, the impact of which is not known. Ganglioside binding site is conserved in all subtypes of BoNT/A. The greatest variability was found in the light (L) chain, mainly between subtypes A3 and A4 (76% identity). The enzymatic site of L chain is conserved, but non-conservative mutations are observed in domains involved in substrate (SNAP-25) recognition. When compared to subtypes A1 and A2, subtypes A3 and A4 show sequence variation in α-exosite and S1′ subsite recognition, respectively, suggesting that these subtypes have a decreased affinity and catalytic efficiency for their substrate (Arndt *et al.*, 2006). Another example of gene variation and toxin activity difference is given by neurotoxins type B. BoNT/B from strain 111 (subtype B2) isolated from infant botulism differs from strain Okra/NT associated with food-borne botulism in Japan by 56 amino acid changes (95.7% identity), from which most occur in the half C-terminal part of the toxin (Ihara *et al.*, 2003). BoNT/B from strain 111 shows an about 10-fold lower specific activity than that of strain Okra/NT, and most of monoclonal antibodies which recognize the C-terminus of BoNT/B Okra/NT do not react with BoNT/B 111. Binding affinity of BoNT/B 111 to the receptor synatotagmin II in the presence of ganglioside GT1b is 4.2 lower than that of BoNT/B Okra/NT. Mutations of 23 residues in the C-terminus of BoNT/B 111 have been attributed to the lower binding affinity of the toxin to its receptor and thus to the lower specific toxicity (Kozaki *et al.*, 1998; Ihara *et al.*, 2003).

Sequence comparison of neurotoxin genes suggests that the BoNT genes have evolved separately in different genomic backgrounds (Hill *et al.*, 2007). BoNT genetic diversity could also reflect a different geographical distribution of strains or their involvement in different epidemiological situations. *C. botulinum* subtype A2 was first identified in infant botulism in Japan and was found to differ from strains involved in food-borne botulism in adults referred as subtype A1 (Tabita *et al.*, 1991; Willems *et al.*, 1993). However, no correlation was evidenced between strains subtype A1 and A2 isolated from the United States and UK and their clinical origin, food-borne or infant botulism (Cordoba *et al.*, 1995; Johnson *et al.*, 2005). But strains A1 are more prevalent in the USA, whereas subtype A2 strains are commonly isolated in Europe. Indeed, all the strains from food-borne botulism in the USA which have been analysed fall into subtype A1, and all 33 *C. botulinum* type A isolated from Italy belong to subtype A2 as well as two strains from infant botulism in the UK (Franciosa *et al.*, 2004; Johnson *et al.*, 2005; Hill *et al.*, 2007). But 18 *C. botulinum* strains type A isolated in France or Europe by Prevot during the period approximately 1950–1960 are of subtype A1 (Hill *et al.*, 2007). Divergent strains of subtype A2 characterized by five amino acid differences in BoNT/A2 and a slightly different botulinum locus organization with a shorter intergenic region between *orfX1* and *botR/A* genes (77 versus 1228 nucleotides) when compared to strain A2 Kyoto-F, have been identified in Italy such as strain associated with consumption of contaminated cheese (Mascarpone) (Fig. 1.1) (Franciosa *et al.*, 2006). Organization of botulinum locus of strain Mascarpone is closely related to that of locus containing *bont/A1* in strain type A(B) NCTC2916 (Henderson *et al.*, 1996; Dineen *et al.*, 2003). Strains Mascarpone and Kyoto-F have probably a common origin and then a distinct evolution including a gene rearrangement in strain Mascarpone with an ancestor of strain NCT2916. Another differential genetic evolution is illustrated by neurotoxigenic *C. butyricum* strains. *C. butyricum* strains producing type E neurotoxin have been isolated from infant botulism and young people in Italy (McCroskey *et al.*, 1986; Fenicia *et al.*, 1999), as well as from botulism associated with consumption of fermented soybean in China (Meng *et al.*, 1997, 1999; Wang *et al.*, 2000). Based on toxin gene sequence, toxigenic *C. butyricum* strains from Italy and China are divided in two distinct subtypes, termed E4 and E5, respectively, indicating an independent evolution of *bont/E* gene after transfer in *C. butyricum* (Hill *et al.*, 2007).

Genomic localization

The genes encoding for the different types of BoNT are present on different genetic elements, including phages, plasmid or chromosomes depending on the species and strain of *Clostridia*. In *C. tetani* and *C. argentinense*, the neurotoxin gene is present within a large plasmid (51 and 76 MDa respectively). Plasmids of various sizes and bacteriophages have been found in *C. botulinum* A, B, E, and F but toxigenicity has not been associated with the presence of these genetic elements (Strom *et al.*, 1984; Weickert *et al.*, 1986; Eklund *et al.*, 1989). The genes encoding for these neurotoxins were assumed to be located on the chromosome. However, it has been recently found that in some strains such as Loch Maree strain (subtype A3) and 657Ba (type B and subtype A4), the botulinum neurotoxin genes are harboured by large plasmids (about 280 kb) (Marshall *et al.*, 2007). In neurotoxigenic *C. butyricum* strains, the location of the BoNT/E encoding gene is still undefined (Hauser *et al.*, 1992; Zhou *et al.*, 1993). In *C. botulinum* C and D, it has been clearly evidenced that BoNT is encoded by bacteriophages (reviewed in Poulain *et al.*, 2006).

Neurotoxin gene transfer

The similarity among the different BoNTs and with TeNT, the fact that different *Clostridium* species can produce BoNT and that some strains contain combinations of *bont* genes (Franciosa *et al.*, 1997; Henderson *et al.*, 1997), strongly suggest that *bont* and *tent* genes derive from a common ancestor and have been transferred between *Clostridium* strains. Bivalent strains producing Ab, Ba, Af, or Bf toxins (Franciosa *et al.*, 1994; Henderson *et al.*, 1997; Santos-Buelga *et al.*, 1998; Barash and Arnon, 2004), as well as strains from type A containing an additional botulinum locus with silent *bont/B* gene, indicate a horizontal gene transfer. Indeed, strains subtypes A2 or A4 with silent *bont/B* gene have the same genomic background, as tested by amplified length polymorphism, suggesting either a horizontal transfer of *bont/B* gene in strains already having *bont/A2* or *bont/A4* genes or that two strains with *bont/B* gene acquired *bont/A2* or *bont/A4* horizontally (Hill *et al.*, 2007). In addition, non-toxigenic derivatives are present in certain toxinotypes such as in *C. botulinum* B (Yamakawa *et al.*, 1997), indicating an instability of a DNA fragment harbouring the *bont* genes. Genetics in *Clostridium* is yet poorly understood, but it can be assumed that toxin gene transfer has been probably mediated by genetically mobile elements (Minton, 1995).

Plasmids

Since the *tent* and *bont*/G genes have been localized on large plasmids in *C. tetani* and *C. argentinense*, of 51 and 76 MDa respectively (Finn *et al.*, 1984; Eklund *et al.*, 1988), their transfer can be achieved by mobilization of the corresponding plasmids. Conjugation and mobilization of large plasmids in *Clostridium* such as *C. perfringens*, have already been reported (Brefort *et al.*, 1977). However, up to now, only non-toxigenic *C. argentinense* and *C. tetani* variants, free of plasmids, have been obtained.

Bacteriophages

Bacteriophages mediate the neurotoxin gene transfer in group III of *C. botulinum*. A pseudolysogenic relationship corresponding to the presence of bacteriophages free within the bacterial cytoplasm, exists between these phages and hosts. Thus, variants free of bacteriophages can be obtained with high frequency using curing reagents such as acridine orange and UV light. Under laboratory culture conditions, a proportion of the bacteria, which depends on the strain and growth conditions (temperature, salinity), are lysed and lose free bacteriophages which can reinfect them. Such lysogeny and reinfection cycles occur probably in the environment (soil, intestinal tract of birds and animals) and account for isolation of non-toxigenic or low toxin producer variants (Eklund and Dowell, 1987). *C. botulinum* C and D strains cured of their phages do not produce BoNT/C or BoNT/D, respectively, while they continue to produce another botulinum toxin, termed C2 toxin, which is not neurospecific and whose encoding gene is chromosomally located. Such bacteria could be converted into neurotoxigenic strains C or D by reinfection with phages obtained from toxigenic *C. botulinum* C or D strains. In addition, BoNT/C1 and BoNT/D genes can be transferred in different *Clostridium* host strains and they determine the toxinotype of these strains (Eklund *et al.*, 1971; Eklund and Poysky, 1974; Eklund *et al.*, 1974).

Recently, the genome sequence of *C. botulinum* type C1 neurotoxin-converting phage was determined (Sakaguchi *et al.*, 2005). It consists of 185,682 bp with 404-bp terminal repeats, and 198 potential orfs including BoNT/C1 and exoenzyme C3. Molecular analysis of the phage genome revealed that this phage exists as a circular plasmid prophage in the lysogen and does not integrate into the host chromosome. This mode of lysogenization appears to be related to the unstable lysogeny of BoNT phages.

In *C. botulinum* A and F, the involvement of bacteriophages has been suggested on the basis of the identification of a gene (*lyc*) in the vicinity of the *bont* genes. *lyc*/A and *lyc*/F genes have been mapped approximately 1 kb downstream from the corresponding *bont* genes, and are partially related to bacteriophage genes encoding lytic enzymes in *Lactobacillus* and *Streptococcus pneumoniae* (Henderson *et al.*, 1997). Since lytic enzymes participate in the bacteriophage life cycle, and the presence of *lyc* gene in the vicinity of *bont*/A gene in NCTC2916 could indicate that the botulinum locus is part of an integrated prophage. Genes encoding two prophages have been identified in the genome of strain Hall, but they are not conserved in two other *C. botulinum* type A1 strains (Sebaihia *et al.*, 2007).

Transposons

Transposable elements have been identified in *C. botulinum* C and D. The exoenzyme C3 gene which codes for an ADP-ribosyltransferase specific for the eukaryotic Rho protein, is also harboured by the same bacteriophages carrying *bont*/C1 and *bont*/D genes. The transformation-defective mutant phage CN shows a deletion of a 21.5-kb fragment containing the C3 gene. This fragment is present in several bacteriophages isolated from type C and type D bacteria, and it is marked by a 6-bp core motif AAGGAG. The DNA sequence on its left end is homologous among the C and D types, whereas the sequences diverge immediately downstream of the core motif. The 21.5-kb fragment appears to be a mobile DNA element responsible for the spreading of the C3 gene in *C. botulinum* C and D, and it has features similar to the site specific Tnl554 transposon family, including (i) asymmetric ends, (ii) absence of either inverted or terminal repeats and (iii) presence of a 6-bp core motif sequence at both insertion junctions and the insertion site itself (Hauser *et al.*, 1993).

Transposable elements analogous to that containing the C3 gene, could account for the different localization of the neurotoxin genes (chromosomal, plasmid, bacteriophage) and subsequently for the gene transfer between *Clostridium* strains. Such elements have not yet been clearly identified. However, nucleotide sequence analysis in *C. botulinum* A suggests their possible existence. A 97-nucleotide stretch downstream of the stop codon of the *bont*/A gene is identical in *C. botulinum* A strains 62 and NTCT2916, whereas the following nucleotides are totally unrelated between both strains. The 97-nucleotide stretch could be part of a mobile DNA element encompassing the *bont*/A and *antp* genes. In addition, the different surrounding sequences indicate that *bont*/A gene has different localizations on the chromosome of strains 62 and of strain NCTC2916. This is confirmed by a different restriction profile of the 3′ part of *bont*/A gene in the two strains (Popoff and Marvaud, 1999). The genomic sequence of *C. botulinum* A strain Hall shows that the *botulinum* locus is flanked by two insertion sequences in the 5′ part and by a transposase gene in the 3′ part (Bennik *et al.*, 2003). In addition, the upstream region of BoNT/A and silent BoNT/B clusters in strain NCTC2916 share marked homology, indicating a similar evolutionary origin, and insertion sequences with multiple internal mutations were identified downstream of the BoNT/A cluster (Dineen *et al.*, 2003).

Neurotoxigenic *C. butyricum* strains probably originated from non-toxigenic *C. butyricum* strains by acquisition of *bont/E* gene from *C. botulinum* E (Poulet *et al.*, 1992). The *bont/E* gene and its flanking regions are absent in non-toxigenic *C. butyricum* strains, suggesting a possible gene transfer mediated by a mobile DNA element (Hauser *et al.*, 1992). This gene has been transferred from a neurotoxigenic *C. butyricum* strain to a non-toxigenic *C. botulinum* E strain following a protocol resembling transduction with a defective phage (Zhou *et al.*, 1993). The precise mechanism of molecular transfer has not been elucidated since DNA/DNA hybridization studies suggest that the *bont/E* gene is localized on chromosomal DNA and not on phage DNA in toxigenic *C. butyricum* (Zhou *et al.*, 1993).

In addition, rearrangement, probably by homologous recombination, appears to have occurred between genes of the botulinum locus of different types of neurotoxigenic bacteria. This is suggested by the identification of atypical strains carrying mosaic genes containing elements derived from *C. botulinum* C and D (Moriishi *et al.*, 1996), or mosaic *ntnh* genes from *C. botulinum* A and C (Kubota *et al.*, 1996) or from proteolytic *C. botulinum* A, B, and F (East *et al.*, 1996; Hutson *et al.*, 1996). Subtype BoNT/A2 also probably results from a recombination event between the 5′ end of the *bont/A1* L chain gene and the 3′ end of the *bont/A3* gene (Hill *et al.*, 2007).

Toxin gene regulation

bont and *antp* genes are transcribed as two operons

Transcriptional analysis has been investigated in *C. botulinum* A and C (Hauser *et al.*, 1995; Henderson *et al.*, 1996). In *C. botulinum* C, a transcription start site was mapped 113 nucleotides upstream of the ATG of *ntnh*/C, and another one 100 nucleotides upstream of the initial codon of *bont*/C1. Both genes are preceded by a consensus ribosome binding site (GGAGG). The analysis of mRNA by reverse transcriptase PCR (RT-PCR) showed that a mRNA overlapped *bont*/C1 and *ntnh*/C1 genes possibly encoding both genes (Hauser *et al.*, 1995). Similar results were found in *C. botulinum* A NCTC 2916 (Henderson *et al.*, 1996). Two transcripts have been identified for *bont*/A gene. The longer (7.5 kb) encompasses the *ntnh-bont*/A genes, and the shorter (4 kb) corresponds to *bont*/A alone. This indicates that *bont*/A and *bont*/C1 are transcribed as a mono or a bicistronic messenger in association with the corresponding *ntnh* gene. A transcription start site was mapped 100 nucleotides upstream of the ATG of *ntnh*/A, but the transcriptional start of *bont*/A was not clearly identified. A transcription start site was determined 20 and 118 nucleotides upstream of the ATG of *bont*/A in *C. botulinum* NCTC 2916 and A62, respectively (Binz *et al.*, 1990; Henderson *et al.*, 1996). However, transcriptional analysis in *C. botulinum* KyotoF type A2 showed that neurotoxin genes are transcribed only as a tricistronic operon, *p47-ntnh-bont*/A2, with a transcriptional origin 28 nucleotides upstream of the ATG of *p47*, and no RNA messenger specific of *bont*/A2 alone was evidenced (Dineen *et al.*, 2004). In *C. tetani*, the transcriptional start site was found 127 nucleotides upstream of the ATG of *tent* (Eisel *et al.*, 1986). Using RT-PCR, it was found that at least one mRNA overlaps the three *ha* genes in *C. botulinum* C (Hauser *et al.*, 1995). By Northern blots in *C. botulinum* A NCTC 2916, a single 3.2 kb transcript encompasses *ha*35, *ha*17 and *ha*70 genes (Henderson *et al.*, 1996). A transcriptional start site was mapped 34 nucleotides upstream of the ATG of *ha*35. Therefore, the *ha* genes seem to form a tricistronic operon. However, the HA35 or HA33 components in *C. botulinum* A and C respectively, were found to be produced in higher amount than the other HA components (Inoue *et al.*, 1996). This suggests that the relative levels of synthesis of the 3 HA components are controlled at the translational level as it was found for other toxin such as cholera toxin (Hirst, 1995). In *C. botulinum* with a subtype A2 locus organization, the transcriptional analysis also revealed that *orfX1*, *orfX2* and *orfX3* genes are transcribed as a tricistronic operon. A 5.5-kb transcript encompassing the three genes was detected and a transcriptional start site was mapped 1179 nucleotides upstream of the ATG of *orfX1* (Dineen *et al.*, 2004).

The genetic organization in the other *C. botulinum* types shows that *bont* and *antp* genes are localized in two clusters which can be considered as polycistronic units by comparison with the findings reported in *C. botulinum* A and C.

BotR and TetR positively regulate neurotoxin gene expression

It was first reported that in *C. botulinum* C, a gene (*botR*/C previously called *orf*22) encodes a 22 kDa protein having the feature of a DNA binding protein: basic pI (10.4) and presence of helix–turn–helix motif (Hauser *et al.*, 1995). *botR*/C is localized upstream of the *ha* genes. A homologous gene (*botR*/A previously called *orf*21) was characterized in *C. botulinum* A (Henderson *et al.*, 1996). *bot*R/A has a different localization than that of *botR*/C, since it is inserted between the *ha* and *ntnh-bont* genes. This gene is also conserved in proteolytic and non-proteolytic *C. botulinum* B, D, F and G, and in *C. tetani* (East *et al.*, 1996; Henderson *et al.*, 1996; Bhandari *et al.*, 1997; Li *et al.*, 1998). However, it has not yet been detected in *C. botulinum* E. *tet*R is the only gene related to a gene from the botulinum locus which is conserved in *C. tetani*. *tetR* is localized immediately upstream of the *tent* gene (Fig. 1.1).

BotRs from the different toxinotypes and TetR have 51–97% identity and are related to other known regulatory proteins such as Uvia (25–28% identity) which regulates the bacteriocin production in *C. perfringens*, Msmr protein (21–26%) which regulates the sugar transport in *Streptococcus mutans* and TcdR (20–24%) in *C. difficile* (previously named TxeR). Interestingly, *tcd*R and *uvi*A have similar localization than that of *bot*R and *tet*R since they lie upstream of the locus of the Toxin A and Toxin B genes (*tcdA* and *tcdB*) and bacteriocin gene (*bcn*), respectively. TcdR was shown to act as a positive regulator as deduced from *in vivo* expression, in *E. coli* and *C. perfringens*, of a reporter gene fused to *tcdA* and *tcdB* gene promoter regions (Moncrief *et al.*, 1997; Mani and Dupuy, 2001). Then, TcdR was recently biochemically identified as a novel RNA polymerase sigma factor that directs transcription from the tox promoters (Mani and Dupuy, 2001). Moreover, TcdR has also been found to stimulate its own synthesis in response to environmental signals, such as glucose and temperature (Mani *et al.*, 2002; Karlsson *et al.*, 2003). We have analysed the function of *bot*R/A by overexpressing this gene in *C. botulinum* A. *bot*R/A was cloned under the control of its own promoter in a high copy number shuttle vector which was transferred in *C. botulinum* A by electroporation. A significant increase of the BoNT/A and ANTPs production and of the corresponding mRNA levels were detected in the recombinant strain. Inversely, partial inhibition of the *botR*/A expression by antisense mRNA resulted in a lower production of BoNT/A and ANTPs. It was concluded that BotR/A is a transcriptional activator of *bont* and *antp* genes (Marvaud *et al.*, 1998b). In the same way, overexpression of *tetR* in *C. tetani* induces an increased level of *tent* mRNA and TeNT production. Interestingly, overexpression of *botR*/A and to a lesser extent *botR*/C in *C. tetani* enhance TeNT synthesis (Marvaud *et al.*, 1998a), thus indicating that BotR and TetR share *in vivo* a similar mechanism of action. This indicates that the regulation of the *bont* and *tent* genes is conserved in *C. botulinum* and *C. tetani*, and this constitutes a further evidence that *bont* and *tent* gene locus derive from a common ancestor (Marvaud *et al.*, 1998a).

BotR and TetR are alternative sigma factors controlling the expression of the neurotoxin and associated protein genes

Preliminary *in vitro* experiments with crude lysates of *C. botulinum* A overexpressing BotR/A indicated that a complex was formed between *ntnh-bont*/A promoter and proteins, which minimally contain BotR/A (Marvaud *et al.*, 1998b). Therefore, purified recombinant BotR/A was used in gel mobility shift assays to investigate the role of this protein. We found that purified BotR/A alone did not induce any retardation in migration of *ntnh-bont*/A promoter DNA fragment. In *C. difficile*, TcdR acts as an alternative sigma factor on toxin gene transcription by interacting specifically with toxin gene promoters in association with *E. coli* or *C. difficile* RNA polymerase core enzyme (Core). Then, we showed that BotR/A bound to *ntnh-bont*/A promoter in a dose-dependent manner when associated with *E. coli* Core, whereas *E. coli* Core alone did not. Similar results were obtained with the second operon promoter of the botulinum locus, *ha*35 (Raffestin *et al.*, 2005). Interaction of BotR/A-Core with the putative promoters of the other botulinum locus genes (*bont*/A, *botR*/A, *ha*17, *ha*70) was also investigated by gel mobility shift assays. BotR/A-Core interact specifically with the operon promoters and also with the putative promoters of the other botulinum locus genes. Similar to BotR/A, purified TetR mixed with Core bind to the *tent* promoter in a dose dependent manner, unlike TetR or Core alone (Raffestin *et al.*, 2005). TetR-Core also bind to the promoter region of its own gene *tetR*, but at a lower extent. Moreover, BotR/A and TetR are able to bind directly to Core in a DNA-independent manner.

To further test that BotR/A and TetR are RNA polymerase sigma factors, *in vitro* run-off transcription reactions were performed. Transcripts were observed when BotR/A-Core was incubated with DNA templates containing *ntnh-bont*/A and *ha*35 promoters. Whereas, BotR/A-Core did not produce any transcript from the DNA templates containing the putative promoters of *bont*/A, *botR*/A, *ha*17 and *ha*70 genes. Similar finding were obtained with TetR. TetR-Core initiated specifically the transcription of *tent*, whereas no transcript was detected from the *tetR* gene promoter. Thereby, BotR/A and TetR are alternative sigma factors required for specific expression of the botulinum locus operons and *tent* gene, respectively.

Footprinting experiments with DNA fragment containing *ntnh-bont*/A promoter indicate that the conserved −10 (GTTATA) and −35 (TTTACA) promoter sequences, are the motif signatures recognized specifically by the sigma factor BotR/A. Sequences of −35 and −10 regions of *ntnh-bont*/A promoter are conserved in *ha*35 promoter of *C. botulinum* A, also in the corresponding promoters of *C. botulinum* B, C, D and G, as well as in *tent* promoter in *C. tetani*. The consensus promoter site sequences are ATTTTagGTTTACAAAA for the −35 and ATGTTATATataa for the −10 regions (Popoff and Marvaud, 1999). The −35 box contains the hexa-motif TTTACA which has also been found in the corresponding promoter sequence recognized by TcdR and UviA. However, the −10 regions of the promoters recognized by TcdR and UviA are divergent and differ from the −10 consensus sequence in *C. botulinum* and *C. tetani*. A −35 consensus sequence was evidenced 119 nucleotides upstream of the start codon of *botR*/A. However, the consensus −10 sequence of the botulinum operons was not found in the corresponding region, and BotR/A-Core was unable to initiate transcription from *botR*/A promoter. *botR*/A might be regulated by an additional factor. Similarly, the *tetR* upstream region does not contain any conserved −35 and −10 motif sequence, and TetR-Core was not functional or insufficient to promote *in vitro* transcription from its own promoter.

BotR and TetR belong to a new subgroup of the σ^{70} family of RNA polymerase sigma factors (group 5)

It has also recently been demonstrated that UviA is required, as alternative sigma factor, for transcription of UV-inducible bacteriocin gene in *C. perfringens* (Dupuy *et al.*, 2005). The majority of the sigma factors in eubacteria belong to the σ^{70} family, which can be divided in four structurally and functionally related groups. BotR, TetR, TcdR and UviA have some similarities with the ECF group (group 4). Because they differ sufficiently in structure and function, it was suggested that they form a new group (group 5) that is distantly related to the σ^{70} family (Helmann, 2002).

Interchangeability between BotR/A, BotR/C and TetR in *C. tetani* was first demonstrated *in vivo* in *C. tetani* (Marvaud *et al.*, 1998a). Then, the four sigma factors, BotR/A, TetR, TcdR and UviA, were found to be functionally interchangeable *in vitro* and *in vivo* (Dupuy *et al.*, 2006). TcdR-related σ factors enable RNA polymerase core enzyme to bind to the toxin and bacteriocin gene promoters. The four sigma factors are able to substitute for each other to initiate *in vitro* transcription from all toxin and bacteriocin gene promoters. Moreover, TcdR-related sigma factors are functionally distinct from the ECF family, at least from *B. subtilis* σ^{W}. These results reinforce the idea that BotR, TetR, TcdR and UviA form a new group of RNA polymerase sigma factors, group 5 of the $\sigma70$ family, distinct from the ECF group. To determine the *in vivo* functional interchangeability of TcdR-related sigma factors, the expression of a reporter gene fused to the promoter regions of *tcd*A, *bcn*, *ntnh-bont* and *tent* genes was investigated in *C. perfringens*. Despite *in vitro* interchangeability between the four clostridial sigma factors, they show a functional specialization *in vivo*. They can be divided in two clusters, one including BotR and TetR, and the other TcdR and UviA. This specialization of function reflects differences in recognition of promoter −10 region by the two subgroups of sigma factors. Promoters recognized by all four sigma factors have very similar −35 region (TTTACA), but vary considerably in their -10 sequences. The specificity of each sigma factor for its cognate promoter site resides principally in two DNA-binding domains located in subregion 2.4 and 4.2, which recognize the −10 and −35 hexanucleotide motifs respectively. While region 4.2 of the group 5 sigma factors is well conserved, 2.4 region is very similar in BotR and TetR (75% identity) but differs significantly in TcdR and UviA (25% identity) (Dupuy *et al.*, 2006).

The functional conservation of sigma factors of group 5 suggests the idea that the member of group 5 were derived from a common ancestor. A common molecular mechanism is involved in the regulation of synthesis of potent *Clostridium* toxins (BoNT, TeNT, TcdA, and TcdB) and bacteriocin.

The expression of *botR*/A and other botulinum locus genes occurs during the transition from the exponential to the stationary growth phase

Expression of *botR*/A, *bont*/A and *antps* genes in *C. botulinum* A strain Hall and NCTC2916 have been monitored by reverse transcription and real-time PCR analysis (Fig. 1.2) (Couesnon *et al.*, 2006). Expression of *botR*/A is concomitant with that of *bont*/A and *antps*. All botulinum locus genes follow the same kinetic of expression in sporulated or non-sporulated cultures. However, the relative expression of *botR*/A in both strains was approximately 100-fold less than those of the other botulinum locus genes. Expression of botulinum locus genes is low during the exponential growth, increases at a maximum level (100-fold increase) at the transition

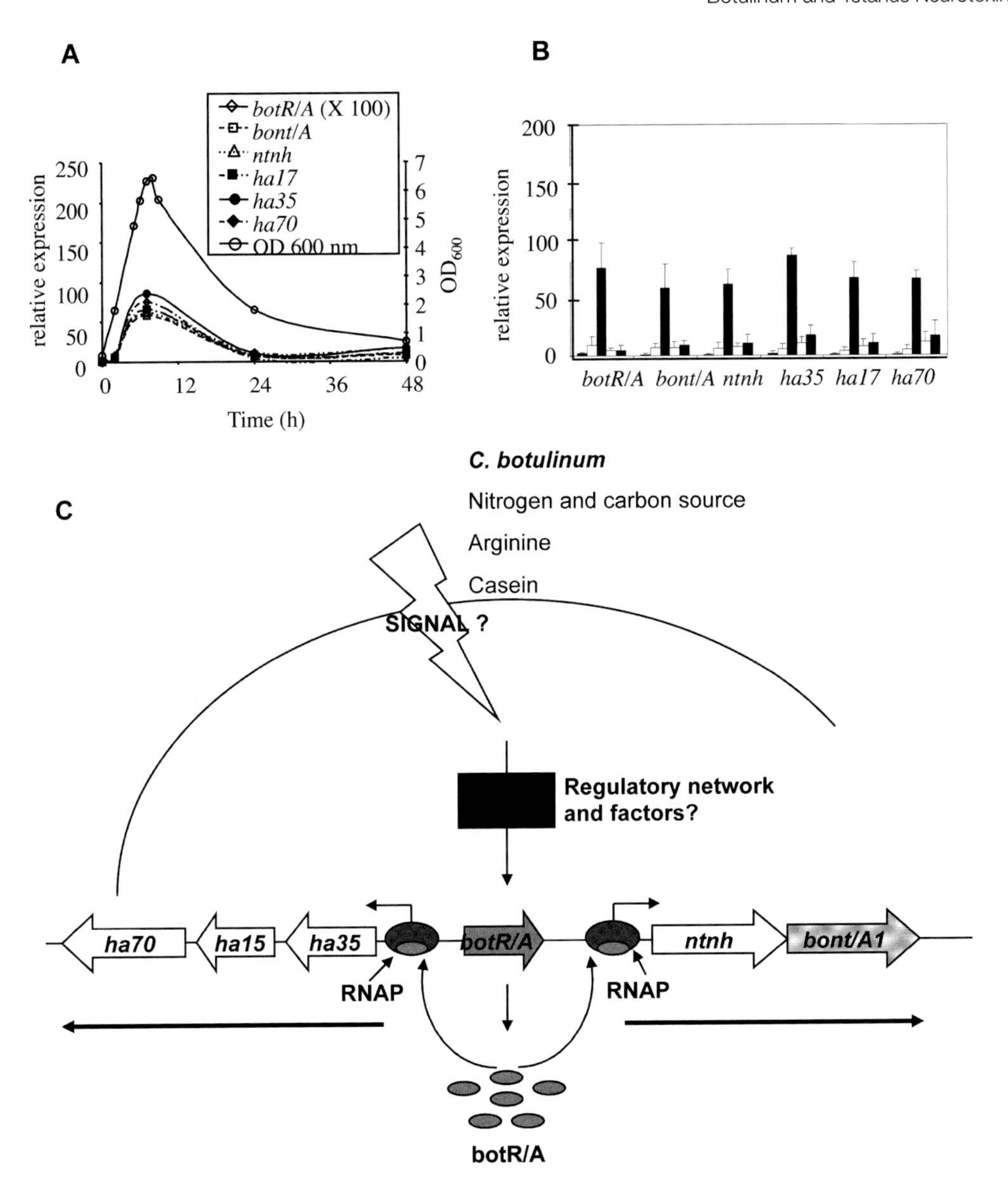

Figure 1.2 (A) Relative expression of botulinum locus genes versus the 16S gene during growth of *C. botulinum* A strain Hall in non-sporulated conditions. (B) Transcript levels from the botulinum locus genes versus the 16SrRNA gene at (from left to right, alternatively black and white) early, mid-exponential, transition and stationary growth phase at 24 and 48h of *C. botulinum* A strain Hall non-sporulated cultures. (C) Model of expression of botulinum locus genes in *C. botulinum* A.

phase between late exponential and stationary growth phase, and then decreases during the stationary phase. After 24 h incubation, relative gene expression decreases to levels found in the early growth phase. The highest toxin concentration (monitored by mouse lethal activity) is observed at 24 h (12 h after the maximal gene expression), and is stable for a period of at least 5 days. These results are in agreement with previous reports for *C. botulinum* A strains A62, Hall and NCTC2916 using analysis of mRNA transcripts by Northern hybridization (Bradshaw *et al.*, 2004). Moreover, in strain NCTC2916, in sporulated or non-sporulated conditions, the relative expression of *bont*/A is approximately twice than that of *ntnh*/A, and *ha*35 transcripts are about twofold higher than those of *ha*17 and *ha*70. In contrast, in non-sporulated culture of

Hall strain, the relative levels of expression of *bont*/A and *antps* are similar, whereas, in sporulated cultures each *ha* gene is expressed twofold higher than *bont*/A and *ntnh*/A (Couesnon *et al.*, 2006).

In C. *botulinum* E, where no *botR* gene was found up to now, the kinetics of BoNT/E production and the relative expression of *bont*/E and *p47* are the same than in C. *botulinum* A. *bont*/E and *p47* are also expressed during the transition phase (Couesnon *et al.*, 2006). All of theses results support the fact that BotR is an alternative sigma factor specific to the transition phase, like TcdR in C. *difficile* (Hundsberger *et al.*, 1997; Dupuy and Sonenshein, 1998; Mani *et al.*, 2002) and other alternative sigma factors. The time correlation between the expression of BotR and the other botulinum locus genes suggests that BotR is the main regulatory factor.

Environmental signals

It is not known whether BotR and TetR are the only regulatory genes or are involved in a regulation cascade and which are the environmental signals which trigger the production of the neurotoxins and ANTPs. In C. *difficile*, a catabolic repression of the expression of Toxin A and Toxin B genes by rapidly metabolizable sugars such as glucose has been evidenced (Dupuy and Sonenshein, 1998). No such regulation has been identified in C. *botulinum* and C. *tetani*. Carbon source, amino-acids, biotin, and growth temperature have been found to regulate the production of Toxin A and B in C. *difficile* (Dupuy and Sonenshein, 1998; Yamakawa *et al.*, 1998; Karlsson *et al.*, 1999; Karlsson *et al.*, 2003). In contrast to C. *difficile*, C. *botulinum* requires glucose for high toxin production. CO_2 seems to stimulate the *bont*/B expression and subsequent BoNT/B production (Lövenklev *et al.*, 2004). Availability of certain nutrients, in particular the nitrogen and carbon sources, is important for toxinogenesis in C. *botulinum* and C. *tetani*. A high level of arginine in minimal medium represses BoNT/A and protease production, whereas casein complementation increases both toxin and protease synthesis (Patterson-Curtis and Johnson, 1989). Moreover, specific short peptides from media containing pancreatic digest of casein are essential for toxin production in C. *tetani* (Porfirio *et al.*, 1997).

Temperature is a common signal involved in the control of alternative sigma factors and subsequent regulated genes expression. In C. *difficile*, a temperature of 42°C prevents transcription of *tcdR* and subsequent expression of *tcdA* and *tcdB* (Karlsson *et al.*, 2003). In C. *botulinum*, high temperature is not an inhibitory signal for toxin gene regulation (Couesnon *et al.*, 2006). Indeed, expression of *botR*/A and other botulinum locus genes was not temperature sensitive, within 37–44°C range. But BoNT/A lethal activity decreases in cultures incubated at 44°C for 24–48 h. Analysis of temperature-dependent stability of BoNT/A and ANTPs showed that NTNH/A and BoNT/A are sensitive to a proteolysis at 44°C, in contrast to HAs proteins. A Ca^{2+}-dependent protease produced by C. *botulinum* A is probably activated at 44°C leading to BoNT/A and NTNH/A degradation, whereas at 37°C this protease only promotes the proteolytic activation of BoNT/A (Couesnon *et al.*, 2006). In contrast, C. *botulinum* E produces no protease activated at 44°C, and BoNT/E as well as NTNH/E are stable in culture supernatant incubated at 44°C (Couesnon *et al.*, 2006).

The environmental signals which trigger the synthesis of neurotoxin in C. *botulinum* and C. *tetani*, and the regulatory network and actors which control the expression of *botR* and *tetR* and subsequent regulated genes, remain to be elucidated (Fig. 1.2).

Structure of clostridial neurotoxins

BoNTs and TeNT share a common structure that is first synthesized as a precursor protein (about 150 kDa), which is inactive or weakly active. The precursor, which does not contain signal peptide, is possibly released from the bacterium by a cell wall exfoliation mechanism. The precursor is proteolytically activated in the extrabacterial medium either by *Clostridium* proteases or by exogenous proteases, such as those naturally found in the intestinal content. The active neurotoxin consists of a light (L) chain (~ 50 kDa) and a heavy (H) chain (~ 100 kDa), which remain linked by a disulphide bridge. Activation of BoNT/A by an endogenous protease from *C. botulinum* cleaves between Lys438-Thr439, thus removing 10 amino acids (Thr439–Lys448) (Krieglstein *et al.*, 1994). In BoNT/E, cleavage occurs at Lys419–Gly420 and Arg422–Lys423, thus excising a Gly420–Ile–Arg422 peptide (Antharavally and DasGupta, 1997). Proteases produced by *C. tetani* cleave TeNT at Glu449 and Ala456 (Krieglstein *et al.*, 1991). For the other BoNTs, cleavage occurs at one or two sites, which include: Arg444/Ser445 and Lys449/Thr450 for BoNT/C; Lys442/Asn443 and Arg445/Asp446 for BoNT/D; as well as Arg435/Lys436 and Lys439/Ala440 for BoNT/F (Sagane *et al.*, 1999). The crystallographic structures of BoNT/A, BoNT/B, as well as the C-terminal domain of TeNT have been solved and reveal three distinct domains: an L-chain containing α-helices and β-strands that include the catalytic zinc-binding motif; the N-terminal part of the H-chain (Hn) forming two unusually long and twisted α-helices; and the C-terminal part of the H-chain (Hc), consisting of two distinct subdomains involved in receptor recognition (Umland *et al.*, 1997; Lacy *et al.*, 1998; Lacy and Stevens, 1999; Emsley *et al.*, 2000; Swaminathan and Eswaramoorthy, 2000; Fotinou *et al.*, 2001).

The binding domain

The Hc binding domain of BoNTs and TeNT consists of two distinct subdomains, the N-terminal (Hc-N) and C-terminal (Hc-C) halves, with few protein–protein contacts between them. Hc-N has two seven-stranded β-segments arranged in a jelly roll motif quite similar to that of lectins. The amino acid sequence of this subdomain is highly conserved among clostridial neurotoxins, thus suggesting a similar folding pattern. The Hc-C subdomain contains a modified β-trefoil folding motif present in several other proteins that possess recognition and binding functions, such as interleukin 1, fibroblast growth factor, and the Kunitz-type trypsin inhibitors. However, the Hc-C sequence is poorly conserved among clostridial neurotoxins. The critical importance of the last 34 residues from Hc-C, and in particular His1293 of TeNT, for binding the oligosaccharide portion of polysialogangliosides was recently demonstrated by photoaffinity labelling (Shapiro *et al.*, 1997). In BoNT/B and BoNT/A, one ganglioside binding site was mapped within the Hc-C, whereas TeNT binds simultaneously to two ganglioside molecules (Swaminathan and Eswaramoorthy, 2000; Fotinou *et al.*, 2001; Rummel *et al.*, 2003, 2004). These data are supportive of a double receptor model of binding for clostridial neurotoxins to the presynaptic membrane. One subdomain binds to a glycoprotein, which differs for the various clostridial neurotoxins, and the other subdomain binds to polysialogangliosides (Chai *et al.*, 2006; Jin *et al.*, 2006; Rummel *et al.*, 2007).

A major difference between the Hc-C of TeNT and BoNT/A resides in the loop structures, thus suggesting that these external segments may bind to unique protein receptors. However, the Hc-N and Hc-C subdomains are required for neurotoxin interactions with neuronal surface receptors. Thereby, antibodies against Hc-N or Hc-C are poorly protective, whereas antibodies raised against the whole Hc domain efficiently neutralize BoNT/A. The whole BoNT/A Hc domain, and especially the native structure of the interfacial region between Hc-N and Hc-C, is necessary to generate efficient protective antibodies (Tavallaie *et al.*, 2004). In addition, the Hc-C domain of TeNT sufficiently mediates binding to neuronal cells (Herreros *et al.*, 2000).

The translocation domain

The Hn regions are highly homologous among the various clostridial neurotoxins, and the conformations of these domains are quite similar in BoNT/A and BoNT/B (Lacy *et al.*, 1998; Swaminathan and Eswaramoorthy, 2000). Hn has a cylindrical shape determined by a pair of unusually long (10 nm), twisted α–helices corresponding to peptide segment 685–827 (BoNT/A numbering), which is overall reminiscent of the α-helical hairpin found in some colicins. At both ends of the helical pair, there is a shorter α-helix, which lies parallel to the main helices and, in addition, several strands pack along the two core helices. The segments involved in ion channel formation at low pH remain unidentified, but the overall structure of Hn resembles that of some viral proteins which undergo an acid-driven conformational change (Bullough *et al.*, 1994).

The catalytic domain

The metalloprotease domain (L chain, 55 Å ~ 55 Å ~ 62 Å) contains both α-helix and β-strand secondary structures, and has little similarity with related enzymes of known structure, apart from the α-helix segment including the zinc binding motif. In addition to the imidazole rings from two histidines of the His-Glu-X-X-His motif and a water molecule bound to the glutamic acid, the zinc atom is coordinated by another glutamic acid conserved among all clostridial neurotoxins (Rigoni *et al.*, 2001; Rossetto *et al.*, 2001). The Glu residue is particularly important because it coordinates the water molecule directly implicated in proteolytic hydrolysis. Its mutation leads to complete inactivation of these neurotoxins (Li *et al.*, 2000). The active site is similar to that of thermolysin and identifies a primary sphere of residues essential for catalysis, which coincides with the zinc coordinating residues. In addition, there is a secondary layer of amino acids more distant from the zinc centre, in which Arg362 and Tyr365 (BoNT/A numbering) are reportedly involved in catalysis by stabilizing the transition state (Rigoni *et al.*, 2001; Rossetto *et al.*, 2001; Binz *et al.*, 2002). The active site of BoNT/A is in a deep recess (20–24 Å) shielded by the H chain; it only becomes accessible to the substrate following reduction of the interchain disulfide bridge, and it appears capable of accommodating a 16-residue segment. Upon reduction at physiological temperature (37°C), BoNT/A loses most of its native tertiary structure and adopts a molten globule conformation that correlates with binding to unstructured SNAP-25 and optimum endopeptidase activity (Cai and Singh, 2001; Kukreja and Singh, 2005). In contrast, the belt region of BoNT/B formed by the Hc-N domain does not completely occlude the active site, which remains accessible to small molecule inhibitors (Swaminathan and Eswaramoorthy, 2000). However, BoNT/B also requires reduction of the disulfide bridge between L and H chains to activate its endopeptidase activity (Herreros *et al.*, 1999; Schiavo *et al.*, 2000). Most of the L chain is required for enzymatic activity. The BoNT/A L chain can only tolerate an eight amino acid truncation at the N-terminus, and a 32 amino acid deletion at the C-terminus; however, in contrast the TeNT L chain is still functional following a 16 or even 65 amino acid deletion at the C-terminus (Kurazono *et al.*, 1992). The L chain C-terminus likely participates in solubility, stability, and catalysis via stabilizing the tertiary structure (Baldwin *et al.*, 2004). Additionally, BoNT/A activity is higher when SNAP-25 is inserted into the membrane or in liposomes containing negatively charged phospholipids, versus soluble SNAP-25, thus indicating that a membrane substrate conformation is better recognized by the toxin and/or phospholipids may facilitate an electrostatic toxin–substrate interaction (Caccin *et al.*, 2003).

Botulinum complexes

BoNTs are associated with non-toxic proteins (ANTPs) to form large complexes (abbreviated Botulinum complex), also known as progenitor toxins. ANTPs encompass a non-toxic and non-haemagglutinin component (NTNH), as well as several haemagglutinin components (HAs) (rev in (Oguma *et al.*, 1999; Quinn and Minton, 2001; Sharma *et al.*, 2003). Botulinum complexes, in which proteins are not covalently linked, are formed in cultures and naturally contaminated food. They are stable at acidic pH, and dissociate at alkaline pH (≥ pH 8). In contrast, TeNT does not yield any complex, and related botulinum *antp* genes are not evident in the *C. tetani* genome (Brüggemann *et al.*, 2003). *C. botulinum* A produces three types of botulinum complex designated LL (19S, 900 kDa), L (16S, 500 kDa), and M (12S, 300 kDa). The 12S complex results from a BoNT/A plus NTNH association, and does not possess any haemagglutinating activity. NTNH has a cleavage site within the N-terminus and is separated into 13- and 106-kDa fragments as assessed by SDS-PAGE. The 16S complex respectively consists of BoNT/A, NTNH, and HAs (HA33, HA17, and HA70) at a presumed 1:1:2 ratio. HA70 is proteolytically cleaved into HA50 and HA20. The 19S complex has the same protein composition as the 16S complex, but the ratio of HA33 is about twice that found in the 16S complex. The 19S complex is likely a dimer of the 16S complex cross-linked by HA33 (Oguma *et al.*, 1999; Sharma *et al.*, 2003). The composition of botulinum complex types A2 and A1 from strain NCTC2916, which contains different proteins like P47, OrfX1, OrfX2, and OrfX3, remains undetermined (Dineen *et al.*, 2003). *C. botulinum* B, C, and D yield two complex types, L and M, the composition of which is equivalent to those of *C. botulinum* A (Oguma *et al.*, 1999; Arimitsu *et al.*, 2003). The complete subunit structure of *C. botulinum* D strain 4947 L complex (650 kDa) was determined as one BoNT/D, two HA70, four HA33, and four HA17b molecules, with intermediate complexes (540 and 610 kDa) constituted of various HA subunits. In addition, an intermediate M complex (410 kDa) contains BoNT/D, NTNH, and two molecules of HA70 (Kouguchi *et al.*, 2002; Mutoh *et al.*, 2003). *C. botulinum* A2, E and F produce only M complexes devoid of haemagglutinating activity, and *C. argentinense* produces only L complex (Oguma *et al.*, 1999). NTNHs from different *C. botulinum* types possess a high identity level (76–83.5%), and are the most conserved proteins in various botulinum complex (Popoff and Marvaud, 1999). NTNH/A, NTNH/C, and NTNH/D contain a cleavage site within their N-terminus, yielding 15-kDa N-terminal and 115-kDa C-terminal fragments. NTNH/A is split into 13 and 106 kDa fragments by cleavage between Pro144/Phe145 (Fujita *et al.*, 1995). NTNH/C and NTNH/D are cleaved at Lys127 by a trypsin-like protease with 7–13 amino acids removed from the N-terminus of the 115-kDa fragment that subsequently results in three proteins starting at Leu135, Val139, or Ser141 (Sagane *et al.*, 2000). NTNH is only cleaved in the 12S (M) complexes from *C. botulinum* types A, C, and D, but not in the L (16S) or LL (19S) complexes. The cleaved NTNH molecules constituted a nicked structure since the two fragments still remain together after NTNH purification (Sagane *et al.*, 2000). In contrast, NTNH/E and NTNH/F show an identical deletion of 33 residues in the corresponding region of NTNH/A, NTNH/C, and NTNH/D encompassing the cutting site, and NTNH/G possesses a slightly different sequence in this region. It is presumed that the processing and additional sequence of NTNH in *C. botulinum* A, C, and D are responsible for forming 12S-, 16S- and 19S-sized complexes. The inability of *C. botulinum* E and F to form L complexes may result from the absence of HA or other related proteins that bind to NTNH, and from the absence of a putative binding site in NTNH/E and NTNH/F (Oguma *et al.*, 1999; Popoff and Marvaud, 1999). HA33-35 is the most abundant haemagglutinin component of the botulinum complex. Type A HA35 binds to oligosaccharides containing galactose-β1-4glucose-*N*-acetyl-D-neuraminic acid (Galβ1-4GlcNAc) (Inoue *et al.*, 2001). Thereby, haemagglutination induced by 16S and 19S type A botulinum complex is mainly mediated through HA35 binding to erythrocyte membrane glycolipids and glycoproteins containing Galβ1-4GlcNAc, such as paragloboside and glycophorin A (Inoue *et al.*, 1999; Inoue *et al.*, 2001). Similarly, HA33 from types C and D botulinum complex bind to paragloboside on Galβ1-4GlcNAc and also sialylglycolipids (GM3), as well as sialoglycoproteins (sialosylparagloboside) on the *N*-acetyl-D-neuraminic acid-α2-3-galactose-β1 motif (Fujinaga *et al.*, 2004). The importance of HA33-35 in haemagglutination is also supported by monoclonal antibody studies. Type C-specific monoclonal antibodies against HA33 inhibit haemagglutination, contrary to those against HA50 and HA17 (Mahmut *et al.*, 2002). However, type C HA70 and its derivative HA50 recognize sialosylparagloboside and GM3 at the *N*-acetyl-D-neuraminic acid-α2-3-galactose-β1motif in erythrocyte membranes, like the corresponding 16S botulinum complex. Thus, HA50 could also be involved in haemagglutination (Fujinaga *et al.*, 2004). HA35 purified from *C. botulinum* A is predominantly a dimeric, β-sheet protein in aqueous solutions. In *C. botulinum* A, five N-terminal amino acids are removed from HA35, but similar posttranslational modification has not been observed in HA33 from *C. botulinum* C. The significance of HA35 processing on its biological activity is not known (Sharma *et al.*, 2006). It was first discovered that the 31 C-terminal amino acids, which contain a predicted carbohydrate recognition site, play an essential role in haemagglutination (Sagane *et al.*, 2001). Structure of type C HA33 shows two β-trefoil domain consisting of a six stranded, antiparallel β-barrel capped on one side by three β-hairpins. Related β-trefoil structures bind to oligosaccharides and are found in other proteins, including various lectins like the ricin B-chain, cytokines, trypsin inhibitor, xylanase, as well as the C-terminal part of BoNTs. Type A HA35 retains a similar structure related to the carbohydrate binding site of ricin, a plant toxin. It is noteworthy that Asp263 and Asn285 of HA35, which are conserved in the lactose-binding site for ricin B chain, are critical for carbohydrate binding (Inoue *et al.*, 2003; Arndt *et al.*, 2005).

Molecular mode of action of clostridial neurotoxins

BoNTs ingested with food or directly synthesized in the intestine by overgrowth of neurotoxigenic *Clostridium* in this site, across the mucosal barrier possibly by transcytosis without altering epithelial cells (Maksymowych *et al.*, 1999; Maksymowych and Simpson, 2004; Couesnon *et al.*, 2007). By blood and/or lymphatic circulation, BoNT is transported to nerve ends. Tetanus toxin formed in wound contaminated by *C. tetani* diffuses locally to nerve ends.

The BoNT and TeNT mode of action consists in four steps: binding, internalization, translocation and intracellular activity (Fig. 1.3). Several reviews detail the mode of action of neurotoxins (Poulain *et al.*, 1997; Herreros *et al.*, 1999; Bigalke and Shoer, 2000; Humeau *et al.*, 2000; Schiavo *et al.*, 2000; Meunier *et al.*, 2002a,b).

Binding to a cell surface receptor

BoNT and TeNT recognize specific receptors on unmyelinated areas of the presynaptic membrane. The precise identity of the clostridial neurotoxin receptors is still unknown. A dual receptor is probably involved consisting in ganglioside of G_{1b} series (G_{D1b}, G_{T1b}) and a glycoprotein. Each neurotoxin recognizes distinct receptor. The specificity of cell surface receptors and also of the intracellular target (see below) explains the different sensitivity of animal species to BoNTs and TeNT. It has been found that synaptotagmin I and II associated to ganglioside G_{T1b} is the receptor of BoNT/B and BoNT/G (Nishiki *et al.*, 1994; Nishiki *et al.*, 1996; Dong *et al.*, 2003; Jin *et al.*, 2006; Rummel *et al.*, 2004; Rummel *et al.*, 2007), whereas BoNT/A recognizes another membrane synaptic vesicle called SV2 (Dong *et al.*, 2006; Mahrhold *et al.*, 2006). This synaptic vesicle membrane protein as a receptor for BoNT could explain that nerve stimulation enhances the intoxication process. The half C-terminal part of BoNT and TeNT H chain is involved in the recognition of the cell surface receptor. This domain consists in two subdomains, a jelly roll folded and a β-trefoil regions (Umland *et al.*, 1997; Lacy *et al.*, 1998; Lacy and Stevens, 1999; Emsley *et al.*, 2000).

Internalization

Neurotoxin bound to its receptor is internalized by receptor-mediated endocytosis. An essential difference between both types of neurotoxins is that BoNTs are directly endocytosed in clathrin coated vesicles which are acidified triggering the translocation of the L chain into the cytosol. Therefore, BoNT L chain is delivered in the peripheral nervous system, in the neuromuscular junctions where it blocks the release of acetylcholine leading to a flaccid paralysis. In contrast, TeNT enters different endocytic vesicles, probably uncoated vesicles which are not acidified and retrogradely transport the toxin, in a microtubule-dependent manner, until the cell body of neurons in the spinal cord (Deinhardt *et al.*, 2006). Then, TeNT carries out a transynaptic migration and reaches the target neurons which are the inhibitory interneurons involved in the regulation of the motoneurons. TeNT enters inhibitory interneurons probably via coated vesicles which are acidified permitting the delivery of L chain into the cytosol where it inhibits the release of glycine and GABA.

Translocation of the light chain

The drop of pH in the endosome is a prerequisite step for the translocation. In acidic conditions, the neurotoxin undergoes a conformational change exposing its hydrophobic domains at the surface. This step is mediated by the half N-terminal part of the H chain which is formed of two long α-helices (Lacy *et al.*, 1998; Lacy and Stevens, 1999). It has been found that H chain oligomerizes, possibly in a tetramer structure, and forms cation-selective channels permeable to small molecules (< 700 Da). The precise

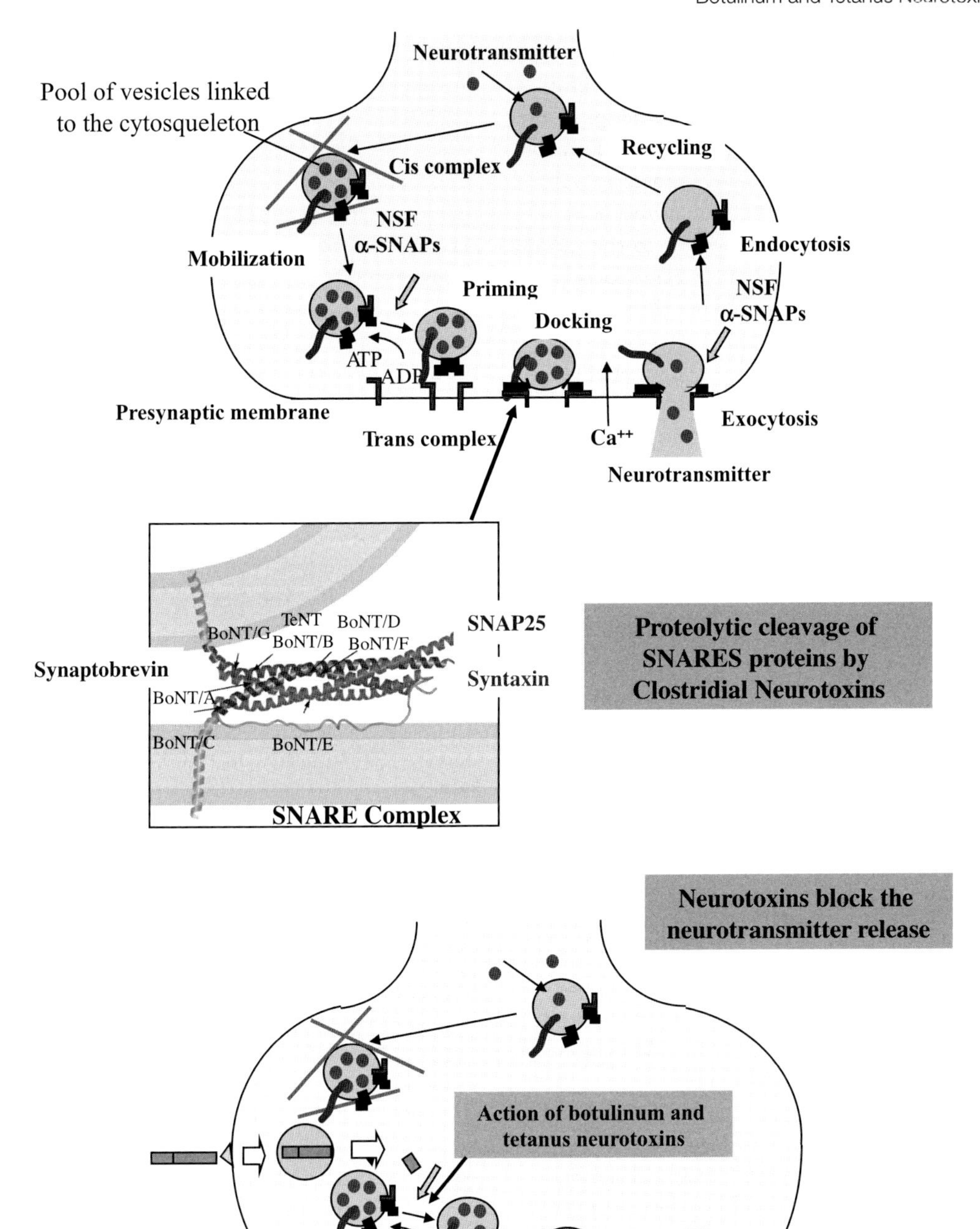

Figure 1.3 Molecular mechanism of neuroexocytosis inhibition by botulinum and tetanus neurotoxins (BoNT and TeNT). BoNTs and TeNT share the same mode of action in the target neuronal cells. They enter the neuronal cells through receptor-mediated endocytosis and translocate the light chain into the cytosol. BoNTs and TeNT specifically cleave one of the three proteins (Synaptobrevin, SNAP-25, Syntaxin) of the SNARE complex which have a key role in the fusion mechanism between synaptic vesicles and presynaptic membrane and in the release of neurotransmitter in the synaptic cleft. SNARE, SNAP-receptor; α-SNAP, soluble accessory protein; NSF, *N*-ethylmaleimide-sensitive factor.

mechanism of translocation remains mysterious. Neurotoxins probably use a mechanism similar to that of diphtheria toxin. In this model, the H chain forms a hydrophilic cleft. The hydrophobic residues of H and partially unfolded L chain face the lipids and the hydrophilic segment of L chain glides on the one of H chain. Then, L chain refolds in the neutral pH of cytosol. Protein chaperones are possibly involved in this mechanism.

Intracellular activity

Molecular genetic studies revealed that clostridial neurotoxins contain a conserved zinc dependent proteolytic site (His-Glu-x-x-His) in the middle of the L chain (Niemann, 1991; Popoff and Marvaud, 1999). It was then demonstrated that clostridial neurotoxins are zinc endopeptidases (inhibition by metalloprotease specific inhibitors and specific antibodies of the zinc binding motif, presence of zinc in neurotoxins, mutations on the zinc motif) and that they cleave previously identified synaptic proteins (Fig. 1.3) (Schiavo *et al.*, 1992). TeNT, BoNT/B, D, F and G cleave at different sites VAMP/synaptobrevin which is a membrane protein from synaptic vesicles. BoNT/A and E cleave SNAP-25, a presynaptic membrane protein, and BoNT/C1 cleaves both SNAP-25 and syntaxin which is also a presynaptic membrane protein. VAMP, SNAP-25 and syntaxin are called SNARE proteins for SNAP receptors. They form a large protein family since many isoforms have been identified in almost all the cells. Each clostridial neurotoxin recognizes their substrate at specific binding sites (SNARE motifs, two in VAMP and syntaxin, four in SNAP-25), and cut at a specific cleavage site. Note that TeNT and BoNT/B cleaves VAMP at the same site. Natural isoforms or recombinant proteins mutated in the SNARE motifs and/or in the cleavage sites are insensitive to neurotoxins (reviewed in Humeau *et al.*, 2000). SNARE proteins are unstructured in solution and assemble in a ternary complex (SNARE complex) consisting of four tightly packed α-helices which lie parallel to the membrane surface (Poirier *et al.*, 1998; Sutton *et al.*, 1998). SNARE complex is very stable to various denaturing agents, and recruits soluble cytosolic proteins such as NSF (*N*-ethymaleimide-sensitive factor) and SNAPs (soluble NSF accessory proteins) forming the 20S SNARE complex which has been recognized as essential in vesicle targeting and fusion (Söllner *et al.*, 1993a,b; Hayashi *et al.*, 1995). The 20S SNARE complex is rapidly disassembled by NSF-dependent hydrolysis of ATP (Barnard *et al.*, 1997). Thus cycling assembly of SNARE proteins in SNARE complex and disassembly is essential in the exocytosis process. Clostridial neurotoxins only cleave SNARE proteins when disassembled, but not in complex form. When an individual SNARE protein is cleaved by a clostridial neurotoxin, the SNARE complex formation is not inhibited but its stability is reduced and the release of neurotransmitter is blocked (Fig. 1.3) (Hayashi *et al.*, 1995; Pellegrini *et al.*, 1995). Even the physiological properties induced by the cleavage of either VAMP, SNAP-25 or syntaxin are not equivalent at the neuromuscular junctions, all the clostridial neurotoxins cause a blockade of the regulated neurotransmission which varies in intensity and duration according to each neurotoxin type (review in Humeau *et al.*, 2000; Meunier *et al.*, 2002a).

The remodelling of neuromuscular junction after clostridial neurotoxin treatment has been mainly studied with BoNT/A. After neuromuscular junction paralysis induced by BoNT/A, outgrowth of intramuscular axons spread out from the node of Ranvier (nodal sprouting) and at motor nerve terminals (terminal sprouting). Sprouts are observed within 24 hours of muscle inactivity and increases in number and complexity for about 50 days after BoNT/A treatment (Juzans *et al.*, 1996). Newly formed sprouts are able to spontaneous and regulated release of acetylcholine and permit the recovery of neuromuscular transmission after sublethal BoNT/A intoxication. Sprouts persist even when muscle recovers normal activity, but are eliminated when the originally intoxicated motor nerve terminal becomes functional (de Paiva *et al.*, 1999). Muscle derived signalling factors such as growth factors IGF-1 and IGF-2 are possibly involved in the induction of sprouting (reviewed in Meunier *et al.*, 2002b). The negative control of sprouting by the original endplate is not well known, neural-cell adhesion molecule (N-CAM) and tenascin-C are possible regulators (reviewed in Meunier *et al.*, 2002a).

Duration of BoNT intoxication is variable according to the toxinotypes. Half lives of inhibition of neurotransmitter release in rat cerebellar neurons for BoNT/A, BoNT/C1, BoNT/B, BoNT/F, and BoNT/E are estimated to >31, >25, 10, 2 and 0.8 days respectively, in agreement with the neuromuscular paralysis times found in mice (Foran *et al.*, 2003). The intoxication process is much longer with BoNT/A than with BoNT/E even both toxins recognize the same intracellular target, SNAP-25. Removal of nine C-terminal residues from SNAP-25 by BoNT/A does not prevent the formation of SNARE complex but its stability is reduced. In contrast, the removal of 17 C-terminal residues from SNAP-25 by BoNT/E totally inhibits the SNARE complex assembly (Hayashi *et al.*, 1995; Eleopra *et al.*, 1998; Raciborska and Charlton, 1999; Bajohrs *et al.*, 2004; Rickman *et al.*, 2004). A first hypothesis is that the most modified form of SNAP-25 by BoNT/E is rapidly degraded and replaced by newly synthesized SNAP-25 allowing a rapid recovery, whereas the limited truncated SNAP-25 by BoNT/A is more slowly recognized by the cell reparation mechanism (Eleopra *et al.*, 1998). However, it has been found that BoNT/A proteolytic activity inside neurons persists for a long period (> 11 weeks in mouse spinal cord neurons), compared with that of BoNT/E (<3 weeks) (Keller *et al.*, 1999). BoNT/A L chain localizes to the plasma membrane with cleaved SNAP-25, whereas BoNT/E L chain has a cytoplasmic distribution. A specific motif in the N-terminus of BoNT/A L chain supports the intracellular localization, and a dileucine motif in the C-terminus is involved in toxin trafficking (Fernandez-Salas *et al.*, 2004). Persistence of BoNTs in neurons is probably the basis of their activity duration.

Concluding remarks

Homologies at the amino acid and protein structure levels between BoNTs and TeNT suggest that their genes have evolved from a common ancestor. Neurotoxin gene localization on mobile structures and/or mobile DNA elements accounts for gene transfer between strains from intra and inter *Clostridium* species. It is noteworthy that TeNT is unique and is produced by a homogeneous *Clostridium* species, whereas *bont* genes show a high degree of variability and are harboured by various *Clostridium* species and strains.

The production of the clostridial neurotoxins is a regulated process, in which BotR and TetR are required. BotR and TetR are alternative RNA polymerase sigma factors from group 5 of the σ^{70} family, containing also TcdR and UviA from *C. difficile* and *C. perfringens*, respectively. Therefore, *C. botulinum* and *C. tetani* probably share a common mechanism of toxin gene expression, which is probably related to that of toxin and bacteriocin synthesis in *C. difficile* and *C. perfringens*, respectively. Additional regulatory factors such as two-component systems in response to unknown signal, and/or environmental or nutritional factors are also possibly involved in the toxinogenesis. Unravelling the mechanism of toxin regulation is essential to optimize the neurotoxin production for vaccine preparation or for therapeutic use, and to develop new strategies to control botulism.

References

Antharavally, B.S., and DasGupta, B.R. (1997). Covalent structure of botulinum neurotoxin type E: location of sulfhydryl groups, and disulfide bridges and identification of C-termini of light and heavy chains. J. Protein Chem. *16*, 787–799.

Arimitsu, H., Inoue, K., Sakaguchi, Y., Lee, J., Fujinaga, Y., Watanabe, T., Ohyama, T., Hirst, R., and Oguma, K. (2003). Purification of fully activated *Clostridium botulinum* serotype B toxin for treatment of patients with dystonia. Infect. Immun. *71*, 1599–1603.

Arndt, E.R., Jacobson, M.J., Abola, E.E., Forsyth, C.M., Tepp, W.H., Marks, J.D., Johnson, E.A., and Stevens, E.S. (2006). A structural perspective of the sequence variability within botulinum neurotoxin subtypes A1–A4. J. Mol. Biol. *362*, 733–742.

Arndt, J.W., Gu, J., Jaroszewski, L., Schwarzenbacher, R., Hanson, M.A., Lebeda, F.J., and Stevens, R.C. (2005). The structure of the neurotoxin-associated protein HA33/A from *Clostridium botulinum* suggests a reoccurring beta-trefoil fold in the progenitor toxin complex. J. Mol. Biol. *346*, 1083–1093.

Bajohrs, M., Rickman, C., Binz, T., and Davletov, B. (2004). A molecular basis underlying differences in the toxicity of botulinum serotypes A and E. EMBO Rep. *5*, 1090–1095.

Baldwin, M.R., Bradshaw, M., Johnson, E.A., and Barbieri, J.T. (2004). The C-terminus of botulinum neurotoxin type A light chain contributes to solubility, catalysis and stability. Protein Express. Purif. *37*, 187–195.

Barash, J.R., and Arnon, S.S. (2004). Dual toxin-producing strain of *Clostridium botulinum* type Bf isolated from a California patient with infant botulism. J. Clin. Microbiol. *42*, 1713–1715.

Barnard, R.J.O., Morgan, A., and Burgoyne, R.D. (1997). Stimulation of NSF ATPase activity by alpha-SNAP is required for SNARE complex disassembly and exocytosis. J. Cell Biol. *139*, 875–883.

Bennik, M.H.J., Mauchline, M., Bosveld, F., Elmore, M., Minton, N.P., Parkhill, J., and Peck, M.W. (2003). The genome sequence of proteolytic *Clostridium botulinum* ATCC3502 (Hall A): some highlights. Clostridial 03 Pathogenesis, Woods Hole, 44.

Bhandari, M., Campbell, K.D., Collins, M.D., and East, A.K. (1997). Molecular characterization of the clusters of genes encoding the botulinum neurotoxin complex in *Clostridium botulinum (Clostridium argentinense)* type G and nonproteolytic *Clostridium botulinum* type B. Curr. Microbiol. 35, 207–214.

Bigalke, H., and Shoer, L.F. (2000). Clostridial neurotoxins. In: Bacterial Protein Toxins, eds. K. Aktories and I. Just, Berlin: Springer, 407–443.

Binz, T., Bade, S., Rummel, A., Kollewe, A., and Alves, J. (2002). Arg(326) and Tyr(365) of the botulinum neurotoxin type A light chain are involved in transition state stabilization. Biochemistry *41*, 1717–1723.

Binz, T., Kurazono, H., Popoff, M.R., Eklund, M.W., Sakaguchi, G., Kozaki, S., Krieglstein, K., Henschen, A., Gill, D.M., and Niemann, H. (1990). Nucleotide sequence of the gene encoding *Clostridium botulinum* neurotoxin type D. Nucleic Acids Res. *18*, 5556.

Bradshaw, M., Dineen, S.S., Maks, N.D., and Johnson, E.A. (2004). Regulation of neurotoxin complex expression in *Clostridium botulinum* strains 62A, Hall A-*hyper*, and NCTC2916. Anaerobe *10*, 321–333.

Brefort, G., Magot, M., Ionesco, H., and Sebald, M. (1977). Characterization and transferability of *Clostridium perfringens* plasmids. Plasmid *1*, 52–66.

Brüggemann, H., Bäumer, S., Fricke, W.F., Wiezr, A., Liesagang, H., Decker, I., Herzberg, C., Martinez-Arias, R., Henne, A., and Gottschalk, G. (2003). The genome sequence of *Clostridium tetani*, the causative agent of tetanus disease. Proc. Natl. Acad. Sci. USA *100*, 1316–1321.

Bullough, P.A., Hughson, F.M., Skehel, J.J., and Wiley, D.C. (1994). Structure of influenza haemagglutinin at the pH of membrane fusion. Nature (London) *371*, 37–43.

Caccin, P., Rossetto, O., Rigoni, M., Johnson, E.A., Schiavo, G., and Montecucco, C. (2003). VAMP/synaptobrevin cleavage by tetanus and botulinum neurotoxins is strongly enhanced by acidic liposomes. FEBS Lett. 542, 132–136.

Cai, S., and Singh, B.R. (2001). Role of the disulfide cleavage induced molten globule state of type A botulinum neurotoxin in its endopeptidase activity. Biochemistry *40*, 15327–15333.

Cato, E.P., George, W.L., and Finegold, S.M. (1986). Genus *Clostridium*. In: Bergey's Manual of Systematic Bacteriology, vol. 2, eds. P.H.A. Sneath, N.S. Mair, M.E. Sharpe, and J.G. Holt, Baltimore, MD: Williams and Wilkins, 1141–1200.

Chai, Q., Arndt, J.W., Dong, M., Tepp, W.H., Johnson, E.A., Chapman, E.R., and Stevens, R.C. (2006). Structural basis of cell surface receptor recognition by botulinum neurotoxin B. Nature *444*, 1096–1100.

Collins, M.D., and East, A.K. (1997). Phylogeny and taxonomy of the food-borne pathogen *Clostridium botulinum* and its neurotoxins. J. Appl. Microbiol. *84*, 5–17.

Cordoba, J.J., Collins, M.D., and East, A.K. (1995). Studies on the genes encoding botulinum neurotoxin type A of *Clostridium botulinum* from a variety of sources. System. Appl. Microbiol. *18*, 13–22.

Couesnon, A., Pereira, Y., and Popoff, M.R. (2008). Receptor-mediated transcytosis of botulinum neurotoxin A through intestinal cell monolayers. Cell. Microbiol. *10*, 375–387.

Couesnon, A., Raffestin, S., and Popoff, M.R. (2006). Expression of botulinum neurotoxins A and E, ans associated non-toxin genes, during the transition phase and stability at high temperature: analysis by quantitative reverse transcription-PCR. Microbiol. *152*, 759–770.

de Paiva, A., Meunier, F., Molgo, J., Aoki, K.R., and Dolly, J.O. (1999). Functional repair of motor endpates after botulinum neurotoxin type A poisoning: biphasic switch of synaptic activity between nerve sprouts and their parent terminals. Proc. Natl. Acad. Sci. USA *96*, 3200–3205.

Deinhardt, K., Berminghausen, O., Willison, H.J., Hopkins, C.R., and Schiavo, G. (2006). Tetanus toxin is internalized by a sequential clathrin-dependent mechanism initiated within lipid microdomains and independent of epsin1. J. Cell Biol. *174*, 459–471.

Dineen, S.S., Bradshaw, M., and Johnson, E.A. (2003). Neurotoxin gene clusters in *Clostridium botulinum* type A strains: sequence comparison and evolutionary implications. Curr. Microbiol. *46*, 342–352.

Dineen, S.S., Bradshaw, M., Karasek, C.E., and Johnson, E.A. (2004). Nucleotide sequence and transcriptional analysis of the type A2 neurotoxin gene cluster in *Clostridium botulinum*. FEMS Microbiol. Lett. *235*, 9–16.

Dong, M., Richards, D.A., Goodnough, M.C., Tepp, W.H., Johnson, E.A., and Chapman, E.R. (2003). Synaptotagmins I and II mediate entry of botulinum neurotoxin B into cells. J. Cell Biol. *162*, 1293–1303.

Dong, M., Yeh, F., Tepp, W.H., Dean, C., Johnson, E.A., Janz, R., and Chapman, E.R. (2006). SV2 is the protein receptor for botulinum neurotoxin A. Science *312*, 592–596.

Dupuy, B., Mani, N., Katayama, S., and Sonenshein, A.L. (2005). Transcription activation of a UV-inducible *Clostridium perfringens* bacteriocin gene by a novel sigma factor. Mol. Microbiol. 55, 1196–1206.

Dupuy, B., Raffestin, S., Matamouros, S., Mani, N., Popoff, M.R., and Sonenshein, A.L. (2006). Regulation of toxin and bacteriocin gene expression in *Clostridium* by interchangeable RNA polymerase sigma factors. Mol. Microbiol. *60*, 1044–1057.

Dupuy, B., and Sonenshein, A.L. (1998). Regulated transcription of *Clostridium difficile* toxin genes. Mol. Microbiol. *27*, 107–120.

East, A.K., Bhandari, M., Stacey, J.M., Campbell, K.D., and Collins, M.D. (1996). Organization and phylogenetic interrelationships of genes encoding components of the botulinum toxin complex in proteolytic *Clostridium botulinum* types A, B, and F: evidence of chimeric sequences in the gene encoding the nontoxic nonhemagglutinin component. Int. J. Syst. Bacteriol. *46*, 1105–1112.

Eisel, U., Jarausch, W., Goretzki, K., Henschen, A., Engels, J., Weller, U., Hudel, M., Habermann, E., and Niemann, H. (1986). Tetanus toxin: primary structure, expression in *E. coli*, and homology with botulinum toxins. EMBO J. 5, 2495–2502.

Eklund, M.W., and Dowell, J. (1987). Avian botulism. Charles C. Thomas: Springfield, IL.

Eklund, M.W., and Poysky, F.T. (1974). Interconversion of type C and D strains of *Clostridium botulinum* by specific bacteriohages. Appl. Microbiol. *27*, 251–258.

Eklund, M.W., Poysky, F.T., and Habig, W.H. (1989). Bacteriophages and plasmids in *Clostridium botulinum* and *Clostridium tetani* and their relationship to production of toxins. In: Botulinum Neurotoxin and Tetanus Toxin., ed. L.L. Simpson, San Diego: Academic Press, 25–51.

Eklund, M.W., Poysky, F.T., Meyers, J.A., and Pelroy, G.A. (1974). Interspecies conversion of *Clostridium botulinum* type C to *Clostridium novyi* type A by bacteriophage. Science *172*, 480–482.

Eklund, M.W., Poysky, F.T., Mseitif, L.M., and S., S.M. (1988). Evidence for plasmid-mediated toxin and bacteriocin production in *Clostridium botulinum* type G. Appl. Environ. Microbiol. *54*, 1405–1408.

Eklund, M.W., Poysky, F.T., Reed, S.M., and Smith, C.A. (1971). Bacteriophage and the toxigenicity of *Clostridium botulinum* type C. Science *172*, 480–482.

Eleopra, R., Tugnoli, V., Rossetto, O., De Grandis, D., and Montecucco, C. (1998). Different time courses of recovery after poisoning with botulinum neurotoxin serotypes A and E in humans. Neurosci. Lett. *256*, 135–138.

Emsley, P., Fotinou, C., Black, I., Fairweather, N.F., Charles, I.G., Watts, C., Hewitt, E., and Isaacks, N.W. (2000). The structures of the Hc fragment of Tetanus Toxin with carbohydrate subunit complexes provide insight into ganglioside binding. J. Biol. Chem. *275*, 8889–8894.

Fenicia, L., Franciosa, G., Pourshaban, M., and Aureli, P. (1999). Intestinal toxemia botulism in two young people, caused by *Clostridium butyricum* Type E. Clin. Infect. Dis. *29*, 381–387.

Fernandez-Salas, E., Steward, L.E., Ho, H., Garay, P.E., Sun, S.W., Gilmore, M.A., Ordas, J.V., Wang, J., Francis, J., and Aoki, K.R. (2004). Plasma membrane localization signals in the light chain of botulinum neurotoxin. Proc. Natl. Acad. Sci. USA *101*, 3208–3213.

Finn, C.W., Silver, R.P., Habig, W.H., Hardegree, M.C., Zon, G., and Garon, C.F. (1984). The structural gene for tetanus neurotoxin is on a plasmid. Science *224*, 881–884.

Foran, P.G., Mohammed, N., Lisk, G.O., Nagwaney, S., Lawrence, G.W., Johnson, E., Smith, L., Aoki, K.R., and Dolly, O.J. (2003). Evaluation of the therapeutic usefulness of botulinum neurotoxin B, C1, E and F compared with the long lasting type A. J. Biol. Chem. *278*, 1363–1371.

Fotinou, C., Emsley, P., Black, I., Ando, H., Ishida, H., Kiso, M., Sinha, K.A., Fairweather, N.F., and Isaacs, N.W. (2001). The crystal structure of Tetanus Toxin Hc fragment complexed with a synthetic GT1b analogue suggests cross-linking between ganglioside receptors and the toxin. J. Biol. Chem. *276*, 3274–3281.

Franciosa, G., Fenicia, L., Pourshaban, M., and Aureli, P. (1997). Recovery of a strain of *Clostridium botulinum* producing both neurotoxin A and neurotoxin B from canned macrobiotic food. Appl. Environ. Microbiol. *63*.

Franciosa, G., Ferreira, J.L., and Hatheway, C.L. (1994). Detection of type A, B, and E botulism neurotoxin genes in *Clostridium botulinum* and other *Clostridium* species by PCR: evidence of unexpressed type B toxin genes in type A toxigenic organisms. J. Clin. Microbiol. *32*, 1911–1917.

Franciosa, G., Floridi, F., Maugliani, A., and Aureli, P. (2004). Differentitation of the gene clusters encoding botulinum neurotoxin type A complexes in *Clostridium botulinum* type A, Ab, and A(B) strains. Appl Environ Microbiol *70*, 7192–7199.

Franciosa, G., Maugliani, A., Floridi, F., and Aureli, P. (2006). A novel type A2 neurotoxin gene cluster in *Clostridium botulinum* strain Mascarpone. FEMS Microbiol. Lett. *261*, 88–94.

Fujinaga, Y., Inoue, K., Watarai, S., Sakaguchi, G., Arimitsu, H., Lee, J.C., Jin, Y., Matsumura, T., Kabumoto, Y., Watanabe, T., Ohyama, T., Nishikawa, A., and Oguma, K. (2004). Molecular characterization of binding subcomponents of *Clostridium botulinum* type C progenitor toxin for intestinal epithelial cells and erythrocytes. Microbiology *150*, 1529–1538.

Fujinaga, Y., Takeshi, K., Inoue, K., Fujita, R., Ohyama, T., Moriishi, K., and Oguma, K. (1995). Type A and B neurotoxin genes in a *Clostridium botulinum* type AB strain. Biochem. Biophys. Res. Commun. *213*, 737–745.

Fujita, R., Fujinaga, Y., Inoue, K., Nakajima, H., Kumon, H., and Oguma, K. (1995). Molecular characterization of two forms of nontoxic-non hemagglutinantinin components of *Clostridium botulinum* type A progenitor toxins. FEBS Lett. *376*, 41–44.

Gurtler, V., Wilson, V.A., and Mayall, B.C. (1991). Classification of medically important clostridia using restriction endonuclease site differences of PCR-amplified 16s rDNA. J. Gen. Microbiol. *137*, 2673–2679.

Hall, J.D., McCroskey, L.M., Pincomb, B.J., and Hatheway, C.L. (1985). Isolation of an organism resembling *Clostridium barati* which produces type F botulinal toxin from an infant with botulism. J. Clin. Microbiol. *21*, 654–655.

Hauser, D., Gibert, M., Boquet, P., and R., P.M. (1992). Plasmid localization of a type E botulinal neurotoxin gene homologue in toxigenic *Clostridium butyricum* strains, and absence of this gene in non-toxigenic *C. butyricum* strains. FEMS Microbiol. Let. *99*, 251–256.

Hauser, D., Gibert, M., Eklund, M.W., Boquet, P., and Popoff, M.R. (1993). Comparative analysis of C3 and botulinal neurotoxin genes and their environment in *Clostridium botulinum* C and D. J. Bacteriol. *175*, 7260–7268.

Hauser, D., Gibert, M., Marvaud, J.C., Eklund, M.W., and Popoff, M.R. (1995). Botulinal neurotoxin C1 complex, Clostridial neurotoxin homology and genetic transfer in *Clostridium botulinum*. Toxicon 33, 515–526.

Hayashi, T., Yamasaki, S., Nauenburg, S., Binz, T., and Niemann, H. (1995). Disassembly of the reconstituted synaptic vesicle membrane fusion complex *in vitro*. EMBO J. *14*, 2317–2325.

Helmann, J.D. (2002). The extracytoplasmic function (ECF) sigma factors. Adv. Microbial. Physiol. *46*, 47–110.

Henderson, I., Davis, T., Elmore, M., and Minton, N. (1997). The genetic basis of toxin production in *Clostridium botulinum* and *Clostridium tetani*. In: The *Clostridia*: Molecular Biology and Pathogenesis, ed. I. Rood, New York: Academic Press, 261–294.

Henderson, I., Whelan, S.M., Davis, T.O., and Minton, N.P. (1996). Genetic characterization of the botulinum toxin complex of *Clostridium botulinum* strain NCTC2916. FEMS Microbiol. Lett. *140*, 151–158.

Herreros, J., Lalli, G., Montecucco, C., and Schiavo, G. (1999). Pathophysiological properties of clostridial neurotoxins. In: The Comprehensive Sourcebook of Bacterial Protein Toxins, vol. 2, eds. J.E. Alouf and J.H. Freer, London: Academic Press, 202–228.

Herreros, J., Lalli, G., and Schiavo, G. (2000). C-terminal half of tetanus toxin fragment C is sufficient for neuronal binding and interaction with a putative protein receptor. Biochem. J. *347*, 199–204.

Hielm, S., Björkroth, J., Hyytiä, E., and Korkeala, H. (1998a). Prevalence of *Clostridium botulinum* in Finnish trout farms: pulse-field gel electrophoresis typing reveals extensive genetic diversity among Type E isolates. Apl. Environ. Microbiol. *64*, 4161–4167.

Hielm, S., Björkroth, K.J., Hyytia, E., and Korkeala, H. (1998b). Genomic analysis of *Clostridium botulinum* group II by pulsed-field gel electrophoresis. Appl Environ Microbiol *64*, 703–708.

Hill, K.K., Smith, T.J., Helma, C.H., Ticknor, L.O., Foley, B.T., Svensson, R.T., Brown, J.L., Johnson, E.A., Smith, L.A., Okinaka, R.T., Jackson, P.J., and Marks, J.D. (2007). Genetic diversity among botulinum neurotoxin-producing clostridial strains. J. Bacteriol. *189*, 818–832.

Hippe, H., Andreesen, J.R., and Gottschalk, G. (1992). The genus *Clostridium*-Nonmedical. In: The Prokaryotes, vol. II, eds. A. Balows, H.G. Trüper, M. Dworkin, W. Harder, and K.H. Scheifer, New York: Springer Verlag, 1800–1866.

Hirst, T.R. (1995). Biogenesis of cholera toxin and related oligomeric toxins. In: Bacterial Toxins and Virulence Factors in Disease, eds. J. Moss, B. Iglewsski, M. Vaughan, and A.T. Tu, New York: Marcel Dekker, 123–184.

Humeau, Y., Doussau, F., Grant, N.J., and Poulain, B. (2000). How botulinum and tetanus neurotoxins block neurotransmitter Biochimie *82*, 427–446.

Hundsberger, T., Braun, V., Weidmann, M., Leukel, P., Sauerborn, M., and von Eichel-Streiber, C. (1997). Transcription analysis of the genes *tcdA-E* of the pathogenicity locus of *Clostridium difficile*. Eur. J. Biochem. *244*, 735–742.

Hutson, R.A., Collins, M.D., East, A.K., and Thompson, D.E. (1994). Nucleotide sequence of the gene coding for non-proteolytic *Clostridium botulinum* type B neurotoxin: comparison with other Clostridial neeurotoxins. Cur. Microbiol. *28*, 101–110.

Hutson, R.A., Zhou, Y., Collins, M.D., Johnson, E.A., Hatheway, C.L., and Sugiyama, H. (1996). Genetic characterization of *Clostridium botulinum* type A containing silent type B neurotoxin gene sequences. J. Biol. Chem. *271*, 10786–10792.

Hyytia, E., Björkroth, K.J., Hielm, S., and Korkeala, H. (1999a). Characterization of *Clostridium botulinum* group I and II by randomly amplified polymorphic DNA analysis and repetitive element sequence-based PCR. Int J Food Microbiol *48*, 179–189.

Hyytia, E., Hielm, S., Björkroth, J., and Korkeala, H. (1999b). Biodiversity of *Clostridium botulinum* Type E strains isolated from fish and fishery products. Appl. Environ. Microbiol. *65*, 2057–2064.

Ihara, H., Kohda, T., Morimoto, F., Tsukamoto, K., Karasawa, T., Nakamura, S., Mukamoto, M., and Kozaki, S. (2003). Sequence of the gene for *Clostridium botulinum* type B neurotoxin associated with infant botulism, expression of the C-terminal half of heavy chain and its binding activity. Biochim. Biophys. Acta *1625*, 19–26.

Inoue, K., Fujinaga, Y., Honke, K., Yokota, K., Ikeda, T., Ohyama, T., Takeshi, K., Watanabe, T., Inoue, K., and Oguma, K. (1999). Characterization of haemagglutinin activity of *Clostridium botulinum* type C and D 16S toxins, and one subcomponent of haemagglutinin (HA1). Microbiol. *145*, 2533–2542.

Inoue, K., Fujinaga, Y., Watanabe, T., Ohyama, T., Takeshi, K., Moriishi, K., Nakajima, H., Inoue, K., and Oguma, K. (1996). Molecular composition of *Clostridium botulinum* type A progenitor toxins. Infect. Immun. *64*, 1589–1594.

Inoue, K., Fujnaga, Y., Honke, K., Arimitsu, H., Mahmut, N., Sakaguchi, G., Ohyama, T., Watanabe, T., Inoue, K., and Oguma, K. (2001). *Clostridium botulinum* type A haemagglutinin positive progenitor toxin (HA^+-PTX) binds to oligosaccharides containing Galβ1–4GlcNAc through one subcomponent of haemagglutinin (HA1). Microbiology *147*, 811–819.

Inoue, K., Sobhany, M., Transue, T.R., Oguma, K., Pedersen, L.C., and Negishi, M. (2003). Structural analysis by X-ray crystallography and calorimetry of a haemagglutinin component (HA1) of the progenitor toxin from *Clostridium botulinum*. Microbiology *149*, 3361–3370.

Jin, R., Rummel, A., Binz, T., and Brunger, A.T. (2006). Botulinum neurotoxin B recognizes its protein receptor with high affinity and specificity. Nature *444*, 1092–1095.

Johnson, E.A., Tepp, W., Bradshaw, M., Gilbert, R.J., Cook, P.E., and McIntosh, E.D. (2005). Characterization of *Clostridium botulinum* strains associated with an infant botulism case in the United Kingdom. J. Clin. Microbiol. *43*, 2602–2607.

Johnson, J.L., and Francis, B.S. (1975). Taxonomy of the Clostridia: ribosomal ribonucleic acid homologies among the species. J. Gen. Microbiol. *88*, 229–244.

Jovita, M.R., Collins, M.D., and East, A.K. (1998). Gene organization and sequence determination of the two botulinum neurotoxin gene clusters in *Clostridium botulinum*. Cur. Microbiol. *36*, 226–231.

Juzans, P., Comella, J.X., Molgo, J., Faille, L., and Angaut-Petit, D. (1996). Nerve terminal sprouting in botulinum type-A treated mouse *Levator auris Longus* muscle. Neuromusc. Disord. *6*, 177–185.

Karlsson, S., Burman, L.G., and Akerlund, T. (1999). Suppression of toxin production in *Clostridium difficile* VPI10463 by amino acids. Microbiology *145*, 1683–1693.

Karlsson, S., Dupuy, B., Mukherjee, K., Norin, E., Burman, L.G., and Akerlund, T. (2003). Expression of *Clostridium difficile* toxins A and B and their sigma factor TcdD is controlled by temperature. Infect. Immun. *71*, 1784–1793.

Keller, J.E., Neale, E.A., Oyler, G., and Adler, M. (1999). Persistence of botulinum neurotoxin action in cultured spinal cord cells. FEBS Lett. *456*, 137–142.

Keto-Timonen, R., Nevas, M., and Korkeala, H. (2005). Efficient DNA fingerprint of *Clostridium botulinum* types A, B, E, and F by amplified fragment length polymorphism analysis. Appl Environ Microbiol *71*, 1148–1154.

Kouguchi, H., Watanabe, T., Sagane, Y., Sunagawa, H., and Ohyama, T. (2002). *In vitro* reconstitution of the *Clostridium botulinum* type D progenitor toxin. J. Biol. Chem. *277*, 2650–2656.

Kozaki, S., Kamata, Y., Nishiki, T., Kakinuma, H., Maruyama, H., Takahashi, H., Karasawa, T., Yamakawa, K., and Nakamura, S. (1998). Characterization of *Clostridium botulinum* type B neurotoxin associated with infant botulism in japan. Infect. Immun. *66*, 4811–4816.

Krieglstein, K.G., DasGupta, B.R., and Henschen, A.H. (1994). Covalent structure of botulinum neurotoxin A: location of sulfhydryl bridges and identification of C-termini of light and heavy chains. J. Protein Chem. *13*, 49–57.

Krieglstein, K.G., Henschenn, A.H., Weller, U., and Habermann, E. (1991). Limited proteolysis of tetanus toxin. Relation to activity and identification of cleavage sites. Eur. J. Biochem. *202*, 41–51.

Kubota, T., Shirakawa, S., Kozaki, S., Isogai, E., Isogai, H., Kimura, K., and Fujii, N. (1996). Mosaic structure of the nontoxic-nonhemagglutinating component gene in *Clostridium botulinum* type A strain isolated from infant botulism in Japan. Biochem. Biophys. Res. Commun. *224*, 843–848.

Kukreja, R., and Singh, B. (2005). Biologically active novel conformational state of botulinum, the most poisonous poison. J. Biol. Chem. *280*, 39346–39352.

Kurazono, H., Mochida, S., Binz, T., Eisel, U., Quanz, M., Grebenstein, O., Wernars, K., Poulain, B., Tauc, L., and Niemann, H. (1992). Minimal essential domains specifying toxicity of the light chains of Tetanus Toxin and Botulinum Neurotoxin type A. J. Biol. Chem. *2667*, 14721–14729.

Lacy, D.B., and Stevens, R.C. (1999). Sequence homology and structural analysis of the clostridial neurotoxins. J. Mol. Biol. *291*, 1091–1104.

Lacy, D.B., Tepp, W., Cohen, A.C., Das Gupta, B.R., and Stevens, R.C. (1998). Crystal structure of botulinum neurotoxin type A and implications for toxicity. Nature Struct. Biol. *5*, 898–902.

Lawson, P.A., Llop-Perez, P., Hutson, R.A., Hippe, H., and Collins, M.D. (1993). Towards a phylogeny of the clostridia based on 16S rRNA sequences. FEMS Microbiol. Let. *113*, 87–92.

Li, B., Qian, X., Sarkar, H.K., and Singh, B.R. (1998). Molecular characterization of type E *Clostridium botulinum* and comparison to other types of *Clostridium botulinum*. Biochim. Biophys. Acta *1395*, 21–27.

Li, L., Binz, T., Niemann, H., and Singh, B.R. (2000). Probing the mechanistic role of glutamate residue in the zinc-binding motif of type A botulinum neurotoxin light chain. Biochemistry *39*, 2399–2405.

Lin, W.J., and Johnson, E.A. (1995). Genome analysis of *Clostridium botulinum* type A by pulsed-field gel electrophoresis. Appl. Environ. Microbiol. *61*, 4441–4447.

Lövenklev, M., Artin, I., Hagberg, O., Borch, E., Holst, E., and Radström, P. (2004). Quantitative interaction effects of carbon dioxide, sodium chloride, and sodium nitrite on neurotoxin gene expression in nonproteolytic *Clostridium botulinum* type B. Appl. Environ. Micrcobiol. *70*, 2928–2934.

Mahmut, N., Inoue, K., Fujinaga, Y., Hughes, L., Arimitsu, H., Sakaguchi, G., Ohtsuka, A., Murakami, T., Yokota, K., and Oguma, K. (2002). Characterization of monoclonal antibodies against haemagglutinin associated with *Clostridium botulinum* type C neurotoxin. J. Med. Microbiol. *51*, 286–294.

Mahrhold, S., Rummel, A., Bigalke, H., Davletov, B., and Binz, T. (2006). The synaptic vesicle protein 2C mediates the uptake of botulinum neurotoxin A into phrenic nerves. FEBS Lett. *580*, 2011–2014.

Maksymowych, A.B., Rienhard, M., Malizio, C.J., Goodnough, M.C., Johnson, E.A., and Simpson, L.L. (1999). Pure botulinum neurotoxin is absorbed from the stomach and small intestine and produces peripheral neuromuscular blockade. Infect. Immun. *67*, 4708–4712.

Maksymowych, A.B., and Simpson, L.I. (2004). Structural features of the botulinum neurotoxin molecule that govern binding and transcytosis across polarized human intestinal epithelial cells. J. Pharmacol. Exp. Ther. *210*, 633–641.

Mani, N., and Dupuy, B. (2001). Regulation of toxin synthesis in *Clostridium difficile* by an alternative RNA polymerase sigma factor. Proc. Ntl. Acad. Sci. USA *98*, 5844–5849.

Mani, N., Lyras, D., Barroso, L., Howarth, P., Wilkins, T., Rood, J.I., Sonenshein, A.L., and Dupuy, B. (2002). Environmental response and autoregulation of *Clostridiium difficile* TxeR, a sigma factor for toxin gene expression. J. Bacteriol. *184*, 5971–5978.

Marshall, K.M., Bradshaw, M., Pellet, S., and Johnson, E.A. (2007). Plasmid encoded neurotoxin genes in *Clostridium botulinum* serotype A subtypes. Biochem. Biophys. Res. Commun. *361*, 49–54.

Marvaud, J.C., Eisel, U., Binz, T., Niemann, H., and Popoff, M.R. (1998a). *tetR* is a positive regulator of the Tetanus toxin gene in *Clostridium tetani* and is homologous to *botR*. Infect. Immun. *66*, 5698–5702.

Marvaud, J.C., Gibert, M., Inoue, K., Fujinaga, V., Oguma, K., and Popoff, M.R. (1998b). *botR* is a positive regulator of botulinum neurotoxin and associated non toxic protein genes in *Clostridium botulinum* A. Mol. Microbiol. *29*, 1009–1018.

McCroskey, L.M., Hatheway, C.L., Fenicia, L., Pasolini, B., and Aureli, P. (1986). Characterization of an organism that produces type E botulinal toxin but which resembles *Clostridium butyricum* from the feces of an infant with type E botulism. J. Clin. Microbiol. *23*, 201–202.

McCroskey, L.M., Hatheway, C.L., Woodruff, B.A., Greenberg, J.A., and Jurgenson, P. (1991). Type F botulism due to neurotoxigenic *Clostridium baratii* from an unknown source in an adult. J. Clin. Microbiol. *29*, 2618–2620.

Meng, X., Karasawa, T., Zou, K., Kuang, X., Wang, X., Lu, C., Wang, C., Yamakawa, K., and Nakamura, S. (1997). Characterization of a neurotoxigenic *Clostridium butyricum* strain isolated from the food implicated in an outbreak of food-borne type E botulism. J. Clin. Microbiol. *35*, 2160–2162.

Meng, X., Yamakawa, K., Zou, K., Wang, X., Kuang, X., Lu, C., Wang, C., Karasawa, T., and Nakamura, S. (1999). Isolation and characterization of neurotoxigenic *Clostridium butyricum* from soil in China. J. Med. Microbiol. *48*, 133–137.

Meunier, F.A., Herreros, J., Schiavo, G., Poulain, B., and Molgo, J. (2002a). Molecular mechanism of action of botulinal neurotoxins and the synaptic remodeling they induce *in vivo* at the skeletal neuromuscular junction. In: Handbook of Neurotoxicology, vol. 1, ed. J. Massaro, Totowa, NJ: Humana Press, 305–347.

Meunier, F.A., Schiavo, G., and Molgo, J. (2002b). Botulinum neurotoxins: from paralysis to recovery of functional neuromuscular trasnmission. J. Physiol. *96*, 105–113.

Minton, N.P. (1995). Molecular genetics of clostridial neurotoxins. Curr. Top. Microbiol. Immunol. *195*, 161–194.

Moncrief, J.S., Barroso, L.A., and Wilkins, T.D. (1997). Positive regulation of *Clostridium difficile toxins*. Infect. Immun. *65*, 1105–1108.

Moriishi, K., Koura, M., Aba, N., Fujii, N., Fujinaga, Y., Inoue, K., and Oguma, K. (1996). Mosaic structures of neurotoxins produced from *Clostridium botulinum* types C and D. Biochim. Biophys. Acta *1307*, 123–126.

Mutoh, S., Kouguchi, H., Sagane, Y., Suzuki, T., Hasegawa, K., Watanabe, T., and Ohyama, T. (2003). Complete subunit structure of the *Clostridium botulinum* type D complex via intermediate assembly with nontoxic components. Biochemistry *42*, 10991–10997.

Nevas, M., Lindström, M., Hielm, S., Björkroth, K.J., Peck, M.W., and Korkeala, H. (2005). Diversity of proteolytic *Clostridium botulinum* strains, determined by a pulse-field gel electrophoresis approach. Appl. Environ. Microbiol. *71*, 1311–1317.

Niemann, H. (1991). Molecular biology of Clostridial neurotoxins. In: Sourcebook of Bacterial Protein Toxins, eds. J.E. Alouf and J. Freer, New York: Academic Press, 299–344.

Nishiki, T., Kamata, Y., Nemoto, Y., Omori, A., Ito, T., Takahashi, M., and Kozaki, S. (1994). Identification of protein receptor for *Clostridium botulinum* type B neurotoxin in rat brain synaptosomes. J. Biol. Chem. *269*, 10498–10503.

Nishiki, T., Tokuyama, Y., Kamata, Y., Nemoto, Y., Yoshida, A., Sato, K., Sekigichi, M., Taakahashi, M., and Kozaki, S. (1996). The high-affinity of *Clostridium botulinum* type B neurotoxin to synaptotagmin II associated with gangliosides G_{T1B}/G_{D1a}. FEBS Lett. *378*, 253–257.

Oguma, K., Inoue, K., Fujinaga, Y., Yokota, K., Watanabe, T., Ohyama, T., Takeshi, K., and Inoue, K. (1999). Structure and function of *Clostridium botulinum* progenitor toxin. J. Toxicol. *18*, 17–34.

Patterson-Curtis, S.I., and Johnson, E.A. (1989). Regulation of neurotoxin and protease formation in *Clostridium botulinum* Okra B and Hall by arginine. Appl. Environ. Microbiol. *55*, 1544–1548.

Pellegrini, L.L., O'Connor, V., Lottspeich, F., and Betz, H. (1995). Clostridial neurotoxins compromise the stability of a low energy SNARE complex mediating NSF activation of synaptic vesicle fusion. EMBO J. *14*, 4705–4713.

Poirier, M.A., Xiao, W., Macosko, J.C., Chan, C., Shin, Y.K., and Bennett, M.K. (1998). The synaptic SNARE complex is a parallel four-stranded helical bundle. Nat. Struct. Biol. *5*, 765–769.

Popoff, M.R. (1995). Ecology of neurotoxigenic strains of *Clostridia*. In: Clostridial neurotoxins, vol. 195, ed. C. Montecucco, Heidelberg: Springer-Verlag, 1–29.

Popoff, M.R., and Marvaud, J.C. (1999). Structural and genomic features of clostridial neurotoxins. In: The Comprehensive Sourcebook of Bacterial Protein Toxins, vol. 2, eds. J.E. Alouf and J.H. Freer, London: Academic Press, 174–201.

Porfirio, Z., Prado, S.M., Vancetto, M.D.C., Fratelli, F., Alves, E.W., Raw, I., Fernandes, B.L., Camargo, A.C.M., and Lebrun, I. (1997). Specific peptides of casein pancreatic digestion enhance the production of tetanus toxin. J. Appl. Microbiol. *83*, 678–684.

Poulain, B., Dousseau, F., Colasante, C., Deloye, F., and Molgo, J. (1997). Cellular and molecular mode of action of botulinum and tetanus neurotoxins. Adv. Organ Biol. *2*, 285–313.

Poulain, B., Stiles, B.G., Popoff, M.R., and Molgo, J. (2006). Attack of the nervous system by clostridial toxins: Physical findings, cellular and molecular actions. In: The Sourcebook of Bacterial Protein Toxins, eds. J.E. Alouf and M.R. Popoff, Amsterdam: Elsevier, Academic Press, 348–389.

Poulet, S., Hauser, D., Quanz, M., Niemann, H., and Popoff, M.R. (1992). Sequences of the botulinal neurotoxin E derived from *Clostridium botulinum* type E (strain Beluga) and *Clostridium butyricum* (strains ATCC43181 and ATCC43755). Biochem. Biophys. Res. Commun. *183*, 107–113.

Quinn, C.P., and Minton, N.P. (2001). Clostridial neurotoxins. In: Clostridia, eds. H. Bahl and P. Dürre, Veinheim: Wiley-VCH, 211–250.

Raciborska, D., and Charlton, M. (1999). Retention of cleaved synaptosome-associated protein of 25 kDa (SNAP25) in neurouscular junctions: a new hypothesis to explain persistence of botulinum A poisoning. Can. J. Physiol. Pharmacol. *77*, 679–688.

Raffestin, S., Dupuy, B., Marvaud, J.C., and Popoff, M.R. (2005). BotR/A and TetR are alternative RNA polymerase sigma factors controlling the expression of the neurotoxin and associated protein genes in *Clostridium botulinum* type A and *Clostridium tetani*. Mol. Microbiol. 55, 235–249.

Rickman, C., Meunier, F.A., Binz, T., and Davletov, B. (2004). High affinity interaction of syntaxin and SNAP-25 on the plasma membrane is abolished by botulinum toxin E. J. Biol. Chem. *279*, 644–651.

Rigoni, M., Caccin, P., Johnson, E.A., Montecucco, C., and Rossetto, O. (2001). Site-directed mutagenesis identifies active-site residues of the light chain of botulinum neurotoxin A. Biochem. Biophys. Res. Commun. *288*, 1231–1237.

Rossetto, O., Caccin, P., Rigoni, M., Tonello, F., Bortoletto, N., Stevens, R.C., and Montecucco, C. (2001). Active-site mutagenesis of tetanus neurotoxin implicates Tyr-375 and Glu-271 in metalloproteolytic activity. Toxicon 39, 1151–1159.

Rummel, A., Bade, S., Alves, J., Bigalke, H., and Binz, T. (2003). Two carbohydrate binding sites in the H_{cc}-domain of tetanus neurotoxin are required for toxicity. J. Mol. Biol. *326*, 835–847.

Rummel, A., Eichner, T., Weil, T., Karnath, T., Gutcaits, A., Mahrhold, S., Sandhoff, K., Proia, R.L., Acharya, K.R., Bigalke, H., and Binz, T. (2007). Identification of the protein receptor binding site of botulinum neurotoxins B and G proves the double-receptor concept. Proc. Natl. Acad. Sci. USA *104*, 359–364.

Rummel, A., Karnath, T., Henke, T., Bigalke, H., and Binz, T. (2004). Synaptotagmins I and II act as nerve cell receptors for botulinum neurotoxin G. J. Biol. Chem. *279*, 30865–30870.

Sagane, Y., Kouguchi, H., Watanabe, T., Sunagawa, H., Inoue, K., Fujinaga, Y., Oguma, K., and Ohyama, T. (2001). Role of C-terminal region, of HA-33 component of botulinum toxin in hemagglutination. Biochem. Biophys. Res. Commun. *288*, 650–657.

Sagane, Y., Watanabe, T., Kouguchi, H., Sunagawa, H., Inoue, K., Fujinaga, Y., Oguma, K., and Ohyama, T. (1999). Dichain structure of botulinum neurotoxin: identification of cleavage sites in types C, D, and F neurotoxin molecules. J. Protein Chem. *18*, 885–892.

Sagane, Y., Watanabe, T., Kouguchi, H., Sunagawa, H., Inoue, K., Fujinaga, Y., Oguma, K., and Ohyama, T. (2000). Characterization of nicking of the nontoxic-nonhemagglutinin components of *Clostridium botulinum* types C and D progenitor toxin. J. Protein Chem. *19*, 575–581.

Sakaguchi, Y., Hayashi, H., Kurokawa, K., Nakayama, K., Oshima, K., Fujinaga, Y., Ohnishi, M., Ohtsubo, E., Hattori, M., and Oguma, K. (2005). The genome sequence of *Clostridium botulinum* type C neurotoxin-converting phage and the molecular mechanisms of unstable lysogeny. Proc. Natl. Acad. Sci. USA *102*, 17472–17477.

Santos-Buelga, J., Collins, M.D., and East, A.K. (1998). Characterization of the genes encoding the botulinum neurotoxin complex in a strain of *Clostridium botulinum* producing type B and F neurotoxins. Cur. Microbiol. *37*, 312–318.

Schiavo, G., Matteoli, M., and Montecucco, C. (2000). Neurotoxins affecting neuroexocytosis. Physiol. Rev. *80*, 717–766.

Schiavo, G., Rossetto, O., Santucci, A., DasGupta, B.R., and Montecucco, C. (1992). Botulinum neurotoxins are zinc proteins. J. Biol. Chem. *267*, 23479–23483.

Sebaihia, M., Peck, M.W., Minton, N.P., Thomson, N.R., Holden, M.T., Mitchell, W.J., Carter, A.T., Bentley, S.D., Mason, D.R., Crossman, L., Paul, C.J., Ivens, A., Wells-Bennik, M.H., Davis, I.J., Cerdeno-Tarraga, A.M., Churcher, C., Quail, M.A., Chillingworth, T., Feltwell, T., Fraser, A., Goodhead, I., Hance, Z., Jagels, K., Larke, N., Maddison, M., Moule, S., Mungall, K., Norbertczak, H., Rabbinowitsch, E., Sanders, M., Simmonds, M., White, B., Whithead, S., and Parkhill, J. (2007). Genome sequence of a proteolytic (Group I) *Clostridium botulinum* strain Hall A and comparative analysis of the clostridial genomes. Genome Res. *17*, 1082–1092.

Shapiro, R.E., Spech, C.D., Collins, B.E., Woods, A.S., Cotter, R.J., and Schnaar, R.L. (1997). Identification of a ganglioside recognition domain of tetanus toxin using a novel ganglioside photoaffinity ligand. J. Biol. Chem. *272*, 30380–30386.

Sharma, S.K., Ferreira, J.L., Eblen, B.S., and Whiting, R.C. (2006). Detection of type A, B, E, and F *Clostridium botulinum* neurotoxins in foods by using an amplified enzyme-linked immunosorbent assay with digoxigenin-labeled antibodies. Appl. Environ. Microbiol. *72*, 1231–1238.

Sharma, S.K., Ramzan, M.A., and Singh, B.R. (2003). Separation of the components of type A botulinum neeurotoxin complex by electrophoresis. Toxicon *41*, 321–331.

Smith, L.D.S. (1975). *Clostridium tetani*. In: The Pathogenic Anaerobic Bacteria, ed. L.D.S. Smith, Springfiel, IL: Charles C. Thomas Publisher, 177–201.

Smith, T.J., Lou, J., Geren, N., Forsyth, M., Tsai, R., La Porte, S.L., Tepp, W.H., Bradshaw, M., Johnson, E.A., Smith, L.A., and Marks, J.D. (2005). Sequence variation within botulinum neurotoxin serotypes impacts antibody binding and neutralization. Infect. Immun. *73*, 5450–5457.

Söllner, T., Bennett, M.K., Whiteheart, S.W., Schelleer, R.H., and Rothmann, J.E. (1993a). A protein assembly-disassembly pathway *in vitro* that may correspond to sequential steps of synaptic vesicle docking, activation, and fusion. Cell 75, 409–418.

Söllner, T., Whiteheart, S.W., Brunner, M., Erdjument-Bromage, H., Geromanos, S., Tempst, P., and Rothman, J.E. (1993b). SNAP receptors implicated in vesicle targeting and fusion. Nature (London) *362*, 318–324.

Strom, M.S., Eklund, M.W., and Poysky, F.T. (1984). Plasmids in *Clostridium botulinum* and related species. Appl. Environ. Microbiol. *48*, 956–963.

Suen, J.C., Hatheway, C.L., Steigerwalt, A.G., and Brenner, D.J. (1988). *Clostridium argentinense* sp. nov.: a genetically homogeneous group composed of all strains of *Clostridium botulinum* toxin type G and some nontoxigenic strains previously identified as *Clostridium subterminale* or *Clostridium hastiforme*. Int. J. Syst. Bacteriol. *38*, 375–381.

Sutton, R.B., Fasshauer, D., Jahn, R., and Brunger, A.T. (1998). Crystal structure of a SNARE complex involved in synaptic exocytosis at 2.4 angstrom resolution. Nature *395*, 347–353.

Swaminathan, S., and Eswaramoorthy, S. (2000). Structural analysis of the catalytic and binding sites of *Clostridium botulinum* neurotoxin B. Nat. Struct. Biol. *7*, 693–699.

Tabita, K.S., Sakaguchi, S., kozaki, S., and Sakaguchi, G. (1991). Distinction between *Clostridium botulinum* type A strains associated with food-borne botulism and those with infant botulism in Japan in intraintestinal toxin production in infant mice and some other properties. FEMS Microbiol. Lett. *79*, 251–256.

Tavallaie, M., Chenal, A., Gillet, D., Pereira, Y., Manich, M., Gibert, M., Raffestin, S., Popoff, M.R., and Marvaud, J.C. (2004). Interaction between the two subdomains of the C-terminal part of the botulinum neurotoxin A is essential for the generation of protective antibodies. FEBS Lett. *572*, 299–306.

Umland, T.C., Wingert, L.M., Swaminathan, S., Furey, W.F., Schmidt, J.J., and Sax, M. (1997). The structure of the receptor binding fragment H_c of tetanus neurotoxin. Nat. Struct. Biol. *4*, 788–792.

Wang, X., Maegawa, T., Karazawa, T., Kozaki, S., Tsukamoto, K., Gyobu, Y., Yamakawa, K., Oguma, K., Sakaguchi, Y., and Nakamura, S. (2000). Genetic analysis of Type E botulism toxin-producing *Clostridium butyricum* strains. Appl. Environ. Microbiol. *66*, 4992–4997.

Weickert, M.J., Chambliss, G.H., and Sugiyama, H. (1986). Production of toxin by *Clostridium botulinum* type A strains cured of plasmids. Appl. Environ. Microbiol. *51*, 52–56.

Willems, A., East, A.K., Lawson, P.A., and Collins, M.D. (1993). Sequence of the gene coding for the neurotoxin of *Clostridium botulinum* type A associated with infant botulism: comparison with other clostridial neurotoxins. Res. Microbiol. *144*, 547–556.

Yamakawa, K., Karasawa, T., Kakinuma, H., Maruyama, H., Takahashi, H., and Nakamura, S. (1997). Emergence of *Clostridium botulinum* typeB-like nontoxigenic organisms in a patient with type B infant botulism. J. Clin. Microbiol. 35, 2163–2164.

Yamakawa, K., Karasawa, T., Ohta, T., Hayashi, H., and Nakamura, S. (1998). Inhibition of enhanced toxin production by Clostridium difficile in biotin-limited conditions. J. Med. Microbiol. *47*, 767–771.

Zhou, Y., Sugiyama, H., and Johnson, E.A. (1993). Transfer of neurotoxigenicity from *Clostridium butyricum* to a nontoxigenic *Clostridium botulinum* type E-like strain. Appl. Environ. Microbiol. 59, 3825–3831.

Improved Understanding of the Action and Genetics of *Clostridium perfringens* Enterotoxin Suggests Potential Applications for Cancer Therapy and Drug Delivery

2

Sameera Sayeed, Susan L. Robertson, Justin A. Caserta and Bruce A. McClane

Abstract

Clostridium perfringens enterotoxin (CPE) causes the intestinal symptoms of a common food-borne illness and ~5–15% of all antibiotic-associated diarrhoea cases. In food poisoning isolates, the enterotoxin gene (*cpe*) is usually present on the chromosome, while *cpe* is carried by conjugative plasmids in antibiotic-associated diarrhoea isolates. CPE action involves its binding to claudin receptors, oligomerization/prepore formation, and prepore insertion to form a functional pore that kills cells by apoptosis or oncosis. The C-terminal half of CPE mediates receptor binding, while its N-terminal half is required for oligomerization. CPE/CPE derivatives are being explored for cancer therapy/diagnosis and improved drug delivery.

Introduction

Clostridium perfringens isolates are classified into types A through E based upon their production of four typing toxins, i.e. alpha, beta, epsilon and iota toxins (Petit *et al.*, 1999; Smedley 3rd *et al.*, 2004). About 1–5% of all *C. perfringens* isolates produce another toxin named *C. perfringens* enterotoxin (CPE), which is a 319 amino acid protein with a unique primary sequence. Most CPE-producing *C. perfringens* isolates classify as type A. Those CPE-positive type A isolates are important human pathogens that cause *C. perfringens* type A food poisoning (McClane, 2007), which is one of the most common food-borne illnesses in developed countries, e.g. ~250,000 cases of *C. perfringens* type A food poisoning occur annually in the USA (Mead *et al.*, 1999). CPE-positive type A isolates are also responsible for between 5% and 15% of all cases of non-food-borne human gastrointestinal (GI) diseases, including antibiotic-associated diarrhoea and sporadic diarrhoea (Carman, 1997; McClane, 2007).

C. perfringens food poisoning develops following the ingestion of CPE-positive type A isolates present in contaminated foods (McClane, 2007). Those isolates then sporulate and produce CPE in the small intestines. After binding to enterocytes and exerting its action, as described later, the enterotoxin then causes diarrhoea and abdominal cramping that typically last for about 12–18 hours before self-resolving in healthy people. In elderly or debilitated individuals, this food poisoning can sometimes be fatal (Bos *et al.*, 2005). In contrast, CPE-associated non-food-borne GI disease may persist for days or weeks and often involves more severe diarrhoea than seen with *C. perfringens* type A food poisoning (Carman, 1997). It has been proposed (Borriello *et al.*, 1984) that CPE-associated non-food-borne GI diseases result from infection rather than from overgrowth of normal GI flora *C. perfringens*. However, several recent surveys have detected the presence of low numbers of type A isolates carrying a *cpe* plasmid in the faeces of healthy people in Japan, Finland and the USA (Birkhead *et al.*, 1988; Heikinheimo *et al.*, 2006; Tanaka *et al.*, 2007). Since type A isolates carrying a *cpe* plasmid cause most cases of CPE-associated non-food-borne GI disease (as discussed later), these findings reopen the possibility that some cases of CPE-associated non-food-borne GI diseases may result from overgrowth of enterotoxigenic strains already present in the GI tract (Heikinheimo *et al.*, 2006).

Considerable epidemiological and experimental evidence supports the importance of CPE in the pathogenesis of both *C. perfringens* type A food poisoning and CPE-associated non-food-borne GI diseases. With respect to epidemiology, CPE is absent from faeces of healthy people, but can be routinely detected in the faeces of patients with *C. perfringens* type A food poisoning or CPE-associated non-food-borne GI diseases (Naik and Duncan, 1978; Saito, 1990). In addition, high numbers of CPE-positive type A isolates are often recovered from the faeces of patients with these diseases (but not from faeces of healthy people) or from food vehicles causing a food poisoning outbreak (McClane, 2007). Human volunteer feeding experiments conducted in the 1970s showed that ingestion of highly purified CPE can reproduce the diarrhoea and abdominal cramps of natural CPE-associated GI disease (Skjelkvale and Uemura, 1977). Perhaps the most compelling support for the importance of CPE in the virulence of CPE-positive type A isolates came from experiments showing that inactivation of the enterotoxin gene in either a human food poisoning isolate or a human non-food-borne GI disease isolate produced isogenic mutants that were avirulent in animal models of intestinal disease (Sarker *et al.*, 1999).

Because of its proven virulence role, CPE has been intensively studied. This chapter will describe the action and genetics of CPE, with a particular focus on recent developments on these two topics. The increased understanding of CPE action and genetics has sparked emerging interest in using CPE or CPE fragments for therapy or drug delivery, so those potential practical applications of CPE or its derivatives will be addressed at the conclusion of this chapter.

CPE genetics

The genetics of *C. perfringens* enterotoxin have now been extensively studied, particularly in type A isolates. Pulsed-field gel mapping studies have shown that, in type A isolates, the enterotoxin gene (*cpe*) can be located either on the chromosome or on large plasmids (Brynestad and Granum, 1999; Brynestad *et al.*, 2001; Brynestad *et al.*, 1997; Collie and McClane, 1998; Cornillot *et al.*, 1995; Miyamoto *et al.*, 2002; Sparks *et al.*, 2001). Interestingly, no type D or E isolates carrying chromosomal *cpe* sequences have yet been identified; instead, the functional *cpe* genes of type D isolates and the silent *cpe* sequences of type E isolates (see below) are plasmid-borne (Li *et al.*, 2007; Sayeed *et al.*, 2007). Type B isolates rarely, if ever, carry the *cpe* gene (Miyakawa *et al.*, 2007), while localization of the *cpe* gene in type C isolates is currently under investigation.

The *cpe* gene locus of type A isolates

Cloning studies revealed that the *cpe* gene of type A isolate NCTC8239 has a 957-basepair open reading frame (ORF) that encodes a 35-kDa protein containing 319 amino acids (Czeczulin *et al.*, 1993). The CPE sequence is unique except for some limited homology, of unknown significance, with proteins made by some strains of *Clostridium botulinum* (Melville *et al.*, 1997). The CPE amino acid sequence is identical among most *cpe*$^{+}$ type A isolates (Czeczulin *et al.*, 1993) and most *cpe*$^{+}$ type C or D isolates (Sayeed *et al.*, 2007); Fisher and McClane, unpublished).

Primer extension analyses, RNase T2 protection assays and deletion mutagenesis clearly identified three *cpe* promoters (named P1, P2 and P3), along with a potential fourth promoter, all located immediately upstream of the *cpe* gene (Melville *et al.*, 1994; Zhao and Melville, 1998). The location and sequences of these *cpe* promoters is largely conserved among type A isolates (Melville *et al.*, 1994). DNA sequencing analysis (Zhao and Melville, 1998) showed P1 is similar to consensus *Sig*K-dependent promoters of *B. subtilis*, while the P2 and P3 sequences are similar to consensus *Sig*E-dependent promoters of *B. subtilis*. This similarity is likely important since, i) *Sig*E and *Sig*K are sporulation-associated sigma factors active in the mother cell compartment of *B. subtilis* sporulating cells and ii) promoter mapping studies (Zhao and Melville, 1998) and Northern blot analyses (Czeczulin *et al.*, 1996) demonstrated that *cpe* mRNA is transcribed almost exclusively during sporulation. Similar *cpe* promoter sequence analyses have not yet been performed for CPE-positive *C. perfringens* isolates belonging to other types.

Downstream of the *cpe* ORF in type A isolates is a stem-loop structure followed by an oligo-T tract (Brynestad *et al.*, 1997; Czeczulin *et al.*, 1993). This region may serve as a *rho*-independent transcriptional terminator for the *cpe* gene. It may also influence CPE expression levels by contributing to message stability.

Association of the *cpe* gene with insertion sequences

Sequencing and Southern blot studies revealed strong associations between the *cpe* gene and insertion sequences (IS) in type A isolates. An IS*200*-like element (Lam and Roth, 1983) named IS*1469* is present immediately upstream of both the chromosomal and plasmid *cpe* genes of type A isolates (Brynestad *et al.*, 1997). In addition, a pair of IS*1470* elements (IS*1470* is a member of the IS*30* family) flank the chromosomal *cpe* gene of *C. perfringens* type A isolate NCTC8239 and most other food poisoning isolates (Fig. 2.1). It has been proposed (Brynestad *et al.*, 1997) that this IS*1470*-IS*1469*-*cpe*-IS*1470* region represents a 6.3-kb transposon that has integrated onto the chromosome between two housekeeping genes, namely, *uapC* and *nadC* (Fig. 2.1). Supporting that possibility, circular intermediates containing the *cpe* gene have been detected in cultures of type A isolates carrying a chromosomal *cpe* gene (Brynestad and Granum, 1999). Those findings suggest the possibility of *cpe* gene movement via IS-element-mediated transposition.

Beyond the conserved IS*1469*-*cpe* region, considerable variation exists between the chromosomal and plasmid-borne *cpe* loci. Flanking IS*1470* elements are absent from the plasmid-borne *cpe* locus of type A isolates (Miyamoto *et al.*, 2002). Instead, those *cpe* plasmids carry either an IS*1151* sequence or a defective IS*1470*-like sequence downstream of the *cpe* gene (Miyamoto *et al.*, 2004). As proposed for the chromosomal *cpe* gene (Brynestad *et al.*, 1997), IS sequences may have mediated mobilization and integration of the *cpe* gene onto plasmids. Whether the nearby IS sequences can still mobilize the plasmid-borne *cpe* is unknown at present.

Plasmids carrying the *cpe* gene can be conjugative

Four *C. perfringens* plasmids have now been shown to transfer via conjugation (Bannam *et al.*, 2006; Brynestad *et al.*, 2001; Parsons *et al.*, 2007), including the tetracycline resistance-encoding plasmid pCW3, the CPE-encoding plasmid pCPF4969, and two epsilon toxin-encoding plasmids (named pJIR3118 and pJIR3119). Complete sequencing of pCW3 and pCPF4969

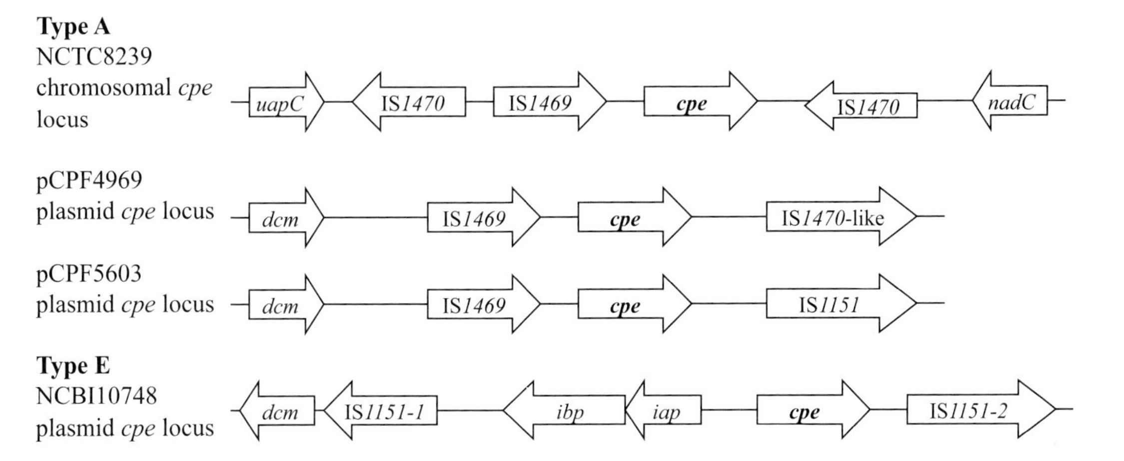

Figure 2.1 Organization of the chromosomal versus plasmid-borne *cpe* locus. The figure shows the organization of the chromosomal *cpe* locus in type A isolate NCTC8239 (top) and the plasmid *cpe* locus in two different type A isolates, F4969 and F5603 (second and third rows). Organization of the silent plasmid *cpe* locus in type E isolate NCBI10748 is shown in the bottom row; note the presence of iota toxin genes (*iap*/*ibp*) near these silent *cpe* sequences. See text for relevant references.

revealed those conjugative plasmids share a locus named *tcp* (transfer clostridial plasmid) comprising 11 genes named *intP* and *tcpA–J* (Bannam *et al.*, 2006; Miyamoto *et al.*, 2006). Several genes of the *tcp* locus encode proteins with similarity to proteins encoded by the conjugative transposon Tn*916* (Clewell *et al.*, 2002), which can conjugatively transfer among most Gram-positive bacteria and can also mobilize plasmids into which it has inserted. Functional genetic studies have demonstrated that several of the *tcp* genes are essential for pCW3 conjugative transfer (Bannam *et al.*, 2006; Parsons *et al.*, 2007). Besides *cpe* plasmids, *tcp* genes have been found on plasmids encoding iota toxin, epsilon toxin and beta toxin of *C. perfringens* (Li *et al.*, 2007; Miyamoto *et al.*, 2006; Sayeed *et al.*, 2007), strongly suggesting that most *C. perfringens* toxin plasmids can transfer conjugatively. If so, horizontal transfer of plasmid-borne virulence genes may be an important contributor to *C. perfringens* virulence evolution and pathogenic diversity.

Complete sequences for two *cpe* gene-carrying plasmids of type A isolates have now been determined (Miyamoto *et al.*, 2006), including the 75.3 kb pCPF5603 (which carries an IS*1151* element downstream of its *cpe* gene) and the 70.5 kb pCPF4969 (which carries IS*1470*-like sequences downstream of its *cpe* gene). These two *cpe* plasmids share a conserved region comprising ~50% of each plasmid, including the *tcp* locus. Given the previously mentioned results from pCW3 functional analyses, the *tcp* locus on *cpe* plasmids probably explains why pCPF4969 and (likely) pCPF5603 are conjugative (Brynestad *et al.*, 2001). The variable region of pCPF5603 carries metabolic genes and a functional *cpb2* gene (encoding beta2 toxin), while the pCPF4969 variable region encodes two bacteriocin loci and a two component regulator related to VirR/VirS. The presence of a functional *cpb2* gene on pCPF5603 opens the possibility of beta2 toxin being an accessory toxin that contributes to the enteric pathogenesis of some non-food-borne human GI disease isolates (Fisher *et al.*, 2005). Surveys have suggested that the *cpe* plasmid present in most type A isolates is closely related to either pCPF5603 or pCPF4969, although a few atypical *cpe* plasmids are now being identified (Heikinheimo *et al.*, 2006; Li *et al.*, 2008).

Type E toxin plasmids carrying a defective *cpe* gene

The *cpe* sequences present in most *C. perfringens* type E isolates are silent due to accumulated mutations in the *cpe* promoter and ORF (Billington *et al.*, 1998). In those type E isolates, iota toxin genes are present immediately upstream of the silent *cpe* sequences (Billington *et al.*, 1998; Li *et al.*, 2007). Furthermore, the iota toxin-silent *cpe* sequences are flanked by IS*1151* sequences (Fig. 2.1). It has been proposed (Billington *et al.*, 1998; Li *et al.*, 2007) that the silent type E *cpe* sequences resulted from insertion of a mobile element carrying the iota toxin genes into the *cpe* promoter region, with subsequent accumulation of additional mutations in the now silenced *cpe* gene.

In type E isolates, the plasmids carrying silent *cpe* genes are very large, ranging from 97 to 130 kb in size (Li *et al.*, 2007). Recent studies suggest this type E toxin plasmid may be conjugative (it carries *tcp*) and that the entire region containing iota toxin-silent *cpe* sequences may be mobilizable from plasmids via adjacent IS sequences (Li *et al.*, 2007).

Recent studies of both healthy human faeces and soil have identified silent *cpe* ORFs in a few type A isolates (Heikinheimo *et al.*, 2006; Li *et al.*, 2008). In some of those atypical isolates, a premature stop codon is present in their *cpe* ORF. In other isolates, an intact *cpe* ORF is present yet the isolate cannot express CPE, possibly due to mutations in their *cpe* promoter region (Li *et al.*, 2008).

The presence of silent *cpe* genes in type E and rare type A isolates has practical implications, i.e. different diagnostic conclusions may be reached using CPE detection assays versus *cpe* PCR assays.

Plasmids carrying functional *cpe* genes in *type D* isolates

As mentioned, the IS*1151*-carrying *cpe* plasmid family found in type A isolates is typically ~75 kb in size, while the IS*1470*-like-carrying *cpe* plasmid family of type A isolates is usually ~70.5 kb (Miyamoto *et al.*, 2006). Plasmids carrying functional *cpe* genes are large in type D isolates, ranging from ~75 to 100 kb in size (Sayeed *et al.*, 2007). Interestingly, a single ~75-kb plasmid present in a few type D isolates can carry the genes encoding CPE, epsilon toxin and beta2 toxin, i.e. it is possible for a single *C. perfringens* plasmid to encode at least three different potent toxins (Sayeed *et al.*, 2007).

CPE expression and secretion

CPE expression is highly regulated at the transcriptional level during sporulation of *C. perfringens* type A isolates (Czeczulin *et al.*, 1996), probably due to the presence of SigE- and SigK-dependent promoters (Melville *et al.*, 1994; Zhao and Melville, 1998). For example, type A isolate NCTC8239 produces at least 1500-fold more CPE during sporulation compared to vegetative growth (Czeczulin *et al.*, 1993). During sporulation, CPE mRNA is transcribed as a monocistronic message (Czeczulin *et al.*, 1996) and CPE protein can represent 10% or more of the total protein present in sporulating cells of some *C. perfringens* type A isolates (Czeczulin *et al.*, 1993). CPE expression is similarly sporulation-dependent in CPE-positive *C. perfringens* type C and D isolates (Fisher *et al.*, 2006; Sayeed *et al.*, 2005).

CPE is not actively secreted from sporulating cells. Instead, it accumulates in the cytoplasm of those cells, where it can localize in paracrystalline inclusion bodies (Labbe, 1989). The toxin is finally released when the mother cell lyses to complete the sporulation process.

CPE mechanism of action

In vivo effects of CPE action

CPE is classified as an enterotoxin since it targets cells and tissues of the gastrointestinal tract to cause fluid and electrolyte loss (effects seen clinically as diarrhoea). In rabbits, it was shown that the small intestine is the major target of CPE *in vivo*, with the ileum being the most CPE-sensitive small intestinal region (McDonel, 1986). In contrast, the rabbit colon showed little response to CPE treatment (McDonel and Demers, 1982). A recent study demonstrated that, *ex vivo*, CPE also binds to human ileal epithelium and induces substantial morphological damage concurrent with reduced transepithelial resistance and net water absorption (Fernandez-Miyakawa *et al.*, 2005). In contrast, CPE had limited affect on human colon. However, there is some clinical indication that CPE might damage the human colon under certain circumstances, e.g. drug-induced constipation, which can allow prolonged contact between the toxin and colonic epithelium (Bos *et al.*, 2005).

When tested in rabbit ileal loop models, CPE induced significant damage within 15–30 minutes (Sherman *et al.*, 1994). This damage includes villus shortening, epithelial desquamation and some inflammation. Development of this histological damage coincides closely with the onset of fluid and electrolyte loss in the CPE-treated ileum, suggesting a cause and effect relationship. Additionally, only those doses of CPE producing histological damage were able to elicit changes in fluid and electrolyte transport in ileal loops (McDonel and Duncan, 1975). A recent study has further supported CPE cytotoxicity/epithelial damage causing enterotoxicity (Smedley 3rd *et al.*, 2008). When two non-cytotoxic rCPE variants were tested in a rabbit ileal loop model of CPE enterotoxicity *in vivo*, both variants bound to enterocytes but failed to produce the histological damage or fluid transport changes seen with wild-type CPE.

Molecular action of CPE

Over the past 20 years there have been detailed analyses of the molecular action of CPE. Fig. 2.2 illustrates the unique, multistep mechanism of action of CPE. Following the release of CPE into the intestinal lumen, the toxin binds to its target cells, i.e. enterocytes. Early studies of CPE action showed that pretreatment of CPE-sensitive cells with proteases prevented the binding and cytotoxicity normally caused by CPE treatment (McClane, 1994). This result indicated the requirement of a proteinaceous receptor for CPE binding. Initial co-immunoprecipitation and affinity chromatography studies (Wieckowski *et al.*, 1994; Wnek and McClane, 1983) suggested this CPE receptor might be an ~45–50 kDa host protein (those results probably detected an aggregate of the true CPE receptor, which is described in the next paragraph).

CPE receptors were definitively identified in the late 1990s by expression cloning approaches, which showed they include certain members of the claudin family of tight junction (TJ) proteins

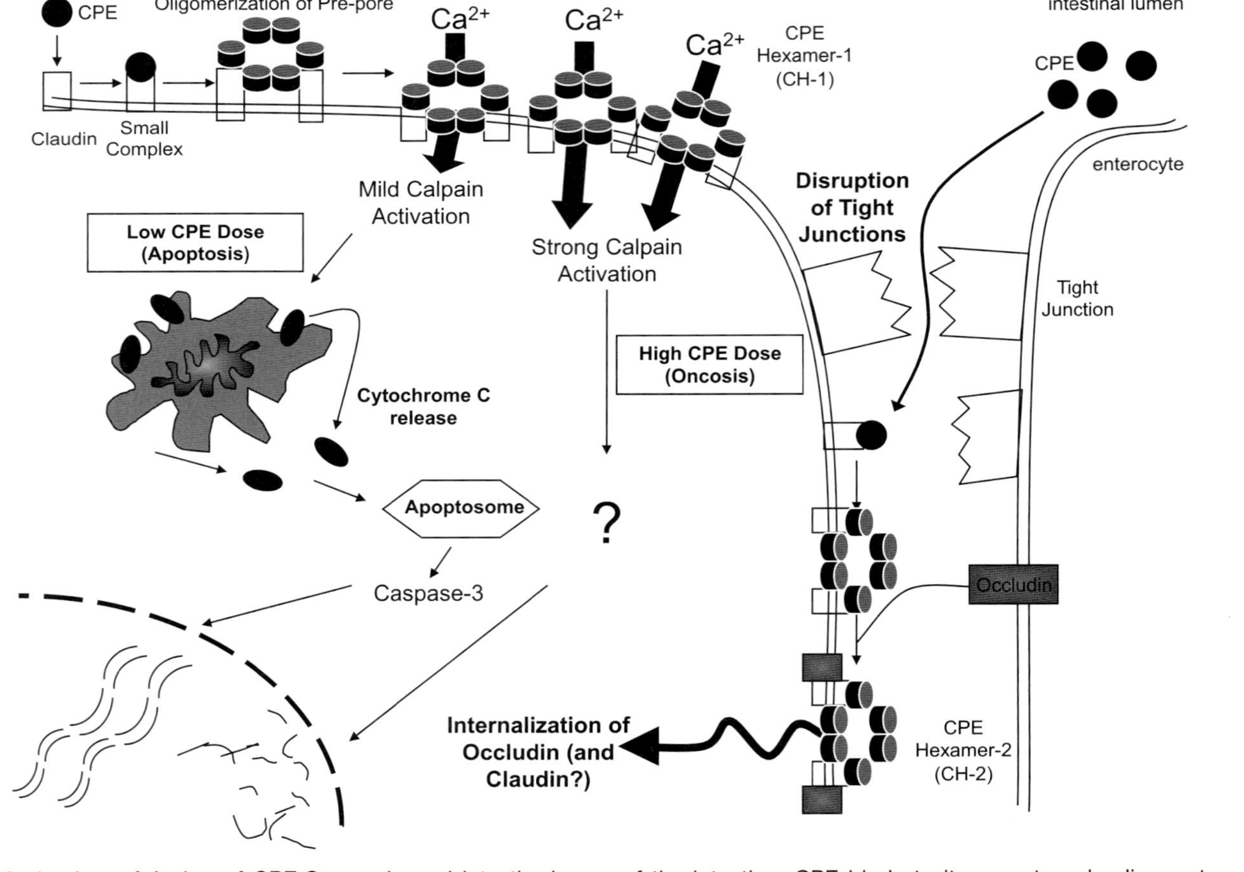

Figure 2.2 Molecular Mechanism of Action of CPE.Once released into the lumen of the intestine, CPE binds to its receptor, claudin, on target enterocytes. This binding event leads to the formation of an ~90 kDa SDS-sensitive complex, termed the small complex (SC). Upon association of other proteins, CPE undergoes an oligomerization event where six CPE molecules come together to form a pre-pore complex which is loosely associated with the target cell membrane. The pre-pore complex, termed the CPE Hexamer-1 (CH-1), rapidly inserts into the membrane resulting in the formation of a pore. This pore leads to an influx in Ca^{2+}, which activates one of two cell death pathways: apoptosis or oncosis. Ca^{2+} influx also results in the disruption of tight junctions between adjacent epithelial cells which allows any unbound toxin to bind to the basolateral side of the cell. Additional SC and CH-1 complexes are able to form, however, a third CPE-containing complex can form when the CH-1 complex associates with occludin. This complex is named the CPE Hexamer-2. Concurrent with the formation of these complexes, the removal of claudin and occludin from the tight junction is observed.

(Fujita *et al.*, 2000; Katahira *et al.*, 1997a,b). Claudins, a family of TJ proteins consisting of >20 members, are important for maintaining the structure and function of the TJ, including creation of a permeability barrier between adjacent epithelial cells (Anderson and Van Itallie, 1999). Transfection studies of CPE-insensitive fibroblasts with different claudins showed that only certain claudins can act as CPE receptors (Fujita *et al.*, 2000). For example, transfection of mouse fibroblasts with claudin-3 or -4 rendered these CPE-insensitive cells sensitive and thus these claudins were concluded to be functional receptors for the enterotoxin. In contrast, transfected fibroblasts expressing claudin-1 or -2 remained insensitive to CPE and did not bind the toxin, so these claudins cannot serve as CPE receptors (Fujita *et al.*, 2000). Additional receptor and non-receptor claudins were subsequently identified (Fujita *et al.*, 2000). For instance, claudins-6, -7, -8 and -14 exhibited CPE-binding capabilities, while claudins-5 and -10 were unable to bind CPE (Fujita *et al.*, 2000). That study (Fujita *et al.*, 2000) also identified the second extracellular loop of claudin as the domain responsible for binding with CPE. Ongoing research is addressing the validity of this preliminary result and further mapping of the claudin region that binds CPE is being investigated.

Binding between CPE and a claudin receptor(s) results in formation of an SDS-sensitive protein complex of ~90 kDa (Wieckowski *et al.*, 1994) that appears to be important, but insufficient, for cytotoxicity (Wieckowski *et al.*, 1994). This ~90 kDa complex, termed the Small Complex (SC), was recently shown to contain both CPE and functional claudin receptors (Robertson *et al.*, 2007). Interestingly, SC also can contain non-receptor claudins (possibly due to interactions between receptor claudins and non-receptor claudins). The stoichiometry of each component (CPE, receptor, and non-receptor) in SC remains to be determined, but is under active investigation.

Under physiologic conditions, the SC represents a transitory step prior to formation of two larger CPE-associated complexes that actually initiate CPE-induced cytotoxicity. The first of these two larger complexes, named CPE hexamer-1 (CH-1), apparently represents a pre-pore complex formed by CPE oligomerization (Smedley 3rd *et al.*, 2007). This pre-pore, which forms on the membrane surface, has recently been shown to contain six copies of CPE (Robertson *et al.*, 2007). Similarly to the SC, the CH-1 complex can contain both receptor and non-receptor claudins. The size of CH-1 was initially thought to be ~155 kDa but it was recently reassessed after CH-1 was found to contain 6 copies of CPE, and at least one receptor and one non-receptor claudin. Using native electrophoresis and gel filtration chromatography approaches, the native CH-1 complex was recently resized to ~450 kDa (Robertson *et al.*, 2007).

Concurrent with CH-1 complex formation, which can occur within minutes of CPE treatment at 37°C, membrane permeability alterations develop in the target host cell due to insertion of the CH-1 pre-pore into plasma membranes (McClane and Wnek, 1990). During this insertion, the CH-1 complex becomes resistant to externally applied proteases and also becomes difficult to dissociate from cells (McClane and Wnek, 1990). Several studies with CPE fragments and mutants support a strong correlation between CH-1 formation and the onset of cytotoxicity (Kokai-Kun *et al.*, 1999; Kokai-Kun and McClane, 1997). Formation of the CPE pore upon CH-1 insertion into membranes renders the host cell permeable to water, ions, and small molecules up to 3000 Da (McClane *et al.*, 1988b). Whole-cell patch clamping experiments indicated the CPE pore is cation-selective (Hardy *et al.*, 1999).

With longer periods of CPE treatment, an additional CPE-containing toxin complex, named CPE hexamer-2 (CH-2), can form in Caco-2 cells (Singh *et al.*, 2000). CH-2 also contains six copies of CPE, receptor claudins and non-receptor claudins (Robertson *et al.*, 2007). In addition to claudins, another TJ protein, occludin, is present within the CH-2 complex (Singh *et al.*, 2000). The size of the CH-2 complex was initially estimated to be ~200 kDa; however, with the recent data indicating six CPE molecules, as well as multiple claudins and occludin, the true molecular weight of this complex was re-evaluated (Robertson *et al.*, 2007). Size exclusion chromatography and Ferguson gel plot analysis reassessed the molecular weight of CH-2 as ~600 kDa. It is unclear at present whether CH-1 is a precursor for CH-2 formation.

The functional significance of CH-2 for CPE action is still unknown because CH-2 complexes have not been formed independently of the CH-1 complex. One effect of CH-2 complex formation may be the removal of occludin and claudins from the TJ (Singh *et al.*, 2001; Sonoda *et al.*, 1999). This may disrupt TJ structure and lead to the increased permeability of water and ions between adjacent epithelial cells, thus contributing to diarrhoea during GI disease.

Activation of cell death pathways by CPE

Formation of the CPE pore triggers death of target cells through recently identified pathways. An early observation was made that the morphological damage and cytotoxicity associated with CPE treatment of Vero or HeLa cells requires the presence of Ca^{2+} in the CPE treatment buffer (Matsuda and Sugimoto, 1979). Ca^{2+} was shown not to be required for CPE receptor binding or for development of CPE-induced pore formation (McClane *et al.*, 1988b). Later studies confirmed a Ca^{2+}-requirement for CPE cytotoxicity against Caco-2 cell cultures and, more importantly, showed that Ca^{2+} is required for activation of cell death pathways in the CPE-treated cell (Chakrabarti and McClane, 2005). This study showed that CPE-treatment increased intracellular Ca^{2+} levels via the CPE pore, with that phenomenon then activating intracellular cell death pathways.

Multiple cell death pathways have been identified in the literature, two of which include apoptosis and oncosis (Fink and Cookson, 2005). Apoptosis, also known as programmed cell death, is characterized by membrane budding, nuclear and cytoplasmic condensation, cleavage of DNA into ladder-like fragments, and the activation of proteases such as caspases. Apoptotic cells are easily phagocytosed, which results in a controlled, non-inflammatory event. On the other hand, oncosis is characterized by cellular swelling and blebbing, random shearing of DNA, and breakdown of the plasma membrane with leads to altered membrane permeability and release of proinflammatory signals. Thus, in contrast to apoptosis, oncosis is a proinflammatory process.

Both apoptosis and oncosis are relevant for CPE-induced cell death. Chakrabarti *et al.* (2003) showed that varying CPE doses have different consequences for cell death pathway activation in host cells. For example, when Caco-2 cells were treated with a relatively low dose of CPE (1 μg/ml), those cells developed morphologic damage (such as cell rounding and detachment) within 1 h. By that time, nuclear condensation and ladder-like fragmentation of ~200 bp was also observed. Furthermore, that low CPE dose, caused caspase-3 activation in Caco-2 cells and, if those cells were pretreated with a caspase 3/7 inhibitor, CPE-induced morphological and nuclear damage was prevented. This caspase 3/7 activation in CPE-treated Caco-2 cells was shown to result from mitochondrial membrane depolarization and the release of cytochrome *c*. Collectively, these low-dose CPE effects were indicative of a classical apoptosis cell death pathway.

The study by Chakrabarti *et al.* (2003) also investigated the effects of higher CPE doses (10 μg/ml) on Caco-2 cells. That higher CPE dose produced morphologic changes faster (i.e. within 30 min of toxin treatment) than the lower CPE dose. In addition, cells treated with the high CPE dose showed little to no nuclear condensation and there was random shearing of their DNA (as opposed to a ladder-like cleavage seen with the low CPE dose). Furthermore, no mitochondrial membrane depolarization, cytochrome *c* release, or caspase 3/7 activation was detected in the higher CPE dose-treated Caco-2 cells. Caspase inhibitors could not inhibit the above effects of high CPE dose treatment. However, glycine, which is an inhibitor of oncosis, did delay these effects. Thus, it was concluded that the cell death pathway responsible for triggering cell death caused by high CPE doses is oncosis, rather than apoptosis.

In terms of pathophysiological relevance, both CPE doses used in Chakrabarti *et al*'s study (Chakrabarti *et al.*, 2003) lie within the range of concentrations of CPE detected in the faeces of patients suffering from CPE-induced diarrhoea (Bartholomew *et al.*, 1985; Birkhead *et al.*, 1988). Additional *in vivo* relevance of the above findings comes from apoptosis not inducing an inflammatory response while oncosis is a proinflammatory process. In animal models, both inflammatory and non-inflammatory damage has been seen within the CPE-treated GI tract and this corresponds, at least in some cases, to the CPE dosage levels used for treatment (McDonel *et al.*, 1978; McDonel and Duncan, 1975; Sarker *et al.*, 1999; Sherman *et al.*, 1994).

As mentioned previously, Ca^{2+} is required for the cytotoxic effects of CPE. A recent study (Chakrabarti and McClane, 2005) showed this requirement reflects a CPE-induced Ca^{2+} influx that increases intracellular Ca^{2+} levels. This increase in intracellular Ca^{2+} then activates a Ca^{2+}-dependent cysteine protease known as calpain. In the case of low CPE doses, where there is limited Ca^{2+} influx, only a mild activation of calpain occurs. In contrast, a strong activation of calpain develops using the high CPE doses that cause the greater influx of Ca^{2+}. In addition, calpain inhibitors can block many of the apoptotic and oncotic events associated with CPE treatment. Inhibitors of calmodulin, a Ca^{2+}-binding protein that regulates many cytoplasmic proteins also block CPE-induced apoptosis and oncosis. Further studies are currently underway to identify the cytoplasmic pathway of CPE-induced host cell death.

CPE structure–function relationship

Dissecting the structure–function relationship of the CPE protein is important to understand the molecular pathogenesis of CPE-associated human and veterinary gastrointestinal illnesses (McClane *et al.*, 1988a) and for improving applied use of CPE or CPE fragments for cancer therapy/diagnosis or drug delivery (discussion later). This section will present current knowledge of how the CPE protein mediates its binding and activities, as described in the preceding section and in the following 'applications' section.

N-terminal CPE region

The earliest studies implicating the N-terminus of CPE in toxicity utilized enzymatic digestion of the enterotoxin with trypsin or chymotrypsin, two proteases typically present in the human GI tract (Granum and Richardson, 1991; Granum *et al.*, 1981). Trypsin was found to remove the 25 N-terminal CPE amino acids, causing a threefold increase in cytotoxicity (Granum *et al.*, 1981). Hanna *et al.*, then showed that this trypsin-induced activation of CPE toxicity did not result from increased CPE binding (Hanna *et al.*, 1992). Later studies using similar recombinant CPE (rCPE) fragments identical to those produced by trypsin digestion of native CPE indicated removal of the first 25 N-terminal amino acids from native CPE enhances formation of CPE large complexes, i.e. CH-1 and, probably, CH-2 (Hanna *et al.*, 1992). Similar digestions with chymotrypsin were shown to remove the first 36 N-terminal amino acids from CPE and the final three amino acids from the CPE carboxy terminus (Granum and Richardson, 1991). This chymotrypsin-generated fragment also possessed increased cytotoxicity and a later study (Kokai-Kun and McClane, 1997) showed an $rCPE_{37-319}$ fragment exhibited enhanced formation of large CPE complexes. These early proteolytic cleavage studies of native CPE were important not only for identifying a CPE region (amino acids 1–36) that is inhibitory for large complex formation and cytotoxicity, but also for suggesting that, during GI disease, proteolysis of native CPE by trypsin or chymotrypsin in the intestines may produce an activated CPE fragment.

Concurrently with the proteolytic cleavage studies described above, Horiguchi and colleagues produced CPE fragments comprising essentially the N-terminal or C-terminal halves of the native toxin (Horiguchi *et al.*, 1987). This was accomplished by chemically cleaving native CPE using 2-nitro-5-thiocyanobenzoic acid, which cleaves CPE at its single cysteine located at residue 186 (Horiguchi *et al.*, 1987). When the resultant CPE fragments were tested for cytotoxicity, Vero cells treated with a fragment comprising essentially the C-terminal half of CPE showed no morphological damage. That result is also consistent with N-terminal CPE sequences playing an important role in cytotoxicity.

Cloning of the *cpe* gene in the early 1990s (Czeczulin *et al.*, 1993) allowed expression of full length rCPE or well-defined rCPE fragments or mutants. As briefly introduced already in this section, the McClane lab has used those rCPE variants to explore the structure–function relationship of CPE in detail. An early study by that group supported the chemical cleavage results of Horiguchi *et al.*, by definitively showing that an $rCPE_{189-319}$ fragment is not cytotoxic despite retaining excellent receptor-binding ability. As mentioned, deletion mutagenesis studies (Kokai-Kun and McClane, 1997) demonstrated that an $rCPE_{37-319}$ fragment resembling the CPE fragment produced by chymotypsin cleavage of native CPE causes increased cytotoxicity due to enhanced large complex formation. That study then went further and showed that this more toxic phenotype is retained if even the N-terminal 45 residues of native CPE are removed. However, an $rCPE_{53-319}$ fragment lost all cytotoxic activity despite retaining normal receptor binding properties.

Since deletion mutagenesis results suggested that the native CPE region from residues 45 to 53 is important for large complex formation and cytotoxicity, a later site-directed study (Smedley 3rd and McClane, 2004) characterized each individual amino acid between CPE residues 45 to 53 by alanine-scanning mutagenesis. Those studies revealed that substituting an alanine for the aspartic acid naturally present at CPE residue 48 produces an rCPE variant that can bind to receptors and form the CPE small complex, but is completely unable to form the large complex or cause a cytotoxic effect in Caco-2 cells. It was also found that substituting an alanine for the isoleucine naturally present at position 51 of native CPE produced an rCPE variant that formed reduced amounts of the CH-1 large complex (Smedley 3rd and McClane, 2004). Other residues in the CPE 47–51 region also appeared to play minor roles for large complex formation and cytotoxicity.

An $rCPE_{162-319}$ C-terminal fragment was non-toxic despite being able to bind to the CPE receptor (Hanna *et al.*, 1991; Hanna *et al.*, 1989). This recombinant C-terminal fragment apparently could not form large complexes (Kokai-Kun and McClane, 1997) or insert into host plasma membranes (Hanna *et al.*, 1991). Since large complex (CH-1) formation was shown (see the preceding paragraph) to involve CPE residues 48 and 51, Smedley *et al.* (2007) later sought to localize the CPE region(s) important for membrane insertion and pore formation. Their study began with a comparison of CPE against other bacterial pore-forming toxins (PFTs). Although CPE does not to share any sequence homology with PFTs, it does exhibit a 25 amino acid region with alternating side chain hydrophobicitiy, which is similar to the transmembrane stem domains (TMDs) of aerolysin, *Clostridium septicum* α-toxin, *Staphylococcus aureus* α-haemolysin, anthrax-protective antigen and the cholesterol-dependent cytolysins (Parker and Feil, 2005). Once PFTs have oligomerized in a pre-pore on the membrane surface, their TMDs then mediate insertion of a β-barrel into the membrane to form the active pore (Parker and Feil, 2005).

The CPE region exhibiting this alternating hydrophobic and hydrophilic pattern typical of TMDs is located between amino acids 81 to 106 (Smedley 3rd *et al.*, 2007). Therefore, Smedley *et al.* characterized three rCPE variants, named D48A, TM1 and TM1-D48A. Those variants correspond, respectively, to an rCPE single point mutant where the natural aspartic acid has been replaced with alanine to block CH-1 formation, an rCPE variant where putative TMD residues 81–106 have been deleted, and an rCPE variant that combines both the point mutation and putative TMD deletion of the first two rCPE variants (Smedley 3rd and McClane, 2004; Smedley 3rd *et al.*, 2007). All three rCPE mutants were found to retain the ability to bind to CPE receptors. The D48A rCPE variant was, as expected from previous studies (Smedley 3rd and McClane, 2004) unable to form CH-1. TM1 showed a novel phenotype where it formed a CH-1-like complex without cytotoxicity. Protease challenge and dissociation studies suggested that the CH-1-like complex formed by TM1 apparently remains on the cell surface in a prepore state. TM1-D48A did not form large complexes or insert into membranes, suggesting that CH-1 formation precedes CPE insertion into membranes. The CH-1-like complexes formed by TM1 were also less stable than CH-1 formed by native CPE, suggesting that the putative TMD region may play roles both in complex insertion and stability.

C-terminal CPE region

Enterotoxin fragments generated by proteolysis, chemical cleavage (Granum, 1982; Granum and Richardson, 1991; Granum *et al.*, 1981; Horiguchi *et al.*, 1987; Smedley 3rd *et al.*, 2007) and deletion or point mutagenesis (Czeczulin *et al.*, 1993; Hanna and McClane, 1991; Hanna *et al.*, 1991; Hanna *et al.*, 1992; Hanna *et al.*, 1989;

Kokai-Kun *et al.*, 1999; Kokai-Kun and McClane, 1997; Smedley 3rd and McClane, 2004) also identified a functional assignment for the C-terminal CPE region. Specifically, CPE fragments lacking sequences in the N-terminal half of native CPE were found to retain similar receptor binding activity as native CPE (Czeczulin *et al.*, 1993; Granum, 1982; Granum and Richardson, 1991; Hanna *et al.*, 1992; Kokai-Kun *et al.*, 1999; Kokai-Kun and McClane, 1997). Hanna *et al.* further localized the C-terminal receptor binding region of CPE by constructing several rCPE fragments containing different regions of the C-terminal half of CPE. rCPE fragments consisting of native CPE amino acids 171–238 or 238–290 did not exhibit any receptor binding activity. However, an rCPE fragment containing the 29 C-terminal residues of native CPE competed strongly against native CPE for binding to Vero cells (Hanna and McClane, 1991). In fact, *E. coli* lysates containing the $CPE_{290-319}$ fragment were able to protect Vero cells from challenge with native CPE (Hanna and McClane, 1991). That same study also showed that a synthetic peptide corresponding to the amino acid 290–319 region of native enterotoxin exhibited similar binding ability as native CPE (Hanna and McClane, 1991).

As mentioned earlier, chymotrypsin reportedly removes not only the N-terminal 36 amino acids from native CPE but also the three extreme C-terminal residues without any detrimental effect on binding (Granum, 1982). Later deletion mutagenesis studies found that deletion of the last five amino acids from CPE prevented receptor binding (Kokai-Kun and McClane, 1997). Combining those results would suggest that the hydrophobic non-polar amino acids leucine and phenylalanine at positions 315 and 316, respectively, may be important for CPE binding but this deduction needs to be experimentally verified.

While those C-terminal CPE studies were ongoing, other efforts were directed towards identifying the CPE receptor. It has been known for over 25 years that CPE binds to a protein receptor causing membrane permeability alterations (McClane and McDonel, 1980; McDonel, 1980), but more recent studies (discussed in detail earlier in the chapter) determined that certain members of the claudin tight junction protein family can function as CPE receptors (Furuse *et al.*, 1993; Katahira *et al.*, 1997a; Singh *et al.*, 2001; Singh *et al.*, 2000; Sonoda *et al.*, 1999). Those studies also showed that C-terminal CPE fragments can remove claudins from tight junctions (Katahira *et al.*, 1997b; Masuyama *et al.*, 2005; Sonoda *et al.*, 1999) and disrupt tight junction strands. Later studies revealed that a rCPE fragment missing amino acids 184–219 could still bind with claudins but cannot disrupt tight junctions (Masuyama *et al.*, 2005). However, a rCPE fragment with the last 16 residues removed from the C-terminus could not even interact with claudin-4, confirming that these C-terminal 16 amino acids of CPE are essential for receptor binding (Takahashi *et al.*, 2005).

While it is now accepted that CPE receptor binding activity localizes to the last 29 C-terminal amino acids of the toxin, the specific CPE amino acids residues mediating this binding are not yet clear. To that end, a recent study introduced single point mutations substituting alanine for the tyrosine residues naturally present at CPE residues 306, 310 and 312. Those single mutations partially reduced binding activity, which could be further reduced by mutating two or all three of those tyrosine residues (Harada *et al.*, 2007). The alanine-substituted mutants lost their ability to disrupt the tight junctions or enhance jejunal absorption in rats (Harada *et al.*, 2007), as described later in this chapter.

Finally, there is at least one neutralizing epitope present in the extreme C-terminal region of the CPE protein (Hanna *et al.*, 1992). The presence of a neutralizing epitope in C-terminal CPE fragments that are not themselves cytotoxic suggested those fragments might represent CPE vaccine candidates. This was explored (Mietzner *et al.*, 1992) using a 30-mer synthetic peptide corresponding to the extreme C-terminal sequence of CPE. That synthetic peptide was then conjugated to a thyroglobulin carrier. When the resultant conjugate was administered i.v. to mice, this immunization produced a very high titre of IgG antibodies capable of neutralizing the cytotoxic activity of native CPE. This result supports the potential of C-terminal CPE fragments for vaccine preparation, if desired. However, a number of challenges would still need to be overcome, e.g. a vaccine protective against CPE-associated

GI disease may need to specifically evoke IgA, in which case CPE vaccine strategies would need to target an IgA response.

Summary of the CPE structure vs. function relationship

The emerging view from the structure/function mapping studies presented above is that binding and activity regions segregate to different halves of the CPE protein, as shown in Fig. 2.3. The CPE binding domain is present in the last 29 C-terminal amino acids of the toxin. The N-terminal half of CPE is also required for the cytotoxic action of CPE as it mediates insertion and large complex formation. However, the first 45 amino acids of native CPE are inhibitory for large complex formation, insertion and cytotoxicity. Removal of those N-terminal sequences enhances CPE cytotoxicity, possibly because this increase exposure of the CPE 45–53 region responsible for CH-1 formation. These N-terminal inhibitory residues can be removed by intestinal proteases, possibly producing activated CPE fragments during GI disease. Within the CPE 45–53 region, residues 47–51 (particularly the aspartic acid at position 48 and the isoleucine at residue 51) are important for CH-1 formation/prepore formation (Kokai-Kun *et al.*, 1999; Smedley 3rd and McClane, 2004). A putative transmembrane stem domain extending from CPE amino acids 81 to 106 may mediate insertion of the CPE prepore oligomer into plasma membranes (Smedley 3rd and McClane, 2004; Smedley 3rd *et al.*, 2007).

Structural information about the CPE protein will be immensely helpful in understanding the structure/function relationship of this protein. Unfortunately, CPE has thus far proven refractory to crystallization but there are now intensive efforts under way to solve that problem.

Practical applications of CPE or CPE-based derivatives

CPE as a cancer therapeutic or diagnostic

The ability of CPE to bind claudins became particularly interesting when it was discovered that claudin expression is upregulated in many cancers (Litkouhi *et al.*, 2007; Long *et al.*, 2001; Michl *et al.*, 2001; Nichols *et al.*, 2004; Santin *et al.*, 2007). Specifically, high-level expression of claudin-3 and/or -4, both functional CPE receptors, has been detected in many pancreatic, prostrate, breast, ovarian, and uterine cancers. Those observations have spawned a series of exciting studies exploring the use of CPE or its derivatives, as potential anti-tumour therapeutics.

To our knowledge, the earliest exploration of CPE as a potential cancer therapeutic occurred in 2001 and involved pancreatic cancer, which is exceptionally difficult to treat with conventional anti-cancer approaches (Michl *et al.*, 2001). RT-PCR analysis revealed that pancreatic cell

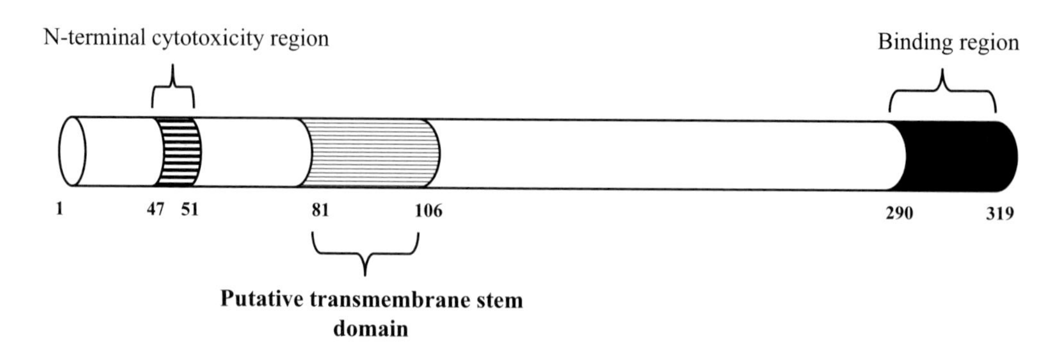

Figure 2.3 CPE structure/function relationship. CPE amino acid residues 1 to 45 have an inhibitory effect on CH-1 formation and cytotoxicity (Kokai-Kun and McClane, 1997). Residues 45 to 53 are important for large complex (CH-1 and CH-2) formation and for cytotoxicity (Kokai-Kun *et al.*, 1999), particularly the aspartic acid at residue 48 and the isoleucine at residue 51 (Smedley 3rd and McClane, 2004). A putative transmembrane stem domain was recently identified between CPE residues 81–106 (Smedley 3rd *et al.*, 2007). This region is thought to be involved in insertion of a pre-pore large complex into membranes and also appears to contribute to CH-1 stability (Smedley 3rd *et al.*, 2007). The receptor binding region is located at the 30 C-terminal amino acids of the toxin (Kokai-Kun and McClane, 1997).

lines often expressed elevated levels of claudin-4 as compared to control, non-cancer cell lines. In cytotoxicity assays, treatment with CPE elicited more cell death in the pancreatic cancer cell lines as compared to controls and the extent of that cell death correlated with how much claudin-4 a cell line expressed. Additionally, that study grew xenografted tumours of claudin-4-expressing, human pancreatic cancer cell Panc-1 on the backs of nude mice. When CPE was injected into those xenografted tumours, the tumour stopped growing and underwent necrosis. When the experiment was similarly repeated with tumours that did not express claudin-4, those tumours increased 2.5 times in size despite CPE treatment and no tumour necrosis was observed. Importantly, mouse survival was unaffected by their CPE injection.

A study by Kominskey *et al.* then examined the effects of CPE treatment of primary breast carcinomas showing high-level expression of claudin-3 and -4 (Kominsky *et al.*, 2004). Those cultured breast cancer cell lines responded to lower CPE doses and exhibited more rapid cytolysis than their non-cancer breast cell line counterparts. This study also examined the effects of CPE treatment *in vivo* using a xenograft approach. When breast cancer cells were grown as tumours on the backs of mice and CPE was then injected into those tumours, necrosis of the tumour and significant tumour size reduction were noted over control tumours grown with other (non-claudin-4 expressing) cells. Again, no systemic toxicity was observed after administration of CPE into the tumours.

An additional study investigated the effects of CPE treatment on chemotherapy-resistant ovarian cancer xenografts that overexpress claudin-3 and -4 (Santin *et al.*, 2005). Similarly to the above reports, those ovarian cancer cell lines *in vitro* were susceptible to CPE-mediated lysis, with sensitivity correlating with the levels of claudin-3 and -4 being expressed by each cell line. This *in vitro* data was complemented with an *in vivo* study where ovarian tumour cells were injected intraperitoneally into mice to allow systemic tumour development. When those tumours had grown, CPE was injected into the tumours. Mice that received this CPE treatment survived significantly longer, and had reduced tumour size, compared with controls receiving a mock injection. No systemic toxicity was associated even with this systemic CPE treatment.

Another intriguing recent use of CPE as a cancer therapy has investigated its potential as a novel, targeted therapeutic for brain metastasis (Kominsky *et al.*, 2007). A major challenge in successful treatment of brain metastasis is to eliminate the tumour without causing critical damage to the CNS. In the recent study by Kominsky *et al., it* was shown that normal brain tissue did not express claudin-3 or -4 and was not susceptible to CPE. However, brain metastases of epithelial cell origin expressing claudin-3 and -4 were quite susceptible to CPE treatment. Using a bioluminescent imaging method (Fig. 2.4), intracranial delivery of CPE produced a 10-fold reduction in tumour size and an increase in survival by 49%. Remarkably, no appreciable local or systemic toxicity was observed with this intracranial delivery method.

A final recent study examined the use of a claudin-targeting, fusion protein between C-terminal CPE sequences (C-CPE) and the protein synthesis inhibitory factor (PSIF) from *Pseudomonas aeruginosa* exotoxin to deliver PSIF to the inside of a cancer cell (Ebihara *et al.*, 2006). This fusion protein, C-CPE-PSIF, was cytotoxic to MCF-7 human breast cancer cells, which express endogenous claudin-4, but was not toxic to mouse fibroblast cells, which do not express any claudins. The cytotoxicity seen with C-CPE-PSIF was attenuated by pre-treating the MCF-7 cells with C-CPE as well as by deleting the binding region of C-CPE. These results show promise for a potential anti cancer drug delivery system that targets claudin-expressing tumour cells.

These promising initial results for use of CPE or CPE derivatives as anticancer agents are now being followed-up by several laboratories. There is also interest in developing CPE derivatives for imaging purposes to improve cancer diagnoses.

Use of CPE derivatives for drug delivery

Another currently exciting area of applied CPE research concerns the construction of a drug delivery vehicle using CPE derivatives. Because CPE interacts closely with the TJ via claudin and occludin interactions, CPE can disrupt the overall structure and function of the TJ barrier

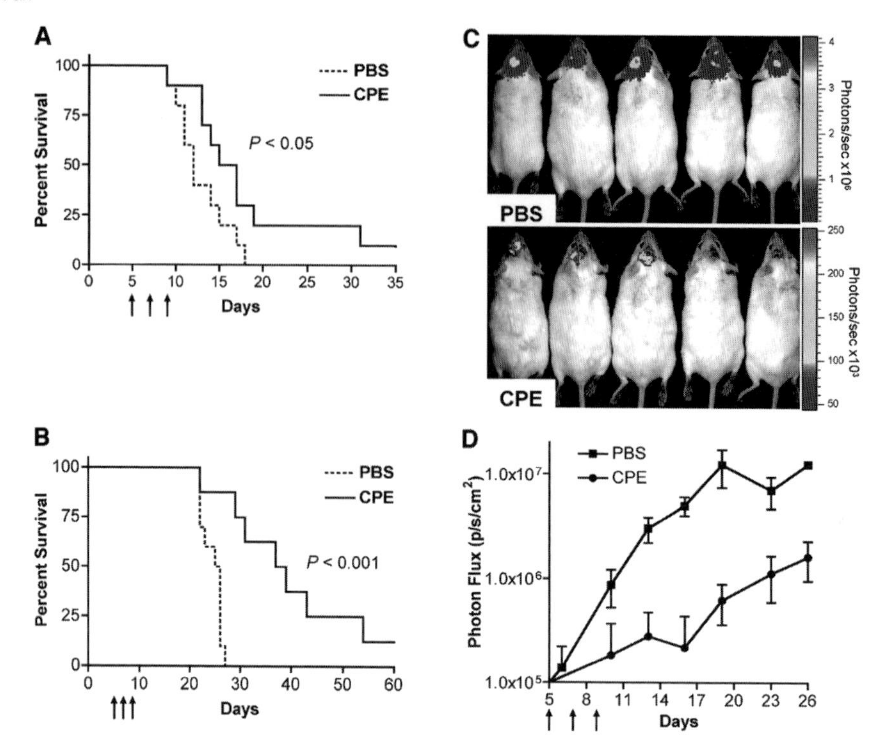

Figure 2.4 Effects of intracranial CPE injection on brain metastasis. (A and B) Brain tumours were established in mice using the human breast cancer cell line MDA-MB-468 (A) and the murine breast cancer cell line NT2.5-luc (B). Tumours were treated by intracranial administration of 0.5 μg CPE or PBS on days 5, 7 and 9. For both cell line metastases, CPE treatment resulted in a statistically significant prolonged per cent survival in mice. (C and D) Five representative mice from each group of the NT2.5-luc tumours were analysed on day 19 using non-invasive bioluminescent imaging. Photon flux was measured to quantitate the bioluminescent signal and revealed that administration of CPE resulted in a significant inhibition of tumour growth.

(Sonoda *et al.*, 1999). This disruption weakens the TJ barrier function to allow larger molecules to pass across the epithelium. C-CPE, a non-toxic fragment containing the binding-proficient C-terminal half of CPE, was then shown to increase transepithelial resistance (Sonoda *et al.*, 1999). This knowledge prompted researchers to investigate the possible application of CPE or its derivatives as a mechanism to increase drug delivery across this epithelial barrier.

An initial study examined the use of C-CPE, corresponding to native CPE residues 184–319, to loosen the TJ barrier and allow movement of large molecules across an epithelial cell layer (Kondoh *et al.*, 2005). In this study, Kondoh *et al.*, treated MDCK cells with C-CPE and demonstrated the removal of claudin-4 from the TJ and a resultant increase in paracellular permeability. Additionally, C-CPE was shown to enhance the intestinal absorption of dextrans up to a molecular weight of 20 kDa. Importantly, C-CPE was 400 times more potent at increasing absorption than currently used clinical absorption enhancers. No intestinal damage was observed with C-CPE treatment of mice in this study.

Summary

The past few years have been extraordinarily productive for CPE research. This recent work has revealed several unusual and interesting features of CPE genetics, structure/function relationships, and action. Some of those CPE features are now being exploited for practical applications, such as cancer therapy/diagnosis or drug delivery. These promising efforts will hopefully tame a powerful toxin so it can be used in the future for beneficial purposes.

Note added in proof

Since submission of this chapter, the structure of the C-terminal, claudin-binding domain

(C-CPE) of *C. perfringens* exterotoxin has been solved (Van Itallie, C.M., Betts, L., Smedley 3rd, J.G., McClane, B.A. and Anderson, J.M. (2008). Structure of the claudin-binding domain of *Clostridium perfringens* enterotoxin. J. Biol. Chem. *283*, 268–274). This domain is a nine strand beta sandwich with similarity to the receptor binding domains of the large Cry family of *Bacillus thuringiensis* toxins. The claudin binding site was suggested to lie on a large surface loop located between strands 8 and 9. This assignment was recently verified experimentally by site-directed mutagenesis approaches (Takahashi, A., Komiya, E., Katutani, H., Yoshida, T., Fujii, M., Horiguchi, Y., Mizuguchi, H., Tsutsumi, Y., Tsunoda, S., Koizumi, N., Isoda, K., Yagi, K., Watanabe, Y., and Kondoh, M. (2008) Domain mapping of a claudin-4 modulator, the C-terminal region of C-terminal fragment of *Clostridium perfringens* enterotoxin, by site-directed mutagenesis, Biochem. Pharm. *75*, 1639–1648).

Acknowledgements

Preparation of this chapter was generously supported by grants R37AI19844 and RO1AI056177 (B.McC.) and T32 AI060525 (S.S.) from the National Institute of Allergy and Infectious Diseases.

References

Anderson, J.M., and Van Itallie, C.M. (1999). Tight junctions: closing in on the seal. Curr. Biol. *9*, R922–924.

Bannam, T.L., Teng, W.L., Bulach, D., Lyras, D., and Rood, J.I. (2006). Functional identification of conjugation and replication regions of the tetracycline resistance plasmid pCW3 from *Clostridium perfringens*. J. Bacteriol. *188*, 4942–4951.

Bartholomew, B.A., Stringer, M.F., Watson, G.N., and Gilbert, R.J. (1985). Development and application of an enzyme linked immunosorbent assay for *Clostridium perfringens* type A enterotoxin. J. Clin. Pathol. *38*, 222–228.

Billington, S.J., Wieckowski, E.U., Sarker, M.R., Bueschel, D., Songer, J.G., and McClane, B.A. (1998). *Clostridium perfringens* Type E animal enteritis isolates with highly conserved, silent enterotoxin gene sequences. Infect. Immun. *66*, 4531–4536.

Birkhead, G., Vogt, R.L., Heun, E.M., Snyder, J.T., and McClane, B.A. (1988). Characterization of an outbreak of *Clostridium perfringens* food poisoning by quantitative fecal culture and fecal enterotoxin measurement. J. Clin. Microbiol. *26*, 471–474.

Borriello, S.P., Larson, H.E., Welch, A.R., Barclay, F., Stringer, M.F., and Bartholomew, B.A. (1984). Enterotoxigenic *Clostridium perfringens*: a possible cause of antibiotic-associated diarrhoea. Lancet *1*, 305–307.

Bos, J., Smithee, L., McClane, B.A., Distefano, R.F., Uzal, F., Songer, J.G., Mallonee, S., and Crutcher, J.M. (2005). Fatal necrotizing colitis following a food-borne outbreak of enterotoxigenic *Clostridium perfringens* type A infection. Clin. Infect. Dis. *40*, e78–83.

Brynestad, S., and Granum, P.E. (1999). Evidence that Tn5565, which includes the enterotoxin gene in *Clostridium perfringens*, can have a circular form which may be a transposition intermediate. FEMS Microbiol. Lett. *170*, 281–286.

Brynestad, S., Sarker, M.R., McClane, B.A., Granum, P.E., and Rood, J.I. (2001). Enterotoxin plasmid from *Clostridium perfringens* is conjugative. Infect. Immun. *69*, 3483–3487.

Brynestad, S., Synstad, B., and Granum, P.E. (1997). The *Clostridium perfringens* enterotoxin genes is on a transposable element in type A human food poisoning strains. Microbiology *143*, 2109–2115.

Carman, R.J. (1997). *Clostridium perfringens* in spontaneous and antibiotic-associated diarrhoea of man and other animals. Rev. Med. Microbiol. *85*, S43–S45.

Chakrabarti, G., and McClane, B.A. (2005). The importance of calcium influx, calpain and calmodulin for the activation of CaCo-2 cell death pathways by *Clostridium perfringens* enterotoxin. Cell. Microbiol. *7*, 129–146.

Chakrabarti, G., Zhou, X., and McClane, B.A. (2003). Death pathways activated in CaCo-2 cells by *Clostridium perfringens* enterotoxin. Infect. Immun. *71*, 4160–4270.

Clewell, D.B., Francia, M.V., Flannagan, S.E., and An, F.Y. (2002). Enterococcal plasmid transfer: sex pheromones, transfer origins, relaxases, and the *Staphylococcus aureus* issue. Plasmid *48*, 193–201.

Collie, R.E., and McClane, B.A. (1998). Evidence that the enterotoxin gene can be episomal in *Clostridium perfringens* isolates associated with non-food-borne human gastrointestinal disease. J. Clin. Microbiol. *36*, 30–36.

Cornillot, E., Saint-Joanis, B., Daube, G., Katayama, S., Granum, P E., Canard, B., and Cole, S.T. (1995). The enterotoxin gene (*cpe*) of *Clostridium perfringens* can be chromosomal or plasmid-borne. Mol. Microbio.l *15*, 639–647.

Czeczulin, J.R., Collie, R.F., and McClane, B.A. (1996). Regulated expression of *Clostridium perfringens* enterotoxin in naturally *cpe*-negative type A, B and C isolates of *C. perfringens*. Infect. Immun. *64*, 3301–3309.

Czeczulin, J.R., Hanna, P.C., and McClane, B.A. (1993). Cloning, nucleotide sequencing, and expression of the *Clostridium perfringens* enterotoxin gene in *Escherichia coli*. Infect. Immun. *61*, 3429–3439.

Ebihara, C., Kondoh, M., Hasuike, N., Harada, M., Mizuguchi, H., Horiguchi, Y., Fujii, M., and Watanabe, Y. (2006). Preparation of a claudin-targeting molecule using a C-terminal fragment of *Clostridium perfringens* enterotoxin. J. Pharmacol. Exp. Ther. *316*, 255–260.

Fernandez-Miyakawa, M.E., Pistone Creydt, V., Uzal, F.A., McClane, B.A., and Ibarra, C. (2005). *Clostridium perfringens* enterotoxin damages the human intestine *in vitro*. Infect. Immun. *73*, 8407–8410.

Fink, S.L., and Cookson, B.T. (2005). Apoptosis, pyroptosis, and necrosis: mechanistic description of dead and dying eukaryotic cells. Infect. Immun. *73*, 1907–1916.

Fisher, D.J., Fernandez-Miyakawa, M.F., Sayeed, S., Poon, R., Adams, V., Rood, J I., Uzal, F.A., and McClane, B.A. (2006). Dissecting the contributions of *Clostridium perfringens* Type C toxins to lethality in the mouse intravenous injection model. Infec.t Immun. *74*, 5200–5210.

Fisher, D.J., Miyamoto, K., Harrison, B., Akimoto, S., Sarker, M R., and McClane, B.A. (2005). Association of beta2 toxin production with *Clostridium perfringens* type A human gastrointestinal disease isolates carrying a plasmid enterotoxin gene. Mol. Microbiol. *56*, 747–762.

Fujita, K., Katahira, J., Horiguchi, Y., Sonoda, N., Furuse, M., and Tsukita, S. (2000). *Clostridium perfringens* enterotoxin binds to the second extracellular loop of claudin-3, a tight junction integral membrane protein. FEBS Lett. *476*, 258–261.

Furuse, M., Hirase, T., Itoh, M., Nagafuchi, A., Yonemura, S., Tsukita, S., and Tsukita, S. (1993). Occludin: a novel integral membrane protein localizing at tight junctions. J. Cell Biol. *123*, 1777–1788.

Granum, P.E. (1982). Inhibition of protein synthesis by a tryptic polypeptide of *Clostridium perfringens* type A enterotoxin. Biochim. Biophys. Acta *708*, 6–11.

Granum, P.E., and Richardson, M. (1991). Chymotrypsin treatment increases the activity of *Clostridium perfringens* enterotoxin. Toxicon *29*, 898–900.

Granum, P.E., Whitaker, J.R., and Skjelkvale, R. (1981). Trypsin activation of enterotoxin from *Clostridium perfringens* type A: fragmentation and some physicochemical properties. Biochim. Biophys. Acta *668*, 325–332.

Hanna, P.C., and McClane, B.A. (1991). A recombinant C-terminal toxin fragment provides evidence that membrane insertion is important for *Clostridium perfringens* enterotoxin cytotoxicity. Mol. Microbiol. *5*, 225–230.

Hanna, P.C., Mietzner, T., Schoolnik, G.K., and McClane, B.A. (1991). Localization of the receptor-binding region of *Clostridium perfringens* enterotoxin utilizing cloned toxin fragments and synthetic peptides. The 30 C-terminal amino acids define a functional binding region. J. Biol. Chem. *266*, 11037–11043.

Hanna, P.C., Wieckowski, E.U., Mietzner, T., and McClane, B.A. (1992). Mapping of functional regions of *Clostridium perfringens* type A enterotoxin. Infect. Immun. *60*, 2110–2114.

Hanna, P.C., Wnek, A.P., and McClane, B.A. (1989). Molecular cloning of the 3′ half of the *Clostridium perfringens* enterotoxin gene and demonstration that this region encodes receptor-binding activity. J. Bacteriol. *171*, 6815–6820.

Harada, M., Kondoh, M., Ebihara, C., Takahashi, A., Komiya, E., Fujii, M., Mizuguchi, H., Tsunoda, S., Horiguchi, Y., Yagi, K., and Watanabe, Y. (2007). Role of tyrosine residues in modulation of claudin-4 by the C-terminal fragment of *Clostridium perfringens* enterotoxin. Biochem. Pharmacol. *73*, 206–214.

Hardy, S.P., Denmead, M., Parekh, N., and Granum., P E. (1999). Cationic currents induces by *Clostridium perfringens* type A enterotoxin in human intestinal CaCo-2 cells. J. Med. Microbiol. *48*, 235–243.

Heikinheimo, A., Lindström, M., Granum, P.E., and Korkeala, H. (2006). Humans as reservoir for enterotoxin gene-carrying *Clostridium perfringens* type A. Emerg. Infect. Dis. *12*, 1724–1729.

Horiguchi, Y., Akai, T., and Sakaguchi, G. (1987). Isolation and function of a *Clostridium perfringens* enterotoxin fragment. Infect. Immun. *55*, 2912–2915.

Katahira, J., Inoue, N., Horiguchi, Y., Matsuda, M., and Sugimoto, N. (1997a). Molecular cloning and functional characterization of the receptor for *Clostridium perfringens* enterotoxin. J. Cell Biol. *136*, 1239–1247.

Katahira, J., Sugiyama, H., Inoue, N., Horiguchi, Y., Matsuda, M., and Sugimoto, N. (1997b). *Clostridium perfringens* enterotoxin utilizes two structurally related membrane proteins as functional receptors *in vivo*. J. Biol. Chem. *272*, 26652–26658.

Kokai-Kun, J.F., Benton, K., Wieckowski, E.U., and McClane, B.A. (1999). Identification of a *Clostridium perfringens* enterotoxin region required for large complex formation and cytotoxicity by random mutagenesis. Infect. Immun. *67*, 5634–5641.

Kokai-Kun, J.F., and McClane, B.A. (1997). Deletion analysis of the *Clostridium perfringens* enterotoxin. Infect. Immun. *65*, 1014–1022.

Kominsky, S.L., Tyler, B., Sosnowski, J., Brady, K., Doucet, M., Nell, D., Smedley, 3rd., J., McClane, B., Brem, H., and Sukumar, S. (2007). *Clostridium perfringens* enterotoxin as a novel-targeted therapeutic for brain metastasis. Cancer Res. *67*, 7977–7982.

Kominsky, S.L., Vali, M., Korz, D., Gabig, T.G., Weitzman, S.A., Argani, P., and Sukumar, S. (2004). *Clostridium perfringens* enterotoxin elicits rapid and specific cytolysis of breast carcinoma cells mediated through tight junction proteins claudin 3 and 4. Am. J. Pathol. *164*, 1627–1633.

Kondoh, M., Masuyama, A., Takahashi, A., Asano, N., Mizuguchi, H., Koizumi, N., Fujii, M., Hayakawa, T., Horiguchi, Y., and Watanbe, Y. (2005). A novel strategy for the enhancement of drug absorption using a claudin modulator. Mol. Pharmacol. *67*, 749–756.

Labbe, R.G. (1989). *Clostridium perfringens*. In *Clostridium perfringens*, D. M.P., ed. (New York, Marcel Decker), pp. 192–234.

Lam, S., and Roth, J.R. (1983). IS*200*: a *Salmonella*-specific insertion sequence. Cell *34*, 951–960.

Li, J., Miyamoto, K., and McClane, B.A. (2007). Comparison of virulence plasmids among *Clostridium perfringens* type E isolates. Infect. Immun. *75*, 1811–1819.

Li, J., Sayeed, S., and McClane, B.A. (2008). The presence of enterotoxingenic *Clostridium perfringens* in Pittsburgh area soils and kitchens. Appl. Environ. Microbiol. 73, 7218–7224.

Litkouhi, B., Kwong, J., Lo, C.M., Smedley 3rd, J.G., McClane, B.A., Aponte, M., Gao, Z., Sarno, J.L., Hinners, J., Welch, W.R. (2007). Claudin-4 overexpression in epithelial ovarian cancer is associated with hypomethylation and is a potential target for modulation of tight junction barrier function using a C-terminal fragment of *Clostridium perfringens* enterotoxin. Neoplasia *9*, 304–314.

Long, H., Crean, C.D., Lee, W.H., Cummings, O.W., and Gabig, T.G. (2001). Expression of *Clostridium perfringens* enterotoxin receptors claudin-3 and claudin-4 in prostate cancer epithelium. Cancer Res. *61*, 7878–7881.

Masuyama, A., Kondoh, M., Seguchi, H., Takahashi, A., Harada, M., Fujii, M., Mizuguchi, H., Horiguchi, Y., and Watanabe, Y. (2005). Role of N-terminal amino acids in the absorption-enhancing effects of the C-terminal fragment of *Clostridium perfringens* enterotoxin. J. Pharmacol. Exp. Ther. *314*, 789–795.

Matsuda, M., and Sugimoto, N. (1979). Calcium-independent and dependent steps in action of *Clostridium perfringens* enterotoxin on HeLa and Vero cells. Biochem Biophys. Res. Commun. *91*, 629–636.

McClane, B.A. (1994). *Clostridium perfringens* enterotoxin acts by producing small molecule permeability alterations in plasma membranes. Toxicon *87*, 43–67.

McClane, B.A. (2007). *Clostridium perfringens* In: Montville, TJ. and M.P. Doyle, eds. Food Microbiology, Fundamentals and Frontiers. 3rd edn (Washington, ASM Press), pp. 423–444.

McClane, B.A., and McDonel, J.L. (1980). Characterization of membrane permeability alterations induced in Vero cells by *Clostridium perfringens* enterotoxin. Biochim. Biophys. Acta *600*, 974–985.

McClane, B.A., and Wnek, A.P. (1990). Studies of *Clostridium perfringens* enterotoxin action at different temperatures demonstrate a correlation between complex formation and cytotoxicity. Infect. Immun. *58*, 3109–3115.

McClane, B.A., Hanna, P.C., and Wnek, A.P. (1988a). *Clostridium perfringens* enterotoxin. Microb. Pathog. *4*, 317–323.

McClane, B.A., Wnek, A.P., Hulkower, K.I., and Hanna, P.C. (1988b). Divalent cation involvement in the action of *Clostridium perfringens* type A enterotoxin. Early events in enterotoxin action are divalent cation-independent. J. Biol. Chem. *263*, 2433–2435.

McClane, B.A., Uzal, F.A., Miyakawa, M.E.F., Lyerly, D., and Wilkins, T. (2006). The Enterotoxic Clostridia., Vol 4 (New York, Springer-Verlag), pp. 688–752.

McDonel, J.L. (1980). *Clostridium perfringens* toxins (type A, B, C, D, E). Pharmacol. Ther. *10*, 617–655.

McDonel, J.L. (1986). The role of toxins in bacterial pathogenesis. In Pharmacology of Bacterial Toxins., D. J. Dorner, ed. (New York, Pergamon Press), pp. 477–518.

McDonel, J.L., and Demers, G.W. (1982). *In vivo* effects of enterotoxin from *Clostridium perfringens* type A in the rabbit colon: binding vs. biologic activity. J. Infect. Dis. *145*, 490–494.

McDonel, J.L., and Duncan, C.L. (1975). Histopathological effect of *Clostridium perfringens* enterotoxin in the rabbit ileum. Infect. Immun. *12*, 1214–1218.

McDonel, J.L., Chang, L.W., Pounds, J.G., and Duncan, C.L. (1978). The effects of *Clostridium perfringens* enterotoxin on rat and rabbit ileum: an electron microscopic study. Lab. Invest. *39*, 210–218.

Mead, P.S., Slutsker, L., Dietz, V., McCaig, L.F., Bresee, J.S., Shapiro, C., Giffin, P.M., and Tauxe, R.V. (1999). Food-related illness and death in the United States. Emerg. Infect. Dis *5*, 607–625.

Melville, S.B., Collie, R.E., and McClane, B.A. (1997). The Clostridia: Molecular biology and pathologenesis. (London, Academic Press.).pp. 471–487.

Melville, S.B., Labbe, R.G., and Sonenshein, A.L. (1994). Expression from the *Clostridium perfringens cpe* promoter in *C. perfringens* and *Bacillus subtilis*. Infect. Immun. *62*, 5550–5558.

Michl, P., Buchholz, M., Rolke, M., Kunsch, S., Lohr, M.M., McClane, B.A., Tsukita, S., Leder, G., Adler, G., and Gress, T.M. (2001). Claudin-4: a new target for pancreatic cancer treatment using *Clostridium perfringens* enterotoxin. Gastroenteology *121*, 678–684.

Mietzner, T.A., Kokai-Kun, J.F., Hanna, P.C., and McClane, B.A. (1992). A conjugated synthetic peptide corresponding to the C-terminal region of *Clostridium perfringens* enterotoxin elicits an enterotoxin-neutralizing antibody response in mice. Infect. Immun. *60*, 3947–3951.

Miyakawa, M.E.F., Fisher, D.J., Poon, R., Sayeed, S., Adams, V., Rood, J., McClane, B.A., and Uzal, F.A. (2007). Both epsilon-toxin and beta-toxin are important for the lethal properties of *Clostridium perfringens* type B isolates in the mouse intravenous injection model. Infect. Immun. *75*, 1443–1452.

Miyamoto, K., Chakrabarti, G., Morino, Y., and McClane, B.A. (2002). Organization of the plasmid *cpe* locus in *Clostridium perfringens* type A isolates. Infect. Immun. *70*, 4261–4272.

Miyamoto, K., Fisher, D.J., Li, J., Sayeed, S., Akimoto, S., and McClane, B.A. (2006). Complete sequencing and diversity analysis of the enterotoxin-encoding plasmids in *Clostridium perfringens* Type A non-food-borne human gastrointestinal disease isolates. J. Bacteriol. *188*, 1585–1598.

Miyamoto, K., Wen, Q., and McClane, B.A. (2004). Multiplex PCR genotyping assay that distinguishes between isolates of *Clostridium perfringens* type A carrying a chromosomal enterotoxin gene (*cpe*) locus, a plasmid cpe locus with an IS*1470*-like sequence, or a plasmid cpe locus with an IS*1151* sequence. J. Clin. Microbiol. *42*, 1552–1558.

Naik, H.S., and Duncan, C.L. (1978). Detection of *Clostridium perfringens* enterotoxin in human fecal samples and anti-enterotoxin in sera. J. Clin. Microbiol. *7*, 337–340.

Nichols, L.S., Ashfaq, R., and Iacobuzio-Donahue, C.A. (2004). Claudin 4 protein expression in primary and metastatic pancreatic cancer: support for use as a therapeutic target. Am. J. Clin. Pathol. *121*, 226–230.

Parker, M.W., and Feil, S.C. (2005). Pore-forming protein toxins: from structure to function. Prog. Biophys. Mol. Biol *88*, 91–142.

Parsons, J.A., Bannam, T.L., Devenish, R.J., and Rood, J.I. (2007). *Tcp*A, an *Fts*K/*Spo*IIIE homolog, is essential for transfer of the conjugative plasmid pCW3 in *Clostridium perfringens*. J.Bacteriol. (in press).

Petit, L., Gilbert, M., and Popoff, M.R. (1999). *Clostridium perfringens*: toxinotype and genotype. Trends Microbiol. *7*, 104–110.

Robertson, S.L., Smedley 3rd, J. G., Singh, U., Chakrabarti, G., Van Itallie, C.M., Anderson, J. M., and McClane, B.A. (2007). Compositional and stoichiometric analysis of *Clostridium perfringens* enterotoxin complexes in Caco-2 cells and claudin 4 fibroblast transfectants. Cell. Microbiol. (in press).

Saito, M. (1990). Production of enterotoxin by *Clostridium perfringens* derived from humans, animals and the natural environment in Japan. J. Food Prot. 53, 115–118.

Santin, A.D., Cane, S., Bellone, S., Palmieri, M., Siegel, E.R., Thomas, M., Roman, J.J., Burnett, A., Cannon, M.J., and Pecorelli, S. (2005). Treatment of chemotherapy-resistant human ovarian cancer xenografts in C.B-17/SCID mice by intraperitoneal administration of *Clostridium perfringens* enterotoxin. Cancer Res. *65*, 4334–4342.

Santin, A.D., Bellone, S., Siegel, E.R., McKenney, J.K., Thomas, M., Roman, J.J., Burnett, A., Tognon, G., Bandiera, E., and Pecorelli, S. (2007). Overexpression of *Clostridium perfringens* enterotoxin receptors claudin-3 and claudin-4 in uterine carcinosarcomas. Clin. Cancer Res. *13*, 3339–3346.

Sarker, M. R., Carman, R.J., and McClane, B.A. (1999). Inactivation of the gene (*cpe*) encoding *Clostridium perfringens* enterotoxin eliminates the ability of two-*cpe*-positive C. *perfringens* type A human gastrointestinal disease isolates to affect rabbit ileal loops. Mol. Microbiol. 33, 946–958.

Sayeed, S., Fernandez-Miyakawa, M.E., Fisher, D. J., Adams, V., Poon, R., Rood, J.I., Uzal, F.A., and McClane, B.A. (2005). Epsilon-toxin is required for most *Clostridium perfringens* Type D vegetative culture supernatants to cause lethality in the mouse intravenous injection model. Intect Immune *73*, 7413–7421.

Sayeed, S., Li, J., and McClane, B.A. (2007). Virulence plasmid diversity in *Clostridium perfringens* type D isolates. Infect. Immun. *75*, 2391–2398.

Sherman, S., Klein, E., and McClane, B.A. (1994). *Clostridium perfringens* type A enterotoxin induces tissue damage and fluid accumulation in rabbit ileum. J Diarrhoeal Dis Res *12*, 200–207.

Singh, U., Mitic, L.L., Wieckowski, E.U., Anderson, J.M., and McClane, B.A. (2001). Comparative biochemical and immunocytochemical studies reveal differences in the effects of *Clostridium perfringens* enterotoxin on polarized CaCo-2 cells versus Vero cells. J. Biol. Chem. *276*, 33402–33412.

Singh, U., Van Itallie, C.M., Mitic, L.L., Anderson, J M., and McClane, B.A. (2000). CaCo-2 cells treated with *Clostridium perfringens* enterotoxin form multiple large complex species, one of which contains the tight junction protein occludin. J. Biol. Chem. *275*, 18407–18417.

Skjelkvale, R., and Uemura, T. (1977). Experimental diarrhea in human volunteers following oral administration of *Clostridium perfringens* enterotoxin. J Appl Bacteriol *46*, 281–286.

Smedley 3rd, J.G., Fisher, D J., Sayeed, S., Chakrabarti, G., and McClane, B.A. (2004). The enteric toxins of *Clostridium perfringens*. Rev Physiol Biochem Pharmacol *152*, 183–204.

Smedley 3rd, J.G., and McClane, B.A. (2004). Fine mapping of the N-terminal cytotoxicity region of *Clostridium perfringens* enterotoxin by site-directed mutagenesis. Infect. Immun. *72*, 6914–6923.

Smedley 3rd, J.G., Saputo, J., Parker, J.C., Fernandez–Miyakawa, M.E., McClane, B.A., Robertson, S. and Hzal, F.A. (2008). Non-cytotoxic *Clostridium perfringens* enterotoxin (CPE) variants localize CPE intestinal binding and demonstrate a relationship between CPE-induced cytotoxicity and enterotoxicity. Infect. Immun. In press.

Smedley 3rd, J. G., Uzal, F.A., and McClane, B.A. (2007). Identification of a prepore large-complex stage in the mechanism of action of *Clostridium perfringens* enterotoxin. Infect. Immun. *75*, 2381–2390.

Sonoda, N., Furuse, M., Sasaki, H., Yonemura, S., Katahira, J., Horiguchi, Y., and Tsukita, S. (1999). *Clostridium perfringens* enterotoxin fragment removes specific claudins from tight junction strands: Evidence for direct involvement of claudins in tight junction barrier. J. Cell Biol. *147*, 195–204.

Sparks, S.G., Carman, R.J., Sarker, M.R., and McClane, B.A. (2001). Genotyping of enterotoxigenic *Clostridium perfringens* fecal isolates associated with antibiotic-associated diarrhea and food poisoning in North America. J. Clin. Microbiol. *39*, 883–888.

Takahashi, A., Kondoh, M., Masuyama, A., Fujii, M., Mizuguchi, H., Horiguchi, Y., and Watanabe, Y. (2005). Role of C-terminal regions of the C-terminal fragment of *Clostridium perfringens* enterotoxin in its interaction with claudin-4. J. Control Release *108*, 56–62.

Tanaka, D., Kimata, K., Shimizu, M., Isobe, J., Watahiki, M., Karasawa, T., Yamagishi, T., Kuramoto, S., Serikawa, T., Ishiguro, F. (2007). Genotyping of *Clostridium perfringens* isolates collected from food poisoning outbreaks and healthy individuals in Japan based on the *cpe* locus. Jpn. J. Infect. Dis. *60*, 68–69.

Wieckowski, E.U., Wnek, A.P., and McClane, B.A. (1994). Evidence that an approximately 50-kDa mammalian plasma membrane protein with receptor-like properties mediates the amphiphilicity of specifically bound *Clostridium perfringens* enterotoxin. J. Biol. Chem. *269*, 10838–10848.

Wnek, A P., and McClane, B.A. (1983). Identification of a 50,000 Mr protein from rabbit brush border membranes that binds *Clostridium perfringens* enterotoxin. Biochem. Biophys. Res. Commun. *112*, 1099–1105.

Zhao, Y., and Melville, S.B. (1998). Identification and characterization of sporulation-dependent promoters upstream of the enterotoxin gene (*cpe*) of *Clostridium perfringens*. J. Bacteriol. *180*, 136–142.

The Cholesterol-dependent Cytolysins and *Clostridium septicum* α-toxin: Pore-forming Toxins of the Clostridia

3

Eileen M. Hotze and Rodney K. Tweten

Abstract

Two classes of pore-forming toxins of the clostridia are represented by the cholesterol-dependent cytolysins (CDCs) and the *Clostridium septicum* α-toxin. The CDCs are found in a wide variety of clostridial species, but are also found in many species from other Gram-positive genera. As a result, various CDCs have evolved specific traits that appear to enhance their ability to complement the pathogenic mechanism of a specific bacterial species. In contrast, closely related toxins to *C. septicum* α-toxin (AT) have not been found in other species of the clostridia, although *C. perfringens* epsilon toxin appears to be distantly related. Remarkably, distant relatives of AT have been found in species of Gram-negative bacteria as well as certain species of mushrooms and the enterolobin tree seed. Although the CDCs appear to be restricted to Gram-positive bacterial pathogens it has recently been shown that the unusual protein fold of their membrane-penetrating domain is present in proteins of the eukaryotic complement membrane attack complex. Both toxins penetrate the membrane by the use of a β-barrel pore but differ significantly in their pore-forming mechanisms. The contribution of both classes of toxins to disease is not yet well understood for the clostridia. It is clear they play important, but probably different, roles in clostridial disease.

Introduction

The species of the genus *Clostridium* are among the most prolific toxin-producing bacteria. The pathogenic clostridia produce toxins that exhibit a variety of mechanisms and contribute in many different ways to pathogenesis. Whereas most of the enzymatically active toxins undertake complex trafficking pathways to deliver the catalytic fragment to the cytosol of the host cell, the pore-forming toxins offer the most direct mechanism by targeting the host membrane directly. There are many different classes of pore-forming toxins including those represented by *Clostridium perfringens* enterotoxin, β-toxin, and epsilon toxin. The most prevalent of these pore-forming toxins are the cholesterol-dependent cytolysins (CDCs) that are found in many clostridial species, as well as in many species from other genera of Gram-positive bacterial pathogens. This review will describe in detail the surprisingly complex mechanism by which the CDCs form pores in eukaryotic membranes. Furthermore, it will become apparent in this review that the CDCs contribute in many ways to pathogenesis, some of which do not apparently depend on pore-formation.

Clostridium septicum produces the pore-forming toxin α-toxin (AT) that has not yet been found in other clostridial species or other Gram-positive bacteria. AT producing bacteria are not as abundant as the CDC producers and its relatives come from widely disparate sources. Interestingly, a distant relative of AT can be found in a Gram-negative bacterial species. *Aeromonas hydrophila* produces aerolysin, which exhibits primary structure and mechanistic similarities to AT. In contrast to the CDCs, AT forms a heptameric complex and so its pore is much smaller than that of the CDCs. As discussed below the cytolytic mechanism of AT exhibits significant differences from the CDCs.

In this review we will survey the work on the pore-forming mechanisms of the CDCs and AT and their roles in pathogenesis. Since the CDCs range across a wide variety of Gram-positive pathogens we will not only discuss the aspects of the clostridia-derived CDCs, but those from other Gram-positive pathogens to provide the reader with a greater sense of the versatility of the CDCs in pathogenesis.

The cholesterol-dependent cytolysins (CDCs)

Overview

The CDCs are a class of pore-forming toxins that act on cholesterol-rich eukaryotic cell membranes. They are unusual among pore-forming toxins as the diameter of the CDC pore measures 200–300 Å, a diameter over 10 times larger than those pores formed by the smaller pore formers such as α-haemolysin from *Staphylococcus aureus* or *Clostridium septicum* AT. The CDCs are among the most widespread virulence factors and are found in more than 25 different species in five Gram-positive genera (*Clostridia, Listeria, Bacillus, Streptococcus* and *Archanobacterium*). It is likely many more will be discovered in the future. To date, the CDC gene has been identified in nine species of Clostridia one of which, perfringolysin O (PFO), from *C. perfringens* has served as the model for studying the pore-forming mechanism of the CDCs (reviewed in Tweten, 2005).

Although the CDCs were discovered over 80 years ago our understanding of the molecular mechanism of pore formation has only been revealed in depth in the last 10–20 years due to technological advances and the solution of the crystal structure of PFO. In contrast, we understand much less about the role these toxins play in disease. While the contribution of the CDCs to the pathogenic mechanisms of a few pathogens is understood to some extent, much remains to be discovered. It is apparent the structures of some of the CDCs have evolved so they better fit the pathogenic mechanism of the pathogen by the introduction of specific traits. Yet, this evolutionary process has not compromised the basic pore-forming mechanism. The ability of the CDC structure to adapt to the various pathogenic mechanisms reflects a plasticity of the CDC structure.

How soluble CDC monomers make the transition to a membrane bound, pore-forming oligomer has been a subject of intense study. Using the clostridial toxin PFO as a model, this chapter will review the work to date that has moulded the current view of the mechanism of pore formation by the CDCs. In addition we will explore how Clostridia, and other CDC producing bacteria, utilize the CDC during infection.

CDC Structure

A sequence alignment of the clostridial CDCs reveals 98% identity between the CDCs isolated from *C. perfringens*, *C. septicum*, and *C. novyi*, suggesting that they have not diverged significantly. This is not unexpected since histotoxic clostridia cause a common disease, gas gangrene. In contrast, the *C. tetani* CDC, tetanolysin, exhibits only 55% sequence identity indicating that it diverged from this group, consistent with the fact that it is not a histotoxic clostridial species. A dendrogram of the primary structures of all of the members of the CDC family (Fig. 3.1) shows that the CDCs have undergone some divergent evolution indicative of selective pressures that have led to changes in their primary structures. Some differences are translated into obvious structural differences in the CDCs. These include the lack of the *sec*-dependent signal sequence on *Streptococcus pneumoniae* pneumolysin (PLY), the addition of about 75 amino acids on the N-terminal region of streptolysin O (SLO) from *Streptococcus pyogenes*, and the addition of a 150 amino acid region which appears to encode for a fucose-binding lectin on the N-terminus of a CDC identified in *Streptococcus mitis*. How these changes in the CDC structure might contribute to the pathogenic mechanism of these and other species remain largely unresolved. There are also less obvious changes in the primary structure of the CDC itself that can affect various features of the mechanism or add to the functionality of the CDC. For example, the pore-forming activity of listeriolysin O (LLO) from *Listeria monocytogenes* exhibits an acidic pH optimum, pneumolysin can activate complement and inter-

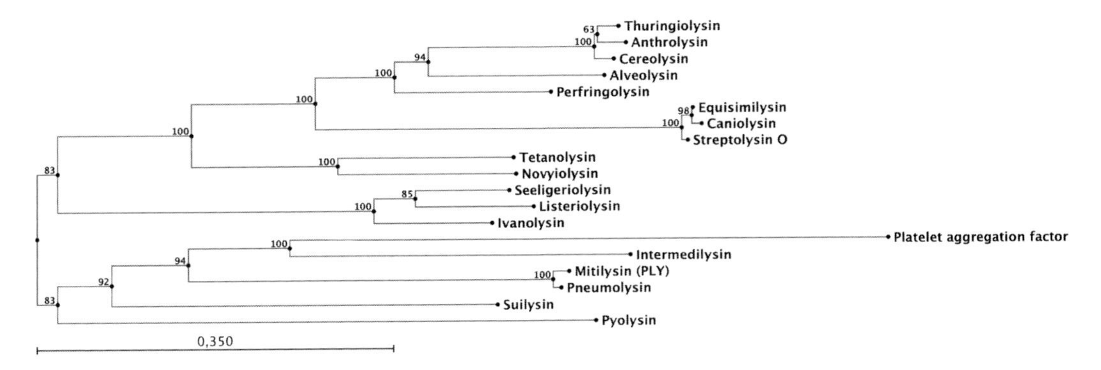

Figure 3.1 CDC phylogenetic tree. Shown is a dendrogram of the sequenced CDCs. Alignment and tree drawing was performed with CLC Free Workbench (CLCbio).

medilysin (ILY) from *Streptococcus intermedius* is human cell specific and binds a human specific receptor rather than directly to cholesterol-rich membranes. These changes appear to result from alterations in the basic CDC structure. The structural differences identified thus far and the implications of these structural changes in disease will be discussed later in the chapter. As yet, no changes in the structures of clostridial CDCs have been identified that appear to encode unique properties.

The crystal structures of two of the more distantly related members of the CDC family, PFO and ILY (Polekhina *et al.*, 2005), have been solved (Fig. 3.2). Not surprisingly the crystal structures revealed two molecules with very similar 3D structures (Fig. 3.2). The crystal structure of the soluble monomer of PFO was the first to be solved and showed that it was an elongated rod-shaped molecule rich in β-sheets. PFO is comprising four discontinuous domains with domain 4 containing the highly conserved undecapeptide or trp-rich sequence (ECTGLAWEWWR). Domain 3 was found to be packed loosely against domain 2 and contained a central β-sheet flanked by two clusters of α-helices that were later identified as the region of the molecule which penetrates the membrane bilayer. Domain 2 consists of four β-strands and is connected to domain 4 through a glycine linker at residue 392. Domain 1 has a highly pronounced curvature and modelling studies predicted that this domain formed extensive contacts with domain 1 in adjacent monomers in the oligomeric complex (Rossjohn *et al.*, 1997), a prediction supported by antibody and electron micrograph data (Darji *et al.*, 1996; de los Toyos *et al.*, 1996).

With an overall sequence identity of 40%, the crystal structure of ILY revealed a fold and topology similar to PFO with the greatest difference in the membrane-binding region of the toxin (domain 4). By comparison with PFO domain 4 of ILY was rotated 20–30° away from the main axis of the molecule (Polekhina *et al.*, 2005). The conformation of the tryptophan-rich loop (the conserved undecapeptide) was also quite different between these two proteins. This is not surprising considering that ILY bears a variant undecapeptide with some pronounced differences (GATGLAWEPWR) from the consensus sequence (ECTGLAWEWWR). The trp-rich loop of ILY adopts an extended conformation compared to the loop of PFO that is folded up against the β-sheet of Domain 4 (Polekhina *et al.*, 2005). It had previously been predicted that cholesterol could in fact bind to PFO on one face of the β-sheet in domain 4 if the trp-rich loop had been displaced. Modelling the extended conformation of the ILY loop on PFO would provide such a binding site. This predicted cholesterol site was not found on ILY in accordance with the fact that ILY does not require cholesterol for membrane binding. Although the majority of CDCs seem to recognize cholesterol-rich membranes no specific binding site had been rigorously identified until recently. As described below the cholesterol-dependent elements have recently been identified, but the contribution of cholesterol to the CDC mechanism is more complex than a simple binding event.

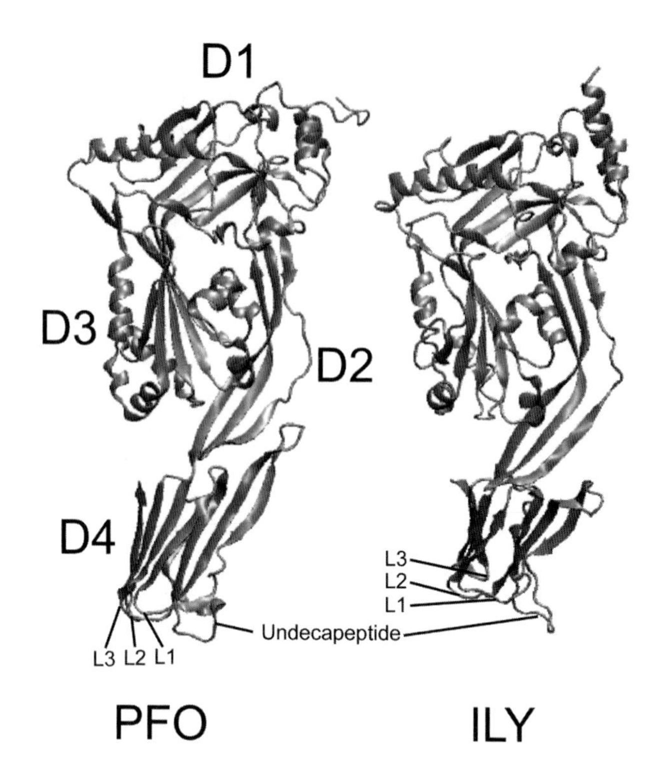

Figure 3.2 Crystal structures of *Clostridium perfringens* PFO and *Streptopcoccus intermedius* ILY. Shown are ribbon representations of the α-carbon backbone structures of PFO (left) and ILY (right) derived from their crystal structures (Polekhina *et al.*, 2005; Rossjohn *et al.*, 1997). The transmembrane region of each toxin is located within D3 and the membrane-binding domain is located in D4. Shown are the locations of the three loops (L1–L3) at the base of D4 that mediate the cholesterol-dependent interaction of both toxins with the membrane (Soltani *et al.*, 2007b). The location of the conserved undecapeptide is also denoted. D1-D4 domains 1–4. The ribbon structures were generated using VMD (Humphrey *et al.*, 1996).

The CDC cytolytic mechanism

Overview

The only pore-forming toxin for which the crystal structures of both the soluble monomer and pore complex have been solved is that of α-haemolysin from *Staphylococcus aureus* (Song *et al.*, 1996). Although the monomer of the α-haemolysin was not actually solved, the closely related leukocidin F (LukF) monomer was solved (Olson *et al.*, 1999). These structures revealed comparatively few conformation changes were required for the transition of the soluble monomer to the membrane complex. In contrast, the CDCs undergo major secondary and tertiary conformational changes during their transition to the pore complex. The soluble monomers of most CDCs do not oligomerize in solution, but instead must interact with cholesterol-rich membranes to initiate pore formation. Binding to the cell membrane alone sets off a cascade of conformational changes that lead to pore formation. PFO has been one of the major models for deciphering the mechanics of pore formation. The crystal structure of the soluble monomer of PFO and fluorescence spectroscopic techniques have been key tools in the study of the CDC pore-forming mechanism. A high-resolution crystal structure of the CDC oligomeric complex is unlikely due to the heterogeneity in the number of monomers incorporated into the pore complex. This has necessitated the use of other approaches to identify these structural transitions. In large part these structural changes have been identified by a combination of fluorescence spectroscopy and cysteine mutagenesis. In addition, other approaches such as planar bilayer analysis, 3D reconstruction of electron microscopic images and atomic force microscopy have contributed to our understanding of the structure of the oligomeric pore complex. Below is a compilation of the structural changes identified thus far during the transition of the soluble CDC monomer to the membrane-embedded pore complex.

Membrane recognition and the cholesterol dependence of the CDCs

The majority of the CDCs are processed and secreted as soluble monomers via the *sec*-dependent secretion pathway, the exception being PLY which lacks a signal peptide and appears to be released by other mechanisms (Walker *et al.*, 1987). Following secretion the soluble CDC monomers then bind to and form pores exclusively on membranes that contain cholesterol. The ability of the CDCs to exhibit such a high degree of selectivity between sterol- and non-sterol-containing membranes is remarkable. This ability targets these toxins to eukaryotic membranes and probably protects the bacterial membrane from the cytolytic activity of the toxin. Early studies in which the toxin was pretreated with cholesterol were shown to abolish cytolytic activity and it was thought that the cholesterol occupied a cholesterol binding site that prevented it from binding to cholesterol containing membranes. These studies led to the assumption that cholesterol acted as a receptor or recognition site for the toxin (reviewed in Alouf, 1999, 2000; Tweten *et al.*, 2001).

Recent reports, however, suggest that the role of cholesterol in the CDC cytolytic mechanism is more complicated and not limited to binding. For example, Jacobs *et al.*, reported that pretreatment of Listeriolysin O (LLO), the CDC from *Listeria monocytogenes*, with cholesterol resulted in a loss of cytolytic activity due a defect in oligomerization rather than cell binding (Jacobs *et al.*, 1998). More recently, Giddings *et al.*, showed that ILY did not bind to cell membranes via cholesterol, but instead utilized CD59 as its receptor (Giddings *et al.*, 2004). In a separate study, Giddings *et al.* showed that SLO and ILY were able to bind and oligomerize on membranes treated with the cholesterol-depleting agent methyl-β-cyclodextrin (MβCD) but their cytolytic activity were severely inhibited by the depletion of cholesterol (Giddings *et al.*, 2003). These studies showed that although ILY did not bind to cells via cholesterol, its mechanism still remained dependent on the presence of membrane cholesterol. Soltani *et al.* subsequently reported that cholesterol was required for conversion of the prepore oligomer to the inserted pore complex in ILY (Soltani *et al.*, 2007a). Currently this collective evidence suggests a multifaceted role for cholesterol in cytolysis.

Early studies of proteolytically derived C-terminal fragments of PFO showed that domain 4 was able to interact with the membrane (Ohno-Iwashita *et al.*, 1988; Tweten, 1988; Tweten *et al.*, 1991). In addition, tryptophan residues within the undecapeptide region of this fragment were shown to penetrate the membrane as shown by an increase in tryptophan emission upon binding (Nakamura *et al.*, 1995) and the fact that the emission of the tryptophans could be quenched by membrane restricted collisional quenchers (Heuck *et al.*, 2003). Since that time fluorescent studies further confirmed that domain 4 of the intact toxin is responsible for membrane recognition. By placing fluorescent probes throughout domain 4 researchers showed that upon initial binding the tip of domain 4 (Fig. 3.2) is superficially embedded in the non-polar interior of the bilayer (Heuck *et al.*, 2000). Recently it was determined that a series of short hydrophobic loops at the tip of domain 4 β-sandwich are also exposed to the bilayer interior. These spectroscopic studies determined that the rest of domain 4 is oriented perpendicular to the membrane surface and is surrounded by water in both the membrane-bound monomer and the oligomeric complex indicating that no appreciable monomer-monomer contacts are made in this region of the toxin (Ramachandran *et al.*, 2002).

It had been long held that the conserved undecapeptide mediated the interaction of the CDCs with cholesterol-rich membranes. This supposition was not unwarranted as many aspects of this peptide region suggested that it did mediate this interaction. Some of these included its highly conserved nature, the fact that it inserts into the membrane surface (Heuck *et al.*, 2000; Nakamura *et al.*, 1995; Sekino-Suzuki *et al.*, 1996) and that early studies suggested chemical modification or mutation of residues in this region affected binding (Iwamoto *et al.*, 1987). Furthermore, the undecapeptide of ILY, the only

CDC known to bind a specific receptor rather than directly to cholesterol-rich membranes, has an altered undecapeptide structure.

Soltani *et al.*, recently provided a detailed investigation of the interaction of domain 4 with cholesterol-rich membranes (Soltani *et al.*, 2007b). Their studies provided an answer to the basis for the cholesterol dependence of the CDCs, the basis for the hallmark sensitivity to oxidation and provided a unifying basis for the cholesterol dependence of PFO-like CDCs that bind directly to cholesterol-rich membranes and those like ILY that first bind to a another receptor. This investigation was prompted by the fact that the pore-forming mechanisms of both ILY and PFO were dependent on membrane cholesterol, yet, as described above, it was established that ILY bound to huCD59 and did not require cholesterol to bind to the membrane. Giddings *et al.* (Giddings *et al.*, 2003) had shown that removal of cholesterol from the membrane trapped ILY in a prepore complex. These observations posed a conundrum: did cholesterol contribute to the pore-forming mechanism of these two CDCs in fundamentally different ways? Soltani *et al.* (Soltani *et al.*, 2007b) first showed that the undecapeptide was, in fact, not the structural element of the CDCs that mediated their interaction with cholesterol-rich membranes. They demonstrated that the three short loops, previously shown by Ramachandran (Ramachandran *et al.*, 2002) to insert into the membrane surface, were responsible for mediating the interaction of the CDCs with cholesterol-rich membranes (Fig. 3.2). They further showed that these same loops in ILY inserted into the membrane in a cholesterol-dependent fashion after it bound to its receptor, huCD59. Hence, for the PFO-like CDCs these loops mediate their direct interaction with cholesterol-rich membranes whereas in ILY these loops insert in a cholesterol-rich fashion after it binds receptor. In both cases the insertion of these loops is necessary to form a functional pore.

A previous study by Soltani *et al.* (Soltani *et al.*, 2007a) showed that mutation of the undecapeptide of ILY could prevent the membrane insertion of the three domain 4 loops. Therefore, although the undecapeptide does not mediate the direct interaction of these toxins with cholesterol-rich membranes, its structure can modulate the interaction of the three loops with the membrane. Presumably, the altered structure of the ILY undecapeptide prevents the direct interaction of its three loops with cholesterol-rich membranes until after it binds to CD59.

Soltani *et al.* (Soltani *et al.*, 2007b) also showed that oxidation of the undecapeptide cysteine of PFO, a treatment known to dramatically reduce the cytolytic activity of CDCs (thus the former name of thiol-activated cytolysins), did not affect the interaction of the PFO domain 4 loops with cholesterol-rich membranes. It did, however, prevent the membrane insertion of the undecapeptide tryptophan residues that trapped PFO in a prepore complex. This key evidence showed that the membrane insertion of the undecapeptide could be uncoupled from the recognition of cholesterol-rich membranes by the D4 loops. The membrane insertion of the undecapeptide therefore occurs after the three domain 4 loops insert into cholesterol-rich membranes, but before the membrane insertion of the transmembrane β-hairpins in domain 3 that form the β-barrel pore. Since oxidation of the conserved undecapeptide cysteine residue blocked the formation of a functional pore but not oligomerization of the monomers into the prepore complex the insertion of the undecapeptide is apparently necessary to the formation of a functional β-barrel, even though the undecapeptide does not directly contribute to the β-barrel structure.

Identification of the transmembrane domain of PFO

The identification of the transmembrane region of PFO that forms the membrane pore by Shepard and Shatursky (Shatursky *et al.*, 1999; Shepard *et al.*, 1998) nullified the original hypothesis that the domain 4 β-sandwich was responsible for penetrating the membrane and forming the pore (Rossjohn *et al.*, 1997). This original idea was based on spectroscopic evidence by Nakamura *et al.*, that some of the domain 4 tryptophans were exposed to the lipid. However, later reports by Palmer *et al.*, utilizing SLO suggested that a region within domain 3 might be associating with the membrane (Palmer *et al.*, 1996). Shepard *et al.*, conclusively identified a region of domain 3 responsible for penetrating the lipid bilayer and its transmembrane structure by using cysteine scanning mutagenesis and several fluorescent techniques exploiting the environmentally sensitive characteristics of the fluorescent dye NBD (*N,N′*-dimethyl-*N*-(iodoacetyl)-*N′*-(7-nitrobenz-2-oxa-1,3-diazolyl)ethylenediamine). They showed that an α-helical bundle located in domain 3 underwent a structural transition to an amphipathic transmembrane β-hairpin (TMH1) in the PFO pore complex (Shepard *et al.*, 1998). The same group subsequently showed that a second α-helical cluster in domain 3 (TMH2) located on the opposite side of the core β-sheet near TMH1 underwent the same structural transition and inserted into and crossed the bilayer (Shatursky *et al.*, 1999). These studies revealed two remarkable features of the CDCs: 1) the transmembrane pore was formed by the contribution of two transmembrane hairpins from each CDC monomer and 2) these TMHs were packed in domain 3 of the soluble monomer as two clusters of three α-helices that underwent a change in their secondary structure to form the transmembrane β-hairpins.

The structural implications of these findings suggested that the α-helical packing motif of the TMHs is a protective mechanism that prevents the exposure of the hydrophobic surface of the amphipathic TMHs in the soluble monomer. Not surprisingly, these hydrophobic residues are mainly found packed against the core of the protein or inside the α-helices with the hydrophilic residues exposed on the surface. This also implies that in order for the α-helices to convert to the TMHs domain 3 must disengage from its interface with domains 1 and 2 in order to extend TMH1 into an amphipathic β-hairpin. Hence, a delicate balancing act is maintained in domain 3 whereby the interaction of domain 3 with domains 1 and 2 is maintained within the monomer, yet can be disrupted upon its interaction with the membrane so that the α-helices can extend into twin TMHs and form the β-barrel pore. This is consistent with the non-complementary nature of the interface between domain 3 and domains 1 and 2 (Rossjohn *et al.*, 1997).

Later it was confirmed that the insertion of the TMHs occurred only after membrane interaction by domain 4 (Heuck *et al.*, 2000). This ordered interaction was thought to prevent the toxin from undergoing possible off pathway conformational changes prior to binding to the membrane. It was shown by Heuck *et al.*, that domains 4 and 3 were conformationally coupled, suggesting that upon the interaction of domain 4 with the membrane conformational signals were communicated to domain 3 that triggered changes that lead to oligomerization and insertion of the β-barrel pore (Heuck *et al.*, 2000). In addition, it was shown that, although domain 4 interacts with the membrane before the TMHs, mutations within domain 3 could affect the rates of insertion of structural elements in both domains 3 and 4. These results demonstrated that although domain 3 and 4 are spatially separated from one another in the toxin, they were conformationally coupled. The structural elements responsible for coupling these two domains have not yet been resolved.

To date no other pore-forming toxin has been shown to exhibit a protein fold similar to that of the CDCs domain 3 TMHs. Recently, however, the crystal structures for the membrane penetrating protein C8α of the complement membrane attack complex (Hadders *et al.*, 2007) and a C9-like molecule from *Photorhabdus luminescens* (MACPF) (Rosado *et al.*, 2007) exhibited a similar fold. These structural findings suggest that these molecules may have a common evolutionary origin.

Formation of the prepore complex

Early studies of the oligomeric pore of PFO had estimated the involvement of approximately 50 monomers. The means by which the pore was created was a significant debate within the field. Rossjohn *et al.* (1997) had hypothesized that the CDCs formed a prepore complex. A prepore complex is a completely oligomerized complex that forms prior to insertion of the β-barrel pore. This hypothesis was not unprecedented as smaller pore-forming toxins such as aerolysin, *S. aureus* α-haemolysin, protective antigen from anthrax, and *C. septicum* α-haemolysin had been shown to assemble into membrane-bound prepore complexes prior to insertion of the pore (Miller *et al.*, 1999; Panchal and Bayley, 1995; Sellman *et al.*, 1997; Van der Goot *et al.*, 1993; Walker *et al.*, 1995). However, these smaller pore-forming toxins formed heptameric prepore complexes, a significantly smaller number of monomers than the CDC complex. Palmer *et al.* (1998) suggested that participation of such a large number of molecules contributing to a prepore complex was difficult to envision and energetically unfavourable due to the amount of membrane to be displaced by the insertion of the large β-barrel pore. Therefore, they proposed a model by which the membrane insertion of the CDCs was initiated with the insertion of the transmembrane domains of a CDC dimer and that monomers were added to the dimer, thus growing the inserted oligomer from a small pore to a larger pore.

Several important studies aided in distinguishing between these two models and resolved the controversy by directly demonstrating the presence of a prepore complex. Shepard *et al.*, examined the assembly of the PFO cytolytic complex to determine whether it formed an oligomeric prepore complex on the membrane prior to the insertion of its membrane-spanning β-sheet (Shepard *et al.*, 2000). They reported that PFO oligomeric complexes could be formed on liposomes at temperatures that did not support pore formation. At low temperature, the processes of oligomerization and membrane insertion could be resolved, and PFO was found to form an oligomer without significant membrane insertion of its β-hairpins. Once the temperature was raised insertion of the β-barrel complex and pore formation proceeded rapidly. These results argued against the suggestion by Palmer *et al.* (1998) that displacement of the large region of membrane by the CDC β-barrel was energetically unfavourable. Since the pore was actually formed more rapidly if PFO was allowed to oligomerize it was clear that formation of the prepore was not energetically unfavourable and, in fact, appeared to facilitate pore formation. Furthermore, PFO was found to increase the conductivity through a planar bilayer by large and discrete stepwise changes in conductance that are consistent with the membrane insertion of a large preassembled pore complex.

Shortly thereafter Hotze *et al.*, showed that PFO could be trapped in the prepore state by engineering a disulfide bond between TMH1 and domain 2 of the toxin, thus preventing the insertion of the TMHs, but not the oligomerization of the monomers (Hotze *et al.*, 2001). Upon reduction of the disulfide, the oligomerized prepore complex rapidly inserted the TMHs into the membrane and formed a pore. Heuck *et al.*, utilized a novel approach that directly determined if small pores could be formed by CDCs. Pore formation was examined for PFO and SLO by encapsulating large and small molecules into liposomes and measuring their simultaneous rate of release under limiting concentrations of each CDC. They showed that both markers were released at the same rate, even at limiting concentrations of PFO and SLO (Heuck *et al.*, 2003). The combined results of these analyses strongly supported the hypothesis that CDCs form a large oligomeric prepore complex on the membrane surface prior to the insertion of its transmembrane β-sheet. These studies are also consistent with the observation that there is a significant reduction in the free energy of partitioning peptide bonds in a membrane if their H-bond potential is satisfied (Wimley *et al.*, 1998). Thus it is also possible that prior to membrane insertion the β-hairpins of the CDCs form into a pre-β-barrel structure.

PFO, like many other CDCs, does not associate into multimeric aggregates even at high concentrations in aqueous solutions suggesting that the monomer–monomer interface may be concealed in the soluble monomer. Since no other cofactors are involved in oligomerization it appears that regulation of oligomerization and formation of the prepore complex is an intrinsic property of the CDC molecule itself. It appears that these surfaces are only exposed once the toxin binds to the membrane. Membrane interaction is the driving force that initiates conformational changes that lead to oligomerization of the monomers, but the details of this process are not completely understood. Ramachandran *et al.* (Ramachandran *et al.*, 2004) identified a key conformational change necessary for oligomerization of monomers. They showed that a probe placed in β-strand 4 (β4, Fig. 3.2) of the core β-sheet of D3 was buried under β5 and was only exposed to the aqueous solvent upon membrane binding. In the crystal structure of the soluble PFO β5 (Fig. 3.2) formed interstrand hydrogen bonds with β-strand 4 (β4) of the core β-sheet of domain 3. They further showed that engineering cysteine substitutions in PFO at amino acid Thr319 of β4 and Val334 in β5 formed a disulfide in the molecule that prevented the dissociation of β5 from β4 and trapped the toxin as a membrane bound monomer and prevented the formation of SDS-resistant oligomers (Ramachandran *et al.*, 2004). Finally, they also showed that two membrane-bound monomers interacted via β1 of one monomer and β4 of the other monomer and that π-stacking of Phe318 of β4 and Tyr181 of β1 was required for the formation of a functional β-barrel. If either of these two residues were changed to a non-aromatic or placed out of register with the other by mutagenesis the PFO monomers were trapped in a prepore state. Thus, these two aromatic residues appear to function to help correctly align β1 and β4 of adjacent monomers so that the correct in-register formation of hydrogen bonds occurs between the two strands. This presumably extends to the correct alignment of the TMHs since they extend from the β-strands of the D3 core β-sheet.

Prepore to pore conversion

The final stage in cytolysis is the formation of the transmembrane β-barrel pore. The process by which this occurs is structurally complex. The pore is formed when the two α-helical bundles located in domain 3 of each monomer extend into the twin amphipathic TMHs and come together to form the β-barrel pore. The mechanics by which 80 TMHs, two per monomer, transition to a transmembrane β barrel was unclear. Insight into this problem was provided by a mutant in PFO in which one of the two aromatics (Tyr-181), necessary for intermolecular alignment of β-strands 1 and 4 (described above), was mutated to alanine. This mutation trapped PFO in the prepore complex by preventing the insertion of the TMHs. When this mutant was combined with native PFO the TMHs from the Tyr-181 mutant were forced to insert suggesting that cooperation between the TMHs was necessary for insertion (Hotze *et al.*, 2002). Hence, cooperative interactions between the monomers of the prepore drive the prepore to pore conversion that results in the formation of the transmembrane β-barrel pore.

The perpendicular orientation of the PFO molecule to the membrane surface proposed by Ramachandran *et al.*, posed a unique problem (Ramachandran *et al.*, 2002). Shepard *et al.*, and Shatursky *et al.*, had conclusively demonstrated that the fully extended TMHs completely spanned the bilayer. Yet, in a perpendicular orientation the extended TMHs appeared to be only long enough to reach the surface of the bilayer, but were not long enough to cross it to form the pore. In essence, the complex needed to collapse approximately 40 Å to position the TMHs so that they could cross the bilayer and form the pore. Atomic force microscopic analysis of the prepore and pore complexes resolved this problem. Czajkowsy *et al.* (2004) showed that a significant difference existed in the height of the prepore and pore complex. Their data showed that the prepore complex was approximately 40 Å higher than the pore complex. Similar results were subsequently obtained by fluorescence resonance energy transfer studies that positioned probes in domain 1 of PFO and the surface bilayer (Ramachandran *et al.*, 2005). Moreover, the monomers in both complexes exhibit nearly identical surface features indicating that the prepore complex appeared to vertically collapse toward the surface, thus positioning the TMHs for insertion across the bilayer. It was proposed that during this transition domain 2, comprised of only 2 long β-strands, was unstable due to the disruption of its interaction with the α-helical bundle of domain 3 as the latter extended to form TMH1. Presumably the domain 2 structure was lost which allowed domains 1 and 3 to move towards the membrane surface, thus bringing the TMHs into proximity of the membrane surface.

Dang *et al.* (Dang *et al.*, 2005) reconstructed two-dimensional (2D) density maps from images of oligomeric PFO. Outer and inner rings of density peaks characterized by these density maps showed that the outer rings of the prepore and pore complexes were reported to be very similar. However, the protein densities that made up the inner ring of the pore complex were more intense and discretely resolved than for the prepore complex. The change in inner-ring protein density was consistent with a mechanism in which the monomers within the prepore complex make a transition from a partially disordered state to a more ordered transmembrane β-barrel in the pore complex.

Tilley *et al.*, performed a 3D EM reconstruction of pneumolysin pore complexes formed on cholesterol-rich lipid layers that were fitted with the PFO crystal structure (Tilley *et al.*, 2005). The densities associated with this 3D construct were also consistent with the structural changes previously demonstrated by other biophysical methods. Their data also showed that a bulge existed on the outer face of the monomer in the pore structure consistent in its location with the predicted bulging of domain 2 (Czajkowsky *et al.*, 2004) as the prepore structure collapsed to the pore complex.

CDCs in disease

We are only beginning to appreciate the many ways in which CDCs contribute to bacterial disease. The dissemination of the CDC gene across so many pathogenic species suggests they are useful in a variety of ways as virulence factors. Appreciation of how the individual CDC has evolved to fulfil a particular function for the pathogen has lagged behind the structural studies. Although this chapter is primarily devoted to the clostridia it is difficult to gain an appreciation of how the CDCs contribute to pathogenesis by studies of the clostridia alone. The fascinating aspect of the evolution of the CDC family is that although the basic mechanism for pore formation has remain conserved, many have undergone evolutionary changes that apparently alters how the pore is used by the pathogen. Below we discuss the structural differences in the CDCs that enables them to better fit the pathogenic lifestyle of the bacterial species.

Perfringolysin O

Clostridium perfringens is a Gram-positive, spore-forming bacillus commonly found in the soil and to a lesser extent in mammalian intestinal tract. The introduction of this bacterium into human tissue via wound or traumatic injury can result in the rapid establishment of clostridial myonecrosis, or gas gangrene. Massive tissue destruction and lack of infiltrating lymphocytes are hallmarks of the disease linked to the proliferation of bacteria and the production of exotoxins. Early histopathological studies of animals infected with either wild-type or isogenic toxin-deficient mutants of *C. perfringens* established that α-toxin, a phospholipase C enzyme, was largely responsible for the tissue damage associated with this disease (Awad *et al.*, 1995). Surprisingly, PFO was shown to play only a minor role in necrosis even though its cytolytic activity suggested that it would play a significant role in tissue destruction. *In vitro* studies with sublytic concentrations with PFO and other CDCs have offered evidence that the toxins have a direct effect on PMNL physiology (Cockeran *et al.*, 2001; Stevens *et al.*, 1997). These studies indicated that PFO is in part responsible for vascular leukostasis by desensitizing the neutrophils. In the mouse model of gas gangrene it appears that the cytolytic function of PFO may be a secondary function in pathogenesis in which it contributes to the hallmark leucostasis. Other studies suggest that the cytolytic activity of PFO may be important in the early stages of disease by contributing to the survival of the bacterium within the phagosome, apparently by assisting in phagosomal escape from macrophages (O'Brien and Melville, 2004). This may allow *C. perfringens* to survive in the host tissue during the early stages of infection when phagocytic cells outnumber the bacteria. Proving this in the mouse model of gangrene may be difficult since relatively high numbers of bacteria are necessary to initiate infection.

Listeriolysin O

Listeria monocytogenes is a Gram-positive pathogen of both humans and animals and is responsible for the life-threatening food-borne illness listeriosis. The pathogenicity of this bacterium is linked to its ability to evade the immune system by establishing an intracellular lifestyle. Replication within the host cell provides protection from the host immune defences and is a rich supply of nutrients for the bacteria. This intercellular niche is established and maintained by the production of several key virulence factors including the production of the CDC listeriolysin O (LLO) (reviewed in Kayal and Charbit, 2006). The pore-forming activity of LLO has been shown to be necessary for phagosomal escape of the bacterium and enables the bacterium to proliferate and spread from cell to cell. However, in order for the bacteria to remain within the protective intracellular environment of the cytosol, LLO has evolved to limit its activity to the phagosomal compartment, thereby preventing damage to the host cell membrane. This trait was shown to be unique to LLO by engineering an LLO deficient strain that expressed the related CDC PFO from *Clostridium perfringens*. This mutant maintained the ability to escape the phagosome but also perforated the cell membrane that resulted in host cell damage (Jones and Portnoy, 1994). The inability to grow intracellular resulted in an avirulent phenotype in the mouse model.

The mechanism by which LLO is able to perforate the phagosomal compartment while maintaining the integrity of the host cell when secreted in the host cytosol has been attributed to several distinct characteristics of this toxin: the negative regulation of translation, the pH dependence of the cytolytic activity, and the degradation of LLO by the N-end rule pathway. The importance of compartmentalizing the activity of LLO is emphasized by the presence of these redundant mechanisms used to inhibit its activity in the cytosol. Although the functionality of these three mechanisms overlaps, they differ by how they are encoded in the soluble structure of LLO.

Traditionally, pH-activated toxins, such as diphtheria toxin and anthrax toxin, remain inactive at neutral pH and are activated at acidic pH after they enter the phagosome. This mechanism does not appear to be operational in LLO, which is also mostly inactive at neutral pH and active at acidic pH. As described above the TMHs in the soluble CDC monomer maintain an α-helical structure, presumably as a protective mechanism to prevent the premature exposure of the amphipathic residues in the solution monomer prior to binding to the membrane. Scheurch *et al.* demonstrated that a triad of acidic amino acid residues located in domain 3 of LLO act as its pH sensor (Schuerch *et al.*, 2005). In the neutral pH environment of the cytosol these residues become deprotonated and charge repulsion presumably destabilizes D3. As indicated above the interface between domains 3 and domains 1 and 2 is not complementary and designed to be disrupted in order to extend the α-helices into the TMHs. It is thought that charge repulsion of these residues carboxylates causes the disruption of this interface and the premature unfurling of the TMHs. Since the carboxylates of these three residues are predicted to point in at each other it would raise the pK_a of the carboxylate group to pH 6–7. Loss of cytolytic activity occurs because premature exposure of the TMHs is an off-pathway conformational change that would normally occur only after monomers bind to and oligomerize on the membrane. The acidic environment of the phagosomal compartment offers a stabilizing environment for the toxin. Under acidic conditions the carboxyl groups of this triad remain protonated, thus reducing charge repulsion and stabilizing domain 3 until such time when the monomer binds to the vacuolar membrane and triggers the unfurling of the TMHs. In the context of the membrane the TMHs of the monomers in the oligomeric complex then cooperate to form the β-barrel pore.

A second important mechanism in compartmentalizing the activity of LLO is the negative regulation of translation of this protein during cytosolic growth. Schnupf *et al.* have recently published evidence that regions within the coding region of the LLO mRNA are responsible for this translational regulation (Schnupf *et al.*, 2006). They reason that negative regulation of translation in the cytosol would offer the bacterium a rapid mechanism by which to shut off LLO activity. They suggest that in addition to regions in the 5′ UTR proposed by Shen and Higgins (2005), the first 52 codons of the LLO mRNA influence the level of production of LLO although the exact mechanism of this control remains unclear. Changes in the environments of the vacuole versus the cytosol presumably affect the secondary structure of the mRNA. Alternatively *trans*-acting elements may play role. Regardless of mechanism, it is clear that *L. monocytogenes* senses the differing environments of the phagosomal compartment and the cytosol to regulate LLO synthesis (Schnupf *et al.*, 2006; Shen and Higgins, 2005).

The ubiquitin-dependent proteolytic pathway rapidly degrades LLO secreted into the host cell cytosol. The N-terminal residue of the substrate appears to influence the half-life of the proteins in the cytosol. Substitution of LLO's N-terminal lysine residue was shown to affect the rate of degradation by this pathway, although it did not strongly affect the intracellular growth or virulence of *Listeria* secreting these mutant toxins. These results suggest that although the N-end rule pathway can function as a back up mechanism to those previously described to inhibit LLO cytolytic activity it does not appear to be a major mechanism for the regulation of cytosolic activity (Schnupf *et al.*, 2007).

LLO also appears to trigger a variety of signalling processes in the cell including the induction of proinflammatory reactions (Rose *et al.*, 2001). It is also involved in generating calcium-dependent and lipid-mediated signalling that provokes host cell changes such as actin rearrangement to aid in bacterial uptake (Gekara and Weiss, 2004). Recently, Hamon *et al.* made a remarkable observation that non-lytic levels of LLO, PFO and PLY could alter histone phosphorylation and acetylation states showing that these toxins can induce epigenetic regulation of genes. These changes led to alterations in the gene expression of several genes including some immunity related genes (Hamon *et al.*, 2007).

Pneumolysin

All clinical isolates of the human pathogen *Streptococcus pneumoniae* produce the CDC pneumolysin and its role as a key virulence factor in pneumonia, bacteraemia, and meningitis has been confirmed by a well established body of evidence. The degree to which pneumolysin plays a role in disease is dependent upon the production of other virulence factors and the background of the particular strain in question. The lytic and non-lytic roles of pneumolysin during the course of disease includes induction of apoptosis, activation of complement, modulation of immune cells, actin remodelling, and widespread cellular and tissue damage (reviewed in Hirst *et al.*, 2004).

All members of the CDC family contain a signal sequence at the N-terminus that targets them for secretion by the *sec*-dependent pathway. This signal peptide is absent in PLY resulting in cytosolic accumulation (Walker *et al.*, 1987). Levels of released PLY have been identified during exponential growth with increasing amounts detected during stationary phase. PLY release has been correlated with the production of LytA, an autolysin responsible for degradation of the cell wall, indicating that lysis of bacteria is essential for PLY release. However, other studies have shown that LytA is not important for PLY release in some strains of *S. pneumoniae* suggesting the presence of an autolysis-independent mechanism of PLY release (Berry *et al.*, 1992). Ultimately, the reason that PLY lacks a signal peptide for secretion remains an enigma.

A second unique characteristic of PLY is its ability to activate complement via the classical pathway independent of antitoxin antibodies (Paton *et al.*, 1984). This activity is independent of its cytolytic activity although both play a role in inducing inflammation and modulating the immune response. Recently it was suggested that the complement activation response was responsible for T cell influx into the lung whereas cytolysis was shown to play a major role in neutrophil recruitment (Jounblat *et al.*, 2003). The ability to activate complement was localized to domain 4 of the toxin and amino acid residues Tyr-384 and Asp-385 (Mitchell *et al.*, 1991). PLY was shown to activate complement by binding the Fc portion of immunoglobulin G (IgG) although both the point mutants deficient in complement activation were still able to bind IgG, albeit to a lesser extent. Clustering of IgG molecules by PLY to activate complement has been attributed to the aggregation properties of PLY in solution (Mitchell *et al.*, 1991). While other CDCs remain soluble at relatively high concentrations, PLY has the capacity to oligomerize to some extent in a membrane-independent fashion. Although point mutants deficient in complement activation maintain wild type cytolytic activity it is unclear if the mutants in domain 4 have an effect on the aggregation of the solution monomer.

Streptolysin O

Streptolysin O (SLO) is produced by the pathogen *Streptococcus pyogenes*. Sequence alignment of the CDCs shows the presence of an additional 60 amino acids at the N-terminus of SLO. Madden *et al.*, have shown in a set of elegant experiments that SLO can be used by *S. pyogenes* to specifically transport the NAD-glycohydrolase (SPN) into the cytosol of keratinocytes (Madden *et al.*, 2001). The transport of SPN increased cell death and was dependent on SLO pore-forming activity. SLO added in *trans* was not sufficient for this cytolysin-mediated transport (CMT). In addition, a mutant of SLO that lacked the additional N-terminal amino acids did not support SPN translocation. The addition of these residues to PFO did not allow it to substitute for SLO in the translocation of SPN, suggesting that other features of SLO are also required for this process. Interestingly, this translocation mechanism does not appear to facilitate the transfer of other secreted proteins into the eukaryotic cell (Meehl and Caparon, 2004). Hence, either the SLO pore or the bacterial cell can select which of the various type II secreted proteins can enter the SLO pore. CMT has been suggested to be the Gram-positive analogue to the type III secretion pathway found in many Gram-negative bacteria, but the mechanism by which SPN translocation occurs remains to be determined.

Intermedilysin

Intermedilysin is produced by *Streptococcus intermedius*, a pathogen often associated with abscess formation in the brain and liver (Nagamune *et al.*, 1996, 1997). Strains isolated from abscess sites were shown to produce elevated levels of ILY compared to strains isolated from oropharynx, suggesting a role in pathogenesis (Nagamune *et al.*, 2000). ILY is the only human cell specific CDC identified to date. The ability of ILY to discriminate between human and non-human cells was attributed to the presence of a cellular receptor, human CD59, required for ILY binding (Giddings *et al.*, 2004). The requirement of a cellular receptor by ILY shifted a paradigm in the CDC field, in which cholesterol was long considered to be the receptor for all CDCs. ILY was the first, and to date the only, CDC to be identified which uses a non-sterol receptor. Human CD59 is a multifunctional GPI-anchored membrane protein found on most cell types. The primary role of CD59 is to protect host cells from the membrane attack complex (MAC) of complement by binding to C8α and C9 (Ninomiya and Sims, 1992; Rollins and Sims, 1990; Rollins *et al.*, 1991). This interaction prevents oligomerization and subsequent pore formation by the MAC. Additionally, CD59 has been shown to be involved in cell signalling such as up-regulation of superoxide production in neutrophils by acting as a co-adapter molecule for exo-calreticulin, a C1q receptor (Ghiran *et al.*, 2003).

The basis for the human cell specificity of ILY appears to be tied up in the structure of CD59. CD59 exhibits homologous restriction and so only the host MAC is typically inactivated by its own CD59 (Rollins *et al.*, 1991). It appears that the structural region of CD59 responsible for its homologous restriction is the same region recognized by ILY (Giddings *et al.*, 2004). It is curious that ILY would evolve to utilize a receptor that was restricted to human cells since this trait does not increase its pore-forming activity over that of other CDCs that bind directly to cells via cholesterol-rich domains. Therefore, why change if pore formation is its primary effect? The use of a specific receptor suggests that there may be other benefits to this change that are not immediately obvious.

One possibility is that recognition of huCD59 may preferentially target ILY to neutrophils during abscess formation. Interestingly, cell surface expression of CD59 is significantly upregulated following activation of PMNs (Gordon *et al.*, 1994), presumably to protect them against elevated levels of activated complement during inflammation. Hence, the neutrophils would provide a target-rich environment for ILY within the abscess. In addition, it is tempting to speculate that ILY may also play a key role in protecting *S. intermedius* from killing by PMNs via other mechanisms. The ILY binding site on CD59 may overlap that of C8α and C9 (Giddings *et al.*, 2004). Therefore, ILY binding CD59 may block the protective effect CD59 affords the host cell from complement. This could increase the host cell susceptibility to MAC within the abscess.

Clostridium septicum α-toxin

Overview

Although *Clostridium septicum* produces several exoproteins the only lethal factor among them appears to be α-toxin (AT) (Ballard *et al.*, 1992). *Clostridium septicum* is a Gram-positive anaerobe that causes a variety of diseases in both humans and animals. It is a major cause of non-traumatic gas gangrene and necrotizing enterocolitis and in many cases the infection is associated with various predisposing diseases ranging from colonic cancer, to leukaemia or diabetes. One of the most horrific diseases caused by *C. septicum* in humans is non-traumatic gas gangrene, a rapidly fatal disease with a mortality rate of 60–100%. AT was originally identified in 1944, but it was not purified to homogeneity until 1992 by Ballard *et al.* (1992). The purified form was the only mouse lethal toxin secreted and was reported to cause shock and death to mice upon injection, similar to the effects seen in human patients with non-traumatic gangrene. Recently Kennedy *et al.*, were able to knock out AT in *C. septicum* and showed that it was absolutely required for virulence in mice (Kennedy *et al.*, 2005).

Although AT is a pore-forming toxin, it is not a member of the CDC family. Rather, AT is a member of the small and unusual class of the aerolysin-like family of pore-forming toxins. Unlike the CDCs, AT is not found in other clostridial species and its gene is not disseminated across bacterial genera. Although AT forms a β-barrel pore in the membrane its mechanism differs from the CDCs in several significant ways. AT forms a comparatively small pore comprising seven monomers, each donating a single β-strand to the β-barrel pore. In contrast to the CDCs, AT binds to a variety of GPI-linked protein receptors and requires proteolytic activation associated with the target cells. The details outlining the cytolytic mechanism and the structural domains involved in the pathway are described below. Similar to the CDCs, our understanding of the mechanism of action of AT exceeds our understanding of the role this toxin plays in *C. septicum* related disease.

Cytolytic mechanism and structure of alpha toxin

The primary structure of α-toxin exhibits similarity with aerolysin, a pore-forming toxin of the Gram-negative pathogen *Aeromonas hydrophila* (Ballard *et al.*, 1995). It is somewhat unusual to find related toxins in a Gram-positive and Gram-negative species. The aerolysin family of toxins is unique; its members appear to range across extremely diverse species and even kingdoms (domains) (Fig. 3.3). Sousa *et al.* (de Sousa and Morhy, 1989) suggested that the haemolytic protein enterolobin from seeds of the Brazilian tree *Enterolobium contortisiliquum* was related to aerolysin. Members of this family may exist in *Photobacterium profundum* (GenBank accession number CAG23788), a psychrotolerant deep sea marine bacterium belonging to the family Vibrionaceae and *Hahella chejuensis*, a marine bacterium isolated from coastal marine sediments (Genbank accession number YP 434730). These proteins exhibit 30–35% identity with aerolysin and/or AT, but whether they are functional, pore-forming toxins of the aerolysin family remains to be determined. Also, a haemolytic lectin from the mushroom *Laetiporus sulphureus*, has been shown to exhibit similarity to AT (Tateno and Goldstein, 2003).

The crystal structure of AT has not been solved but due to the high degree of similarity to aerolysin (73%), molecular modelling studies have generated a structural model based on the crystal structure of aerolysin (Parker *et al.*, 1994). Aerolysin is a bi-lobal protein rich in β-sheets that is composed of four domains. The smaller lobe comprises domain 1 and the remaining domains (2–4) are located in the larger lobe. AT, on the other hand, is a single-lobed protein lacking the smaller lobe of aerolysin (Melton *et al.*, 2004) (Fig. 3.4).

AT has been shown to bind to the target cell membrane by various to GPI-anchored cell surface receptors. AT receptors include the human folate receptor (hFR), members of the SAG family on the surface of *Toxoplasma gondii* and Thy-1 as well as the neuronal surface molecule contactin (Gordon *et al.*, 1999; MacKenzie *et al.*, 1999; Wichroski *et al.*, 2002). Both Thy-1 and contactin are common receptors shared with

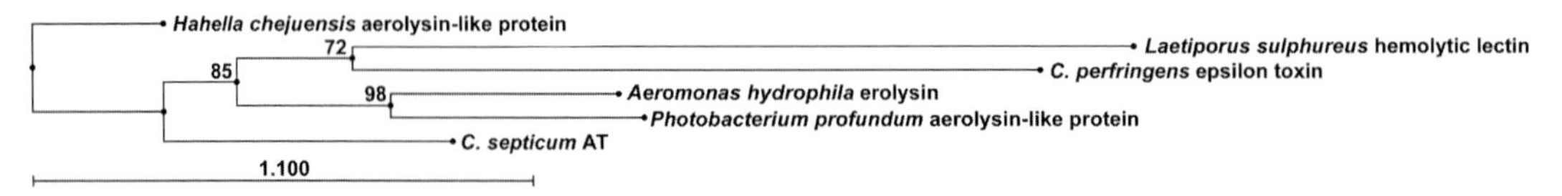

Figure 3.3 Aerolysin-like toxins phylogenetic tree. Shown is a dendrogram of known and putative members of the aerolysin-like family of toxins (from GenBank). Cytolytic activity has only been shown for aerolysin, AT, epsilon toxin and enterolobin, the remaining members have not been characterized.

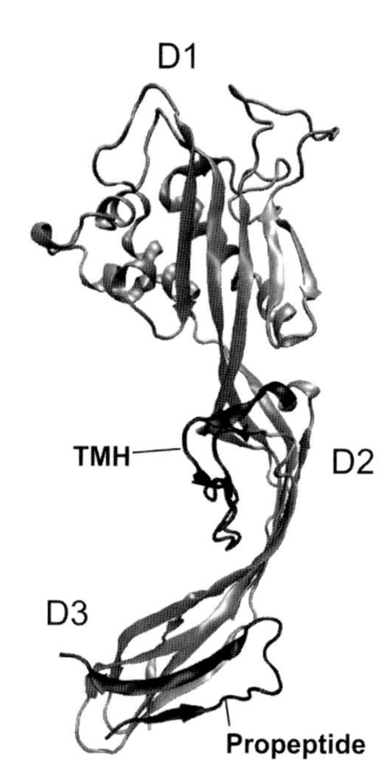

Figure 3.4 Structural model of *C. septicum* AT. Show is the aerolysin based ribbon representation of the structural model of AT with locations of the membrane spanning β-hairpin and propeptide (both highlighted in black) denoted. The receptor-binding domain is located at the top of the molecule in D1.

aerolysin. The smaller lobe of aerolysin allows for recognition of additional receptors unique to aerolysin, fusion of this lobe to AT resulted in a more aerolysin-like molecule in terms of its receptor repertoire (Diep *et al.*, 1999).

Melton-Witt *et al.*, performed alanine and cysteine scanning mutagenesis to map the functional domains of AT (Melton *et al.*, 2006). Their studies revealed that the receptor binding function was restricted to amino acid residues located within domain 1 of the molecule (Fig. 3.4). This region was shown to be particularly rich in aromatic residues consistent with the observation that aromatic residues are frequently involved in interactions between carbohydrate moieties and proteins. Interestingly, none of the protein components of the AT receptors share sequence homology other than their common GPI anchor indicating that the GPI anchor plays the major role in toxin recognition. A cytolytically inactive mutant of AT was proposed as a diagnostic tool identifying paroxysmal nocturnal haemoglobinuria (PNH) patients, a hematopoietic stem cell disorder caused by the loss of glycosylphosphatidylinositol (GPI)-anchored proteins on the cell membrane (Shin *et al.*, 2006; Shin *et al.*, 2004).

Alpha toxin is secreted as a protoxin and requires activation by cell surface proteases. When AT was originally purified most preparations contained two antibody-reacting species, one that was 48 kDa and one migrating at 44 kDa. It was later shown that this discrepancy in size was due to the proteolytic cleavage of a C-terminal propeptide and that the 48-kDa protein represented an inactive protoxin (Ballard *et al.*, 1993). Subsequently, it was shown that furin and other furin-like proteases found on the cell surface were able to process the protoxin into the active form (Gordon *et al.*, 1997). The presence of the propeptide was reported to act as an intramolecular chaperone that prevented premature oligomerization of the toxin (Sellman and Tweten, 1997). Therefore activation by proteolytic cleavage was required prior to oligomerization. Sellman *et al.* (Sellman and Tweten, 1997) showed that activating AT in solution in the absence of membranes led to the formation of an inactive AT oligomer that was significantly less cytolytically active than monomeric toxin when added to erythrocytes. It was also shown that displacement of the propeptide did not occur upon cleavage. Rather, dissociation of the propeptide occurred during the oligomerization process. In the molecular model (Fig. 3.4), the propeptide is located in domain 3 of the toxin. It appears that the propeptide protects an interfacial domain located in this region of the toxin. Not surprisingly, the saturation mutagenesis studies performed on AT revealed that domain 3 was involved in oligomerization (Melton *et al.*, 2006). However, unlike the receptor-binding domain, the oligomerization sites were not limited to one domain of the toxin and additional sites were located in domain 1 of the toxin as well. These results showed that multiple contact sites exist between monomers of AT in the pore oligomer.

Similar to the CDCs, AT also forms a prepore structure prior to the formation of the pore complex. Like the CDCs, the AT prepore complex may organize and coordinate the insertion of the membrane-spanning β-barrel. Sellman *et al.*, showed that AT was able to bind to erythrocyte membranes at 4°C, a temperature that inhibits formation of the pore (Sellman *et al.*,

1997). Conversion of the prepore to the inserted pore was observed by shifting the temperature to 37°C. The mechanism by which the prepore complex transitions into the inserted β-barrel pore has not been resolved. The mutagenesis studies of Melton-Witt *et al.* (2006) reported the identification of a loop in domain 1 near the receptor binding domain that appears to control the prepore to pore conversion.

AT was classified as a β-pore-forming toxin upon the identification of an amphipathic β-hairpin shown to form a membrane-spanning β-barrel. The membrane-spanning domain of AT was located in domain 2 of AT and shown to comprise amino acid residues F200 to S235 (Melton *et al.*, 2004). Deletion mutants in this region demonstrated a complete loss of cytolytic activity yet they retained the ability to bind to receptor and oligomerize into the prepore complex. Fluorescent techniques exploiting the environmentally sensitive fluorescent dye NBD employed in the identification of the TMHs of the CDC PFO were used to demonstrate a direct interaction of this region with the membrane. In addition, collisional quenchers of NBD located in the hydrophobic region of the membrane were used to further verify the location of the residues within the membrane. This span of amino acids in AT as well as aerolysin in the primary structure is a region of alternating hydrophobic and hydrophilic side chains, typical of an amphipathic β-strand. It was subsequently shown that this region in aerolysin also formed its membrane spanning amphipathic β-hairpin structure (Iacovache *et al.*, 2006).

Role in disease

Clostridial disease caused by *C. septicum* is more common in livestock than in humans. In cattle it causes a malignant oedema, also known as blackleg. This disease is often fatal within 12 hours of onset of symptoms. Fatal bacteraemia in lambs or sheep, know as braxy, is also caused by *C. septicum*. Fortunately, clostridial myonecrosis in humans is a rare, with only 1000–3000 cases reported in the United States. Approximately 10% of these cases are classified as non-traumatic or 'spontaneous' clostridial infections, with *C. septicum* being the major causative agent. This form of the disease results in a higher mortality rate than wound-related infections because of its rapid onset and difficulty diagnosing it in the early stages of the infection. The period of time from contamination to myonecrosis ranges from between 6 and 48 h, with death typically occurring within 24–48 h of disease onset. In addition *C. septicum* has been associated with necrotizing enterocolitis (Bignold and Harvey, 1979), meningitis (Dirks *et al.*, 2000), osteomyelitis (Neimkin and Jupiter, 1985), pericarditis (Brahan and Kahler, 1990), and as a superinfection of *E. coli* O157:H7-induced haemolytic uraemic syndrome (Barnham and Weightman, 1998). In the majority of cases, a predisposing disease or malignancy is associated with *C. septicum* disease. In fact, in one study a colonic malignancy was detected in 81–85% of the patients with a *C. septicum* infection. Often an infection with *C. septicum* is the first clinical sign of an occult colonic carcinoma. It is thought that these underlying conditions could lead to multiplication of the organism in the anaerobic environment of the tumour and its escape across the intestinal mucosa into the bloodstream from where it could cause gangrene.

Until recently the relationship between AT and *C. septicum* pathogenesis remained difficult to prove. It was thought that, since AT is the only lethal factor secreted by *C. septicum* (Ballard *et al.*, 1992), it was a critical virulence factor produced by this pathogen. In addition, immunization with formalin-inactivated purified AT protected guinea pigs and mice challenged with *C. septicum* (Amimoto *et al.*, 2002; Ballard *et al.*, 1992). The technical difficulty involved in genetically manipulating *C. septicum* prevented the creation of an AT-deficient strain necessary for critical genetic experiments. This technical issue was overcome by Kennedy *et al.*, who reported the development of a protocol to genetically engineer a strain of *C. septicum* in which the AT gene was knocked out (Kennedy *et al.*, 2005). In addition, they successfully complemented this deficient strain with the wild-type gene. Virulence testing of these mutant strains revealed that the development of fulminant myonecrosis in mice was absolutely dependent on the production of AT. They also found that the inhibition of leucocyte infiltration into the infection site was also dependent on the presence of AT. These results provided strong evidence that AT is the primary pathogenesis factor of *C. septicum*.

Perhaps the most unusual aspect of this disease is that it is essentially the same in all respects to the gangrene caused by *C. perfringens*, yet the main pathogenic factors responsible for the major manifestations of the disease exhibit very different mechanisms of action. In the case of *C. septicum*, AT is a proteolytically activated pore-forming toxin whereas in *C. perfringens* alpha toxin is a phospholipase C. Whether each could substitute for the other would be an interesting experiment in the study of the evolution of factors associated with the pathogenesis of these organisms.

Summary and perspectives

Clostridium septicum α-toxin and *C. perfringens* PFO (and related CDCs) are examples of two different pore-forming toxin families. Both classes of toxins form the pore using a β-barrel motif that is comprising several amphipathic β-hairpins. Apart from these features, however, their mechanisms appear to be quite different, although it is possible that as we understand more of the fine details of their pore-forming mechanisms we may identify other common features. The contribution of each class of toxin to pathogenesis is not fully understood, although it seems that cell lysis by pore formation may not be the end of the story, at least for the CDCs. The future of AT research will likely be somewhat limited due to the fact that it is one of only a few distantly related toxins of this toxin family and *C. septicum* causes comparatively few infections each year. On the other hand, work on the molecular mechanism of the CDCs and their contribution to pathogenesis will persist for the foreseeable future since they are ubiquitous in Gram-positive pathogens and have evolved specific features that are pathogen specific. Throughout the evolution of the CDCs the basic pore-forming mechanism has been preserved, emphasizing the importance of pore formation in disease. It has, however, become increasingly apparent that that the function of the CDC is not limited to cell death by lysis and that sublytic doses of the CDC can have significant effects on the host cell. In the future it is likely that other CDCs will be discovered with novel features and these features will likely be important to the pathogenic mechanism of the organism. Future studies on the molecular biology of the CDC cytolytic mechanism will undoubtedly reveal additional aspects to the remarkable mechanism of pore formation. The study of the roles of these toxins in pathogenesis and how evolutionary changes in their structures contribute to this process is where much of the future work will be accomplished.

References

Alouf, J.E. (1999). Introduction to the family of the structurally related cholesterol-binding cytolysins ('sulfhydryl-activated toxins'). In Bacterial Toxins: A Comprehensive Sourcebook, J. Alouf, and J. Freer, eds. (London, Academic Press), pp. 443–456.

Alouf, J.E. (2000). Cholesterol-binding cytolytic protein toxins. Int. J. Med. Microbiol. *290*, 351–356.

Amimoto, K., Ohgitani, T., Sasaki, O., Oishi, E., Katayama, S., Isogai, M., and Ota, S. (2002). Protective effect of *Clostridium septicum* alpha-toxoid vaccine against challenge with spores in guinea pigs. J. Vet. Med. Sci. *64*, 67–69.

Awad, M.M., Bryant, A.E., Stevens, D.L., and Rood, J.I. (1995). Virulence studies on chromosomal alpha-toxin and theta-toxin mutants constructed by allelic exchange provide genetic evidence for the essential role of alpha-toxin in *Clostridium perfringens*-mediated gas gangrene. Mol. Microbiol. *15*, 191–202.

Ballard, J., Bryant, A., Stevens, D., and Tweten, R.K. (1992). Purification and characterization of the lethal toxin (alpha toxin) of *Clostridium septicum*. Infect. Immun. *60*, 784–790.

Ballard, J., Crabtree, J., Roe, B.A., and Tweten, R.K. (1995). The primary structure of *Clostridium septicum* alpha toxin exhibits similarity with *Aeromonas hydrophila* aerolysin. Infect. Immun. *63*, 340–344.

Ballard, J., Sokolov, Y., Yuan, W.-L., Kagan, B.L., and Tweten, R.K. (1993). Activation and mechanism of *Clostridium septicum* alpha toxin. Mol. Microbiol. *10*, 627–634.

Barnham, M., and Weightman, N. (1998). *Clostridium septicum* infection and hemolytic uremic syndrome. Emerg. Infect. Dis. *4*, 321–324.

Berry, A.M., Paton, J.C., and Hansman, D. (1992). Effect of insertional inactivation of the genes encoding pneumolysin and autolysin on the virulence of *Streptococcus pneumoniae* type 3. Microb. Pathog. *12*, 87–93.

Bignold, L.P., and Harvey, H.P. (1979). Necrotising enterocolitis associated with invasion by *Clostridium septicum* complicating cyclic neutropaenia. Aust. NZ J.Med. *9*, 426–429.

Brahan, R., and Kahler, R.C. (1990). *Clostridium septicum* as a cause of pericarditis and mycotic aneurysm. J. Clin. Microbiol. *28*, 2377–2378.

Cockeran, R., Steel, H.C., Mitchell, T.J., Feldman, C., and Anderson, R. (2001). Pneumolysin potentiates production of prostaglandin E(2) and leukotriene

B(4) by human neutrophils. Infect. Immun. *69*, 3494–3496.

Czajkowsky, D.M., Hotze, E.M., Shao, Z., and Tweten, R.K. (2004). Vertical collapse of a cytolysin prepore moves its transmembrane β-hairpins to the membrane. EMBO J *23*, 3206–3215.

Dang, T., Hotze, E.M., Rouiller, I., Tweten, R.K., and Wilson-Kubalek, E.M. (2005). Prepore to Pore Transition of a Cholesterol-Dependent Cytolysin Visualized by Electron Microscopy. J Struct Biol *150*, 100–108.

Darji, A., Niebuhr, K., Hense, M., Wehland, J., Chakraborty, T., and Weiss, S. (1996). Neutralizing monoclonal antibodies against listeriolysin: mapping of epitopes involved in pore formation. Infect. Immun. *64*, 2356–2358.

de los Toyos, J.R., Mendez, F.J., Aparicio, J.F., Vazquez, F., Del Mar Garcia Suarez, M., Fleites, A., Hardisson, C., Morgan, P.J., Andrew, P.W., and Mitchell, T.J. (1996). Functional analysis of pneumolysin by use of monoclonal antibodies. Infect. Immun. *64*, 480–484.

de Sousa, M.V., and Morhy, L. (1989). Enterolobin, a hemolytic protein from *Enterolobium contortisiliquum* seeds (Leguminosae--Mimosoideae). Purification and characterization. An. Acad. Bras. Cienc. *61*, 405–412.

Diep, D.B., Nelson, K.L., Lawrence, T.S., Sellman, B., Tweten, R.K., and Buckley, J.T. (1999). Expression and properties of an aerolysin-*Clostridium septicum* alpha toxin hybrid protein. Mol. Microbiol. *31*, 785–794.

Dirks, C., Horn, H., Christensen, L., and Pedersen, C. (2000). CNS infection with *Clostridium septicum*. Scand. J. Infect. Dis. *32*, 320–322.

Gekara, N.O., and Weiss, S. (2004). Lipid rafts clustering and signalling by listeriolysin O. Biochem. Soc. Trans. *32*, 712–714.

Ghiran, I., Klickstein, L.B., and Nicholson-Weller, A. (2003). Calreticulin is at the surface of circulating neutrophils and uses CD59 as an adaptor molecule. J. Biol. Chem. *278*, 21024–21031.

Giddings, K.S., Johnson, A.E., and Tweten, R.K. (2003). Redefining cholesterol's role in the mechanism of the cholesterol-dependent cytolysins. Proc. Natl. Acad. Sci. USA *100*, 11315–11320.

Giddings, K.S., Zhao, J., Sims, P.J., and Tweten, R.K. (2004). Human CD59 is a receptor for the cholesterol–dependent cytolysin intermedilysin. Nat Struct Mol Biol *12*, 1173–1178.

Gordon, D.L., Papazaharoudakis, H., Sadlon, T.A., Arellano, A., and Okada, N. (1994). Upregulation of human neutrophil CD59, a regulator of the membrane attack complex of complement, following cell activation. Immunol Cell Biol *72*, 222–229.

Gordon, V.M., Benz, R., Fujii, K., Leppla, S.H., and Tweten, R.K. (1997). *Clostridium septicum* alpha toxin is proteolytically activated by furin. Infect. Immun. *65*, 4130–4134.

Gordon, V.M., Nelson, K.L., Buckley, J.T., Stevens, V.L., Tweten, R.K., Elwood, P.C., and Leppla, S.H. (1999). *Clostridium septicum* alpha toxin uses glycosylphosphatidylinositol-anchored protein receptors. J. Biol. Chem. *274*, 27274–27280.

Hadders, M.A., Beringer, D.X., and Gros, P. (2007). Structure of C8alpha-MACPF reveals mechanism of membrane attack in complement immune defense. Science *317*, 1552–1554.

Hamon, M.A., Batsche, E., Regnault, B., Tham, T.N., Seveau, S., Muchardt, C., and Cossart, P. (2007). Histone modifications induced by a family of bacterial toxins. Proc. Natl. Acad. Sci. USA *104*, 13467–13472.

Heuck, A.P., Hotze, E., Tweten, R.K., and Johnson, A.E. (2000). Mechanism of membrane insertion of a multimeric b-barrel protein: Perfringolysin O creates a pore using ordered and coupled conformational changes. Molec Cell *6*, 1233–1242.

Heuck, A.P., Tweten, R.K., and Johnson, A.E. (2003). Assembly and topography of the prepore complex in cholesterol-dependent cytolysins. J. Biol. Chem. *278*, 31218–31225.

Hirst, R.A., Kadioglu, A., O'Callaghan, C., and Andrew, P.W. (2004). The role of pneumolysin in pneumococcal pneumonia and meningitis. Clin Exp Immunol *138*, 195–201.

Hotze, E.M., Heuck, A.P., Czajkowsky, D.M., Shao, Z., Johnson, A.E., and Tweten, R.K. (2002). Monomer-monomer interactions drive the prepore to pore conversion of a beta-barrel-forming cholesterol-dependent cytolysin. J. Biol. Chem. *277*, 11597–11605.

Hotze, E.M., Wilson-Kubalek, E.M., Rossjohn, J., Parker, M.W., Johnson, A.E., and Tweten, R.K. (2001). Arresting pore formation of a cholesterol-dependent cytolysin by disulfide trapping synchronizes the insertion of the transmembrane beta-sheet from a prepore intermediate. J. Biol. Chem. *276*, 8261–8268.

Humphrey, W., Dalke, A., and Schulten, K. (1996). VMD: visual molecular dynamics. J Mol Graph *14*, 33–38, 27–38.

Iacovache, I., Paumard, P., Scheib, H., Lesieur, C., Sakai, N., Matile, S., Parker, M.W., and van der Goot, F.G. (2006). A rivet model for channel formation by aerolysin-like pore-forming toxins. Embo J *25*, 457–466.

Iwamoto, M., Ohno-Iwashita, Y., and Ando, S. (1987). Role of the essential thiol group in the thiol-activated cytolysin from *Clostridium perfringens*. Eur. J. Biochem. *167*, 425–430.

Jacobs, T., Darji, A., Frahm, N., Rohde, M., Wehland, J., Chakraborty, T., and Weiss, S. (1998). Listeriolysin O: cholesterol inhibits cytolysis but not binding to cellular membranes. Mol. Microbiol. *28*, 1081–1089.

Jones, S., and Portnoy, D.A. (1994). Characterization of *Listeria monocytogenes* pathogenesis in a strain expressing perfringolysin O in place of listeriolysin O. Infect. Immun. *62*, 5608–5613.

Jounblat, R., Kadioglu, A., Mitchell, T.J., and Andrew, P.W. (2003). Pneumococcal behavior and host responses during bronchopneumonia are affected differently by the cytolytic and complement-activating activities of pneumolysin. Infect. Immun. *71*, 1813–1819.

Kayal, S., and Charbit, A. (2006). Listeriolysin O: a key protein of *Listeria monocytogenes* with multiple functions. FEMS Microbiol Rev *30*, 514–529.

Kennedy, C.L., Krejany, E.O., Young, L.F., O'Connor, J.R., Awad, M.M., Boyd, R.L., Emmins, J.J., Lyras, D., and Rood, J.I. (2005). The alpha-toxin of *Clostridium septicum* is essential for virulence. Mol. Microbiol. *57*, 1357–1366.

MacKenzie, C.R., Hirama, T., and Buckley, J.T. (1999). Analysis of receptor binding by the channel-forming toxin aerolysin using surface plasmon resonance. J. Biol. Chem. *274*, 22604–22609.

Madden, J.C., Ruiz, N., and Caparon, M. (2001). Cytolysin-mediated translocation (CMT): a functional equivalent of type III secretion in gram-positive bacteria. Cell *104*, 143–152.

Meehl, M.A., and Caparon, M.G. (2004). Specificity of streptolysin O in cytolysin-mediated translocation. Mol. Microbiol. *52*, 1665–1676.

Melton, J.A., Bentsen, L.M., and Tweten, R.K. (2006). Mutational analysis of *Clostridium septicum* alpha toxin reveals residues critical for receptor-binding, oligomerization and control of membrane insertion. Biochemistry *45*, 14347–14354.

Melton, J.A., Parker, M.W., Rossjohn, J., Buckley, J.T., and Tweten, R.K. (2004). The identification and structure of the membrane-spanning domain of the *Clostridium septicum* alpha toxin. J. Biol. Chem. *279*, 14315–14322.

Miller, C.J., Elliot, J.L., and Collier, R.L. (1999). Anthrax protective antigen: prepore-to-pore conversion. Biochemistry *38*, 10432–10441.

Mitchell, T.J., Andrew, P.W., Saunders, F.K., Smith, A.N., and Boulnois, G.J. (1991). Complement activation and antibody binding by pneumolysin via a region of the toxin homologous to a human acute-phase protein. Mol. Microbiol. *5*, 1883–1888.

Nagamune, H., Ohnishi, C., Katsuura, A., Fushitani, K., Whiley, R.A., Tsuji, A., and Matsuda, Y. (1996). Intermedilysin, a novel cytotoxin specific for human cells secreted by *Streptococcus intermedius* UNS46 isolated from a human liver abscess. Infect. Immun. *64*, 3093–3100.

Nagamune, H., Ohnishi, C., Katsuura, A., Taoka, Y., Fushitani, K., Whiley, R.A., Yamashita, K., Tsuji, A., Matsuda, Y., Maeda, T. (1997). Intermedilysin. A cytolytic toxin specific for human cells of a *Streptococcus intermedius* isolated from human liver abscess. Adv. Exp. Med. Biol. *418*, 773–775.

Nagamune, H., Whiley, R.A., Goto, T., Inai, Y., Maeda, T., Hardie, J.M., and Kourai, H. (2000). Distribution of the intermedilysin gene among the anginosus group streptococci and correlation between intermedilysin production and deep-seated infection with *Streptococcus intermedius*. J. Clin. Microbiol. *38*, 220–226.

Nakamura, M., Sekino, N., Iwamoto, M., and Ohno-Iwashita, Y. (1995). Interaction of theta-toxin (perfringolysin O), a cholesterol-binding cytolysin, with liposomal membranes: change in the aromatic side chains upon binding and insertion. Biochemistry *34*, 6513–6520.

Neimkin, R.J., and Jupiter, J.B. (1985). Metastatic non-traumatic *Clostridium septicum* osteomyelitis. J. Hand Surg. [Am] *10*, 281–284.

Ninomiya, H., and Sims, P.J. (1992). The human complement regulatory protein CD59 binds to the alpha-chain of C8 and to the 'b'domain of C9. J. Biol. Chem. *267*, 13675–13680.

O'Brien, D.K., and Melville, S.B. (2004). Effects of *Clostridium perfringens* alpha-toxin (PLC) and perfringolysin O (PFO) on cytotoxicity to macrophages, on escape from the phagosomes of macrophages, and on persistence of C. perfringens in host tissues. Infect. Immun. *72*, 5204–5215.

Ohno-Iwashita, Y., Iwamoto, M., Mitsui, K., Ando, S., and Nagai, Y. (1988). Protease nicked θ-toxin of *Clostridium perfringens*, a new membrane probe with no cytolytic effect, reveals two classes of cholesterol as toxin-binding sites on sheep erythrocytes. Eur. J. Biochem. *176*, 95–101.

Olson, R., Nariya, H., Yokota, K., Kamio, Y., and Gouaux, E. (1999). Crystal structure of Staphylococcal LukF delineates conformational changes accompanying formation of a transmembrane channel. Nat. Struct. Biol. *6*, 134–140.

Palmer, M., Saweljew, P., Vulicevic, I., Valeva, A., Kehoe, M., and Bhakdi, S. (1996). Membrane-penetrating domain of streptolysin O identified by cysteine scanning mutagenesis. J. Biol. Chem. *271*, 26664–26667.

Palmer, M., Vulicevic, I., Saweljew, P., Valeva, A., Kehoe, M., and Bhakdi, S. (1998). Streptolysin O: A proposed model of allosteric interaction between a pore-forming protein and its target lipid bilayer. Biochemistry *37*, 2378–2383.

Panchal, R.G., and Bayley, H. (1995). Interactions between residues in staphylococcal alpha-hemolysin revealed by reversion mutagenesis. J. Biol. Chem. *270*, 23072–23076.

Parker, M.W., Buckley, J.T., Postma, J.P.M., Tucker, A.D., Leonard, K., Pattus, F., and Tsernoglou, D. (1994). Structure of the *Aeromonas* toxin proaerolysin in its water-soluble and membrane-channel states. Nature *367*, 292–295.

Paton, J.C., Rowan, K.B., and Ferrante, A. (1984). Activation of human complement by the pneumococcal toxin pneumolysin. Infect. Immun. *43*, 1085–1087.

Polekhina, G., Giddings, K.S., Tweten, R.K., and Parker, M.W. (2005). Insights into the action of the superfamily of cholesterol-dependent cytolysins from studies of intermedilysin. Proc. Natl. Acad. Sci. USA *102*, 600–605.

Ramachandran, R., Heuck, A.P., Tweten, R.K., and Johnson, A.E. (2002). Structural insights into the membrane-anchoring mechanism of a cholesterol-dependent cytolysin. Nat. Struct. Biol. *9*, 823–827.

Ramachandran, R., Tweten, R.K., and Johnson, A.E. (2004). Membrane-dependent conformational changes initiate cholesterol-dependent cytolysin oligomerization and intersubunit β-strand alignment. Nat. Struct. Mol. Biol. *11*, 697–705.

Ramachandran, R., Tweten, R.K., and Johnson, A.E. (2005). The domains of a cholesterol-dependent cytolysin undergo a major FRET-detected rearrangement during pore formation. Proc. Natl. Acad. Sci. USA *102*, 7139–7144.

Rollins, S.A., and Sims, P.J. (1990). The complement-inhibitory activity of CD59 resides in its capacity to block incorporation of C9 into membrane C5b-9. J Immunol *144*, 3478–3483.

Rollins, S.A., Zhao, J., Ninomiya, H., and Sims, P.J. (1991). Inhibition of homologous complement by CD59 is mediated by a species-selective recognition conferred through binding to C8 within C5b-8 or C9 within C5b-9. J. Immunol. *146*, 2345–2351.

Rosado, C.J., Buckle, A.M., Law, R.H., Butcher, R.E., Kan, W.T., Bird, C.H., Ung, K., Browne, K.A., Baran, K., Bashtannyk-Puhalovich, T.A. (2007). A common fold mediates vertebrate defense and bacterial attack. Science *317*, 1548–1551.

Rose, F., Zeller, S.A., Chakraborty, T., Domann, E., Machleidt, T., Kronke, M., Seeger, W., Grimminger, F., and Sibelius, U. (2001). Human endothelial cell activation and mediator release in response to *Listeria monocytogenes* virulence factors. Infect. Immun. *69*, 897–905.

Rossjohn, J., Feil, S.C., McKinstry, W.J., Tweten, R.K., and Parker, M.W. (1997). Structure of a cholesterol-binding thiol-activated cytolysin and a model of its membrane form. Cell *89*, 685–692.

Schnupf, P., Hofmann, J., Norseen, J., Glomski, I.J., Schwartzstein, H., and Decatur, A.L. (2006). Regulated translation of listeriolysin O controls virulence of *Listeria monocytogenes*. Mol. Microbiol. *61*, 999–1012.

Schnupf, P., Zhou, J., Varshavsky, A., and Portnoy, D.A. (2007). Listeriolysin O secreted by *Listeria monocytogenes* into the host cell cytosol is degraded by the N-end rule pathway. Infect. Immun. *75*, 5135–5147.

Schuerch, D.W., Wilson-Kubalek, E.M., and Tweten, R.K. (2005). Molecular basis of Listeriolysin O pH-dependence. Proc Natl Acad Sci *102*, 12537–12542.

Sekino-Suzuki, N., Nakamura, M., Mitsui, K.I., and Ohno-Iwashita, Y. (1996). Contribution of individual tryptophan residues to the structure and activity of theta-toxin (perfringolysin o), a cholesterol-binding cytolysin. Eur. J. Biochem. *241*, 941–947.

Sellman, B.R., Kagan, B.L., and Tweten, R.K. (1997). Generation of a membrane-bound, oligomerized pre-pore complex is necessary for pore formation by *Clostridium septicum* alpha toxin. Mol. Microbiol. *23*, 551–558.

Sellman, B.R., and Tweten, R.K. (1997). The propeptide of *Clostridium septicum* alpha toxin functions as an intramolecular chaperone and is a potent inhibitor of alpha toxin-dependent cytolysis. Mol. Microbiol. *25*, 429–440.

Shatursky, O., Heuck, A.P., Shepard, L.A., Rossjohn, J., Parker, M.W., Johnson, A.E., and Tweten, R.K. (1999). The mechanism of membrane insertion for a cholesterol dependent cytolysin: A novel paradigm for pore-forming toxins. Cell *99*, 293–299.

Shen, A., and Higgins, D.E. (2005). The 5′ untranslated region-mediated enhancement of intracellular listeriolysin O production is required for *Listeria monocytogenes* pathogenicity. Mol. Microbiol. *57*, 1460–1473.

Shepard, L.A., Heuck, A.P., Hamman, B.D., Rossjohn, J., Parker, M.W., Ryan, K.R., Johnson, A.E., and Tweten, R.K. (1998). Identification of a membrane-spanning domain of the thiol-activated pore-forming toxin *Clostridium perfringens* perfringolysin O: an α-helical to β-sheet transition identified by fluorescence spectroscopy. Biochemistry *37*, 14563–14574.

Shepard, L.A., Shatursky, O., Johnson, A.E., and Tweten, R.K. (2000). The mechanism of assembly and insertion of the membrane complex of the cholesterol-dependent cytolysin perfringolysin O: Formation of a large prepore complex. Biochemistry *39*, 10284–10293.

Shin, D.J., Cho, D., Kim, Y.R., Rhee, J.H., Choy, H.E., Lee, J.J., and Hong, Y. (2006). Diagnosis of paroxysmal nocturnal hemoglobinuria by fluorescent *Clostridium septicum* alpha toxin. Journal of molecular microbiology and biotechnology *11*, 20–27.

Shin, D.J., Lee, J.J., Choy, H.E., and Hong, Y. (2004). Generation and characterization of *Clostridium septicum* alpha toxin mutants and their use in diagnosing paroxysmal nocturnal hemoglobinuria. Biochem. Biophys. Res. Commun. *324*, 753–760.

Soltani, C.E., Hotze, E.M., Johnson, A.E., and Tweten, R.K. (2007a). Specific protein-membrane contacts are required for prepore and pore assembly by a cholesterol-dependent cytolysin. J. Biol. Chem. *282*, 15709–15716.

Soltani, C.E., Hotze, E.M., Johnson, A.E., and Tweten, R.K. (2007b). Structural elements of the cholesterol-dependent cytolysins that are responsible for their cholesterol-sensitive membrane interactions. Proc. Natl. Acad. Sci. USA *104*, 20226–20231.

Song, L.Z., Hobaugh, M.R., Shustak, C., Cheley, S., Bayley, H., and Gouaux, J.E. (1996). Structure of staphylococcal alpha-hemolysin, a heptameric transmembrane pore. Science *274*, 1859–1866.

Stevens, D.L., Tweten, R.K., Awad, M.M., Rood, J.I., and Bryant, A.E. (1997). Clostridial gas gangrene: Evidence that alpha and theta toxins differentially modulate the immune response and induce acute tissue necrosis. J. Infect. Dis. *176*, 189–195.

Tateno, H., and Goldstein, I.J. (2003). Molecular cloning, expression, and characterization of novel hemolytic lectins from the mushroom *Laetiporus sulphureus*, which show homology to bacterial toxins. J. Biol. Chem. *278*, 40455–40463.

Tilley, S.J., Orlova, E.V., Gilbert, R.J., Andrew, P.W., and Saibil, H.R. (2005). Structural basis of pore formation by the bacterial toxin pneumolysin. Cell *121*, 247–256.

Tweten, R.K. (1988). Cloning and expression in *Escherichia coli* of the perfringolysin O (theta-toxin) gene from *Clostridium perfringens* and characterization of the gene product. Infect. Immun. *56*, 3228–3234.

Tweten, R.K. (2005). The cholesterol-dependent cytolysins; a family of versatile pore-forming toxins. Infect. Immun. *73*, 6199–6209.

Tweten, R.K., Harris, R.W., and Sims, P.J. (1991). Isolation of a tryptic fragment from *Clostridium perfringens* θ-toxin that contains sites for membrane binding and self-aggregation. J. Biol. Chem. *266*, 12449–12454.

Tweten, R.K., Parker, M.W., and Johnson, A.E. (2001). The Cholesterol-Dependent Cytolysins. In Pore-

Forming Toxins, G. van der Goot, ed. (Heidelberg, Springer-Verlag), pp. 15–33.

Van der Goot, F.G., Pattus, F., Wong, K.R., and Buckley, J.T. (1993). Oligomerization of the channel-forming toxin aerolysin precedes insertion into lipid bilayers. Biochemistry *21*, 2636–2642.

Walker, B., Braha, O., Cheley, S., and Bayley, H. (1995). An intermediate in the assembly of a pore-forming protein trapped with a genetically-engineered switch. Chem Biol. *2*, 99–105.

Walker, J.A., Allen, R.L., Falmagne, P., Johnson, M.K., and Boulnois, G.J. (1987). Molecular cloning, characterization, and complete nucleotide sequence of the gene for pneumolysin, the sulfhydryl-activated toxin of *Streptococcus pneumoniae*. Infect. Immun. *55*, 1184–1189.

Wichroski, M.J., Melton, J.A., Donahue, C.G., Tweten, R.K., and Ward, G.E. (2002). *Clostridium septicum* Alpha-Toxin Is Active against the Parasitic Protozoan *Toxoplasma gondii* and Targets Members of the SAG Family of Glycosylphosphatidylinositol-Anchored Surface Proteins. Infect. Immun. *70*, 4353–4361.

Wimley, W.C., Hristova, K., Ladokhin, A.S., Silvestro, L., Axelsen, P.H., and White, S.H. (1998). Folding of beta-sheet membrane proteins: a hydrophobic hexapeptide model. J. Mol. Biol. *277*, 1091–1110.

Binary Bacterial Toxins: Evolution of a Common, Intoxicating Theme

4

Bradley G. Stiles and Michel R. Popoff

Abstract

Several proteins from Gram-positive, spore-forming bacilli use a synergistic binary mechanism for intoxicating eukaryotic cells. These toxins include *Clostridium botulinum* C2 toxin, *Clostridium difficile* toxin (CDT), *Clostridium perfringens* iota (ι) toxin, and *Clostridium spiroforme* toxin (CST). Furthermore, closely related *Bacillus* species such as *Bacillus anthracis*, *Bacillus cereus*, and *Bacillus thuringiensis* produce strikingly similar binary toxins. As per existing literature, these latter proteins have provided a 'model' for the clostridial binary toxins. Each of these clostridial and bacillus binary toxins consists of distinct enzymatic 'A' and binding 'B' proteins that work in concert. Conservation of a basic intoxication theme between different genera clearly suggests retention of an evolutionarily successful mechanism promoting bacterial survival and dissemination throughout Nature. In particular, this chapter highlights the existing literature linked to *Clostridium* binary toxins and their role in pathogenesis, structure, function, as well as current/future use as protein shuttles.

Introduction

The *Clostridium* and *Bacillus* genera represent ubiquitous, spore-forming bacilli commonly found throughout the world in soil, water, and gastrointestinal tracts of animals as well as humans. Many clostridial and bacillus species are deemed harmless, but there are a number of notable exceptions that will be discussed throughout this chapter. Anaerobic clostridia (or archaic relatives of) likely represent some of the first bacteria on Earth, with perhaps closely related aerobic bacilli evolving after the Earth's atmosphere had become more oxygenated via the metabolic efforts of other life forms (Fox *et al.*, 1980; Novichkov *et al.*, 2004). *Clostridium* and *Bacillus* species both grow in low-oxygen environments, but the clostridia are better adapted for anaerobic life with varying levels of aerotolerance evident among a large number of species. The genome sequencing of *C. botulinum* (Sebaihia *et al.*, 2007), *C. difficile* (Sebaihia *et al.*, 2006), *C. perfringens* (Myers *et al.*, 2006), and *B. anthracis* (Read *et al.*, 2003) will inevitably reveal additional similarities between these, as well as other, microorganisms and their ways of conducting life throughout our shared biosphere.

When compared to other microbes, various *Clostridium* and *Bacillus* species have developed unique mechanisms for survival within, and outside of, numerous types of hosts. This is evidenced by the various diseases caused by these microorganisms that are often linked to protein toxins and spores. *C. botulinum*, *C. difficile*, *C. perfringens*, as well as *C. spiroforme* are associated with a multitude of animal and human diseases/intoxications such as gas gangrene, food poisoning, antibiotic-associated diarrhoea, pseudomembranous colitis and enterotoxaemia. Maladies attributed to *B. anthracis* also occur in both animals plus humans, and they include three forms of anthrax (cutaneous, intestinal and inhalational). As subsequently described, the binary toxins from *Clostridium* and *Bacillus* species are evident in diverse diseases. This fact further accentuates the differences that exist within this toxin group and especially among the bacteria that produce them.

Historically, the protein components of these related clostridial and bacillus binary toxins have been considered by many to not bind cells as a pre-formed enzymatic 'A' – binding 'B' complex (Table 4.1). The 'A' and 'B' proteins of clostridial and bacillus binary toxins are released from the bacterium as separate proteins that then assemble upon the surface of a targeted eukaryotic cell. However, this paradigm has recently been challenged as per the *C. botulinum* C2 (Kaiser *et al.*, 2006) and *B. anthracis* toxins (Chvyrkova *et al.*, 2007). Many other protein toxins synthesized by diverse bacterial genera differ from the clostridial and bacillus binary toxins, as they engage cells as a preformed 'A–B' package composed of single, or multiple, proteins (Table 4.2). The adaptation of bacterial proteins that first bind to unique, eukaryotic cell-surface receptors and then influence targeted cells via diverse intracellular mechanisms is a remarkable testimony to 'simple' prokaryotic life and its uncanny ability to evolve. Ultimately, such protein-based virulence factors promote bacterial survival throughout different ecological niches found around the world.

Intoxication by C2, CDT, CST, ι, or related *Bacillus* binary toxins initially involves specific, receptor-mediated binding of 'B' components to a targeted cell (Barth *et al.*, 2004). This occurs as 'B' monomers that form homoheptamers on the cell surface or in solution prior to cell interaction. In either scenario, the 'B' oligomers are generated only after proteolysis of 'B' precursor molecules by serine-type proteases. The 'B' heptamer – receptor complex then acts as a docking platform that subsequently translocates an enzymatic 'A' component(s) into the cytosol via acidified endosomes. Once inside the cytosol, 'A' components from this binary group can inhibit normal cell functions by different processes that include: (1) mono-ADP-ribosylation of globular actin (G actin) causing cytoskeletal disarray and cell death; (2) proteolysis of mitogen-activated protein kinase kinases (MAPKK) disrupting cell signal pathways; or (3) increasing intracellular levels of cyclic AMP leading to oedema and immunosuppression.

Table 4.1 *Clostridium* and *Bacillus* binary toxins

Toxin and components (molecular weight in kDa)	Gene location
C. perfringens ι toxin	Plasmid
Ia (45)	(140 kilobase pairs)
Ib (94, precursor form; 81, activated form)	
C. spiroforme CST	Chromosome
Sa (44)	
Sb (92, precursor form; 76, activated form)	
C. difficile CDT	Chromosome
CDTa (48)	
CDTb (99, precursor form; 75, activated form)	
C. botulinum C2 toxin	Chromosome
C2I (49)	
C2II (81, precursor form; 60, activated form)	
B. anthracis oedema and lethal toxins	Plasmid pXO1
EF (89)	(182 kilobase pairs)
LF (90)	
PA (83, precursor form; 63, activated form)	
B. thuringiensis VIP	Uncertain
VIP2 (49)	
VIP1 (86, precursor form; 70, activated form)	

Table 4.2 Examples of bacterial toxins consisting of single and multiple proteins

Toxin	Structure
Clostridium botulinum neurotoxins A–G	Single protein
Clostridium difficile toxins A and B	Single protein
Clostridium novyi alpha toxin	Single protein
Clostridium sordellii haemorrhagic and lethal toxins	Single protein
Clostridium tetani neurotoxin	Single protein
Corynebacterium diphtheriae diphtheria toxin	Single protein
Pseudomonas aeruginosa exotoxin A	Single protein
Bordetella pertussis pertussis toxin	Multiple proteins
Escherichia coli heat-labile enterotoxin	Multiple proteins
Shigella dysenteriae and other Enterobacteriaceae-produced shiga and shiga-like toxins	Multiple proteins
Vibrio cholerae cholera toxin	Multiple proteins

We will now bring to light some important aspects of *Clostridium*, as well as the closely related *Bacillus*, binary toxins regarding their genetics, structure, function, and use as protein-based delivery tools for targeted cells. The following section briefly introduces each binary toxin and the responsible microbe.

Clostridium and related *Bacillus* binary toxins

Clostridium perfringens iota (ι) toxin

C. perfringens, also known as *Bacillus aerogenes capsulatus* or *Clostridium welchii* in some older literature, was first described in late 1891 by William Welch *et al.*, at the newly opened Johns Hopkins Hospital in Baltimore, Maryland (Welch *et al.*, 1896; Lucey, 2004). In particular, the bacterium was isolated following autopsy of a 38-year-old man previously suffering from various maladies that included cancer and repeated haemorrhages. Of note was a rather profuse, unusually explosive, formation of gas bubbles within the cadaver's blood vessels and organs only eight hours after death. It can be argued that this discovery of a 'novel' bacterium was one of Dr Welch's most significant during a long, illustrious career at Johns Hopkins.

There are five serotypes (A–E) of *C. perfringens* that are classically based upon the production of four lethal, dermonecrotic toxins (alpha, beta, epsilon, and iota). These 'major' toxins are neutralized by type-specific antisera in mouse-lethal and guinea-pig dermonecrotic assays. Today, rapid genetic methods involving multiplex polymerase chain reactions are more commonly used by many diagnostic laboratories for typing *C. perfringens* isolates (Gkiourtzidis *et al.*, 2001). The binary ι toxin is exclusively produced by type E strains, implicated in sporadic diarrheic outbreaks among calves as well as lambs, and interestingly linked on the same plasmid to a highly conserved (yet curiously silent) enterotoxin gene flanked by insertion sites (Bosworth, 1940; Billington *et al.*, 1998; Li *et al.*, 2007). Although *C. perfringens* ι toxin was initially described in 1940 by Bosworth, its binary nature was first elucidated forty-five years later by exploiting the neutralization of ι toxin with cross-reacting *C. spiroforme* antiserum (Stiles and Wilkins, 1986). The two proteins that comprise ι toxin were then designated as iota a or Ia (slower moving) and iota b or Ib (faster moving), based upon electrophoretic mobility in crossed-immunoelectrophoresis. Ia or Ib are each non-toxic, as is the case for individual components from any *Clostridium* and *Bacillus* binary toxin described in this chapter. However, the combination of Ia and Ib forms a potent cytotoxin that becomes lethal to mice, dermonecrotic in guinea pigs, induces cell rounding *in vitro*, and elicits fluid accumulation within rabbit ileal loops. Later studies revealed that Ia is a mono-ADP-ribosyltransferase (Simpson *et al.*, 1987) specific for actin (Schering *et al.*, 1988). Although Ib lacks any discernible enzymatic activity, this molecule binds to a cell-surface protein(s) and subsequently translocates Ia into the cytosol of a targeted cell via lipid rafts (Stiles *et al.*, 2000; Hale *et al.*, 2004; Nagahama *et al.*, 2004).

Clostridium spiroforme toxin (CST)

Similar to the classic rod-shaped *C. perfringens* and enteric-acting ι toxin, the distinctly coiled *C. spiroforme* also causes diarrheic deaths that are spontaneous or antibiotic induced in rabbits (Kaneuchi *et al.*, 1979; Carman and Borriello, 1982, 1984; Borriello and Carman, 1983; Carman and Evans, 1984; Borriello *et al.*, 1986) and perhaps humans (Babudieri *et al.*, 1986). A major virulence factor produced by *C. spiroforme* is a binary ι-like toxin called CST. For rabbits, they are most susceptible to *C. spiroforme*-induced diarrhoea during stress that includes lactation, old age, weaning, and an altered diet (Carman and Evans, 1984). The Sa and Sb components of CST are respectively analogous to Ia and Ib of *C. perfringens* ι toxin, as first determined by crossed-immunoelectrophoresis and neutralization studies with *C. perfringens* type E antiserum (Borriello and Carman, 1983; Popoff *et al.*, 1989). During the late 1970s it was erroneously thought by different researchers that *C. perfringens* type E represented the causative agent of various diarrheic outbreaks within rabbit colonies. This was evidently so as type E antiserum neutralized the *in vitro* cytotoxic effects of caecal contents from enterotoxaemic animals (Katz *et al.*, 1978; Carman and Borriello, 1982). However, *C. perfringens* type E was never isolated and the real breakthrough came in 1982 when Carman and Borriello revealed a strong correlation between disease in rabbits with intestinally residing *C. spiroforme*. As described a few years later in further detail, this bacterium is not commonly associated with the normal intestinal flora of lagomorphs (Carman and Evans, 1984).

Clostridium difficile toxin (CDT)

The ι toxin and CST are closely related to yet another ι-like toxin, designated as CDT, which is produced by *C. difficile* (Popoff *et al.*, 1988; Perelle *et al.*, 1997a; Goncalves *et al.*, 2004). *C. difficile* was first recognized as a major pathogen in the 1970s as per its role in pseudomembranous colitis and antibiotic-induced diarrhoea (Bartlett, 1994). This bacterium still causes many serious problems to this day, especially in hospital settings throughout the world (Loo *et al.*, 2005; McDonald *et al.*, 2005). Initial discovery/isolation of *C. difficile* is credited to Hall and O'Toole in 1935 following their studies of the intestinal flora among healthy infants. Unlike adults, large populations of *C. difficile* associated with high levels of high-molecular-weight toxins A and B inside the infant intestinal tract do not portend a diseased state.

The CDT consists of CDTa and CDTb components that respectively share amino acid sequence identity (80%) with *C. perfringens* Ia and Ib. Such relatedness is also reflected in the fact that protein components of CDT, CST, and ι toxin are interchangeable; therefore, mix-match studies generate biologically active chimeras and further demonstrate the conserved functionality evident among these clostridial species (Popoff *et al.*, 1989; Perelle *et al.*, 1997b; Gülke *et al.*, 2001). The CDT, CST, and ι toxin form one family of clostridial binary toxins (iota family) based upon sequence homology, immunological cross-reactivity, and interchangeable components. Interestingly, *C. difficile*, *C. perfringens*, and *C. spiroforme* are all associated with gastrointestinal diseases in humans and/or animals (Bosworth, 1940; Borriello and Carman, 1983; Braun *et al.*, 2000). The synthesis of common binary toxins reveals a shared evolutionary path for these ubiquitous pathogens in a common niche. It is not likely a random coincidence that these intestinal pathogens possess very similar protein toxins.

For *C. difficile*, there have been many recent studies among hospital patients suggesting that CDT is linked to emerging epidemic strains which are particularly virulent (Geric *et al.*, 2003, 2006; Loo *et al.*, 2005; McDonald *et al.*, 2005; Blossom and McDonald, 2007); however, the contribution that CDT makes towards *C. difficile*-associated pathogenesis has been rightly questioned by the medical research community. A most recent, and rather extensive, epidemiological study from France suggests that strains of *C. difficile* expressing CDT are more virulent than those not expressing the binary toxin (Barbut *et al.*, 2007). Part of the uncertainty linked to CDT and disease severity is due to no efficient and standardized means, at a clinical or research level, for readily detecting the CDT proteins in patient's stool samples. Do high levels of CDT in stool samples correlate with a particularly nasty case of *C. difficile* enterocolitis? As per the enteric/systemic effects of other clostridial (and bacillus) binary toxins upon various mammals, it is quite likely that CDT contributes in some synergistic or additive fashion to *C. difficile*-associated disease among humans and animals.

Clostridium botulinum C2 toxin

C. botulinum, initially identified as *Bacillus botulinus*, was first described in 1895 by Emile van Ermengem following a funeral dinner consisting of contaminated ham (Devriese, 1999). Like *C. perfringens*, the types (A–G) of *C. botulinum* are classically determined by mouse lethal assays with neurotoxin-specific antisera. However, the binary C2 enterotoxin produced by toxin types C and D lacks neurotoxicity but is implicated in a fatal enteric disease of waterfowl. This toxin is cytopathic for many different cell types and induces vascular permeability, necrotic–haemorrhagic lesions, as well as a lethal fluid accumulation in the lungs and intestinal tracts of various animals (Ohishi *et al.*, 1980; Simpson, 1982; Ohishi, 1983a,b; Ohishi and Miyake, 1985; Ohishi and Okada, 1986; Kurazono *et al.*, 1987). The C2 toxin is synthesized by *C. botulinum* during sporulation (Nakamura *et al.*, 1978) which is a characteristic unlike any of the other clostridial binary toxins that are produced during vegetative growth. Additionally, C2 toxin is quite distinct in other ways from the iota family of clostridial binary toxins and thus represents the sole member of the C2 toxin family.

Investigators have found that there are physical (molecular weight and epitopes) as well as functional (cytotoxicity) variations discerned between C2I (enzymatic 'A') and C2II (binding 'B') components produced by different *C. botulinum* strains (Ohishi and Okada, 1986; Ohishi and Hama, 1992). Similar comparison data are lacking for the protein components of other *Clostridium* and *Bacillus* binary toxins, with one notable exception being the highly conserved protective antigen from *B. anthracis* (Price *et al.*, 1999). Like the other clostridial binary toxins, C2I possesses ADP-ribosyltransferase activity targeting an arginine residue (R_{177}) that is readily exposed in monomeric G-actin (Aktories *et al.*, 1986b). This finding represents the discovery of a new group of bacterial mono-ADP-ribosylating proteins that fatally disrupt the eukaryotic cytoskeleton.

Bacillus anthracis oedema and lethal toxins

In contrast to *C. botulinum* or *C. perfringens*, the phylogenetic relative called *B. anthracis* is not composed of serologically distinguishable toxin types. However, *B. anthracis* produces similarly structured binary toxins involved in animal and human cases of anthrax (Petosa *et al.*, 1997; Han *et al.*, 1999; Leppla, 2006). A review of the literature quickly reveals much more information available for the *B. anthracis* binary toxins versus those produced by clostridia. In many ways, the *B. anthracis* toxins have historically served as 'templates' for investigators of the clostridial binary toxins. There are though major differences existing between these protein toxins that will be highlighted throughout this chapter.

Overall, there is remarkably limited genetic variation among *B. anthracis* isolates throughout the world (Keim and Smith, 2002). Some investigators suggest that *B. anthracis* strains are merely a genetic variant of *Bacillus cereus* and *Bacillus thuringiensis* that possess plasmids encoding for unique toxins and capsule (Helgason *et al.*, 2000; Pannucci *et al.*, 2002; Ivanova *et al.*, 2003; Read *et al.*, 2003; Rasko *et al.*, 2004). The natural disease elicited by *B. anthracis* mainly occurs among herbivores after spore transmission from contaminated soil. An anthrax infection can naturally occur via ingestion, inhalation, soiling of an open skin wound, or even biting flies. The nefarious use of *B. anthracis* as a biological weapon was first associated with espionage and targeting of horses, mules, as well as cattle during the First World War (Turnbull, 2002).

The binary toxins of *B. anthracis* were the first multicomponent bacterial toxins ever described in the literature (Smith *et al.*, 1956). They consist of three synergistically acting proteins called oedema factor (EF), lethal factor (LF), and protective antigen (PA) (Stanley and Smith, 1961; Smith, 2002). From a historical perspective, this discovery of a multicomponent toxin in 1956 by Harry Smith's laboratory is quite appropriate.

Both Robert Koch and Louis Pasteur initially used *B. anthracis* to prove profound scientific concepts that have forged the microbiology and immunology we practice today. The rather versatile PA molecule ('B' component) combines with EF and/or LF ('A' components) on the cell surface to respectively form oedema and/or lethal toxins. These proteins represent virulence factors that work synergistically towards bacterial survival and dissemination (Cataldi *et al.*, 1990; Popov *et al.*, 2002; Mock and Mignot, 2003; Tournier *et al.*, 2005; Leppla, 2006). Unlike the protein components of CDT, CST, and ι toxin (the aforementioned ι toxin family), those for *C. botulinum* C2 or *B. anthracis* oedema/lethal toxins do not form biologically active chimeras with other binary toxins (Perelle *et al.*, 1997b; Stiles *et al.*, 2002a). It is also believed that these latter three toxins can uniquely form intact 'A-B' toxins in solution before the 'B' components interact with cell membrane receptors (Mogridge *et al.*, 2002ab; Kaiser *et al.*, 2006; Chvyrkova, 2007).

Other *Bacillus* binary toxins

Some *B. cereus* and *B. thuringiensis* strains synthesize soluble, binary toxins during vegetative growth. *B. cereus* is associated with certain types of food poisoning (Michelet *et al.*, 2006) and resides as a normal intestinal inhabitant among various insects like roaches, sow bugs, as well as termites (Margulis *et al.*, 1998). *B. thuringiensis* is a common soil microbe that has been used for many years as a natural pesticide. Certain coleopteran-specific proteins produced by these bacilli are called vegetative insecticidal proteins (VIPs) and consist of a non-enzymatic, cell-binding component (VIP1) plus an ADP-ribosyltransferase (VIP2) that targets actin (Warren, 1997; Han *et al.*, 1999; Shi *et al.*, 2004). In particular, the VIPs are effective against Northern and Western corn rootworms, as well as the European corn borer. Although VIP1 and VIP2 share significant identity at the amino acid level with clostridial equivalents, the VIPs form yet another distinct family of toxins. Like other 'B' components among the *Clostridium* and *Bacillus* binary toxins, recent studies show that VIP1 forms oligomers and subsequent channels in membranes (Leuber *et al.*, 2006). The unique channels generated by VIP1 are anion selective versus the cation-selective structures induced by *C. perfringens* Ib, *C. botulinum* C2II, and *B. anthracis* PA. An additional characteristic of the VIP1 channels is that they are not blocked by 4-aminoquinolones, unlike C2II and PA (Bachmeyer *et al.*, 2001; Orlik *et al.*, 2004, Leuber *et al.*, 2006)

Biochemistry and genetics

The protein components of *C. botulinum* C2 toxin (Ohishi *et al.*, 1980), *C. difficile* CDT (Perelle *et al.*, 1997a), *C. perfringens* ι toxin (Stiles and Wilkins, 1986), *C. spiroforme* CST (Popoff *et al.*, 1989), as well as *B. anthracis* oedema and lethal toxins (Smith *et al.*, 1956; Stanley and Smith, 1961) are each produced as separate 'A' and 'B' molecules not commonly associated in solution. Table 4.1 lists the gene locations and molecular weights of these bacterial binary toxins. The cell-binding components are evidently, by all accounts, enzymatically inert and produced by the bacterium as monomeric, precursor molecules. These 'B' components are activated by various serine-type proteases like *C. perfringens* lambda toxin, furin, chymotrypsin, or trypsin derived from the bacterium, host, or exogenous addition *in vitro* (Fernie *et al.*, 1984; Klimpel *et al.*, 1992; Gibert *et al.*, 2000). The resultant loss of an N-terminal peptide (approximately 20 kDa) from a 'B' precursor induces conformational changes that facilitate homoheptamerization, either in solution or on the cell surface. This 'B' complex acts as a base that subsequently docks with an 'A' component(s).

Like other multi-component toxins, the enzymatic and binding proteins of all *Clostridium* and *Bacillus* binary toxins are encoded by distinct genes that possess 27–31% G+C content (Popoff, 2000). Amongst the clostridia, 'A' and 'B' component genes are transcribed in the same orientation from a common operon. One mRNA strand contains the transcript for both 'A' and 'B' molecules. The 'A' gene is located 40–50 nucleotides upstream of the 'B' gene, with an exception being the open reading frames for C2 toxin that are separated by 247 nucleotides (Perelle *et al.*, 1993; Fujii *et al.*, 1996; Gibert *et al.*, 1997; Kimura *et al.*, 1998). There is also another significant genetic difference between these toxins as *C. botulinum* C2 toxin, *C. difficile* CDT, and *C. spiroforme* CST are chromosome-encoded versus the plasmid-localized *C. perfringens* ι toxin (Li *et al.*, 2007). The *B. anthracis* binary toxins are also coded on a plasmid called pXO1 (Okinaka *et al.*, 1999; Leppla, 2006).

Each 'A' and 'B' component of the *Clostridium* and *Bacillus* binary toxins, but not those that compose C2, is respectively synthesized with a signal peptide consisting of 29–49 and 39–47 residues (Popoff, 2000). These findings correlate with proteins secreted during logarithmic growth; however, the C2 toxin is only evident from sporulating *C. botulinum* after sporangium lysis (Nakamura *et al.*, 1978). Production of another sporulation-linked protein, *C. perfringens* enterotoxin, employs sigma factors (SigK and SigE) that are also likely used (yet not proven) for the synthesis of *C. botulinum* C2 toxin (Zhao and Melville, 1998). It has also been shown that an extracellular chaperon protein, PrsA, is important for efficient folding and secretion of PA83 from *B. anthracis* via a Sec-dependent route (Williams *et al.*, 2003). It is unknown if similar chaperon molecules play any role with binary toxin components produced by the other bacilli.

Among the *Clostridium* and *Bacillus* binary toxins, most genetic-based work has been done with *B. anthracis*. *B. anthracis* toxin production is controlled by the positive regulatory gene *atxA* (anthrax toxin activator), also located on pXO1, via protein binding approximately 110 base pairs upstream of the ATG start codon. Importance of the atxA protein in pathogenesis is further evidenced by less virulent strains lacking the *atxA* gene (Dai *et al.*, 1995). The 56-kDa protein encoded by *atxA* is rather unique as it shares little sequence homology with other known transcriptional activators. In addition to the genes for oedema and lethal toxins, *atxA* also regulates (in a positive or negative fashion) the expression of other genes found on the pXO1 plasmid (Hoffmaster and Koehler, 1997), the pXO2 plasmid (93 kilobase pairs) that encodes for capsule (Guignot *et al.*, 1997), and the chromosome (Bourgogne *et al.*, 2003; Mignot *et al.*, 2003). There are two transcription start sites for *atxA* (designated P1 and P2) that are respectively situated 58 and 26 base pairs upstream of the start codon. The P2 site induces approximately 10-fold higher levels of messenger RNA than P1. Environmental factors such as bicarbonate concentrations, carbon dioxide levels, and temperature also represent environmental factors governing synthesis of *B. anthracis* oedema and lethal toxins (Mignot *et al.*, 2003; Leppla, 2006). The gene products or environmental factors that may affect production of other bacterial binary toxins described in this chapter are largely unknown, although divalent cations play a role in *C. spiroforme* CST synthesis by an undefined mechanism (Carman *et al.*, 1987). A recent study by Stare *et al.* (2007) shows that the CDT genes for 146 strains of *C. difficile* come in two chromosomal forms that are either full-length (4.3 kilobase pairs) and functional, or truncated (2.3 kilobase pairs) and dysfunctional.

Amino acid sequences among the *Clostridium* and *Bacillus* binary toxins quite evidently reveal similar evolutionary paths. Interchangeable protein components of the ι-toxin family (CDT, CST, and ι toxin) share 80–85% identity, but the signal peptide sequences among them are less conserved (40–61% identity). There is 31–40% identity between C2 and iota family toxins while 26–30% identity exists between PA and clostridial 'B' components. Overall, such data suggest that these toxin genes are derived from a common ancestor. 'B' molecules have throughout time become rather proficient in transporting unique 'A' components into cells. There are striking differences that exist between 'A' components such as *B. anthracis* EF (an adenylcyclase) and LF (a metalloprotease), which are structurally and enzymatically distinct, versus the rather homogeneous ADP-ribosyltransferases evident within the C2 or ι-toxin families. Although unproven, it is very plausible that the binary toxin genes originated from an ancestral form of clostridia or anaerobic progenote. These genes were perhaps horizontally transferred between *Clostridium* and *Bacillus* species via plasmids capable of inserting themselves into the bacterial chromosome.

Proteolytic activation

C. perfringens Ibp and Ia

Helen Ross and colleagues initially reported in 1949 that ι toxin requires proteolytic activation, as determined by mouse lethal and guinea pig dermonecrotic assays (Ross *et al.*, 1949). It took

35 more years before additional clues revealed that an Ib precursor molecule (designated as Ibp) was the likely target of exogenously added, or culture-derived, serine-type proteases (Barth *et al.*, 2004). Ia and Ibp produced by early log phase (less than 10 h) cultures of *C. perfringens* type E were separated by ion-exchange chromatography. It was then found that trypsin proteolysis of fractions containing Ibp or Ia markedly increased ELISA readings for Ibp, but not Ia. These findings suggest a conformational change of 'activated' Ibp that unveils cryptic epitopes recognized by Ib-specific antibodies. In addition to the ELISA results, there were dramatic increases in guinea-pig dermonecrotic activity and mouse lethality of the Ibp fraction when combined with untreated Ia. Subsequent cloning and sequencing of the ι-toxin gene showed that proteolytic activation of Ibp into Ib occurs at residue A_{211} (Perelle *et al.*, 1993). Proteolysis of Ibp facilitates Ib oligomerization into SDS-stable heptamers on cell membranes and a much less stable form in solution (Blöcker *et al.*, 2001; Nagahama *et al.*, 2002; Stiles *et al.*, 2002b).

An Ia molecule(s) readily docks with the Ib oligomer but not Ibp monomer (Stiles *et al.*, 2000). There are also voltage-dependent, ion-permeable channels formed in membranes by the Ib oligomer versus Ibp monomer (Knapp *et al.*, 2002). The latter molecule does not form a homo-oligomer in solution or on the surface of cells unless there is prior proteolytic activation (Stiles *et al.*, 2002b). Ib oligomers formed in solution are structurally fragile and do not promote Ia-induced cytotoxicity versus the solution-generated oligomers of PA63 or C2IIa. Additionally, Ib heptamers formed in solution do not induce potassium release and are readily digested by pronase after binding to Vero cells at 37°C (Nagahama *et al.*, 2002). This latter effect suggests that solution-generated Ib oligomers, once bound to cells, remain exposed and do not insert into cell membranes. Like C2II, which also lacks a furin cleavage site, Vero cell-bound Ibp is not activated over time *in vitro* with or without exogenous trypsin or chymotrypsin (Stiles *et al.*, 2002b). These data suggest that the proteolytic activation site of Ibp, once the molecule has bound to the cell membrane, is sterically encumbered as per conformational change and/or membrane insertion. Of note is that the cell-binding site for Ib/Ibp localized within the C-terminus is quite distal from the N-terminal activation site. To date, extensive proteolytic activation studies similar to those for C2II, Ib, and PA have not been conducted with *C. difficile* CDTb or *C. spiroforme* Sb.

There are other proteases like pepsin, proteinase K, subtilisin, alpha-chymotrypsin, and thermolysin that activate Ibp more efficiently than trypsin. Besides Ibp, Ia is also proteolytically activated by these same enzymes with a resultant loss of 9–13 amino acids from the N-terminus (Gibert *et al.*, 2000). It is still uncertain to date whether proteolysis of Ia increases: (1) docking efficiency to cell-bound Ib; (2) translocation efficiency into cells; and/or (3) ADP-ribosyltransferase activity. As per current knowledge in the field, proteolytic activation of Ia is quite unique among the 'A' components that compose other clostridial or bacillus binary toxins.

Another distinct group of 'A–B' toxins is composed of heterologous proteins formed as holotoxins once they are released by the bacterium. These molecules are quite different structurally than the *Clostridium*/*Bacillus* binary toxins described throughout this chapter. As just a few examples, these other 'A-B' toxins consist of the similar *Escherichia coli* heat labile, *Shigella dysenteriae* shiga, and *Vibrio cholerae* cholera enterotoxins (Sandvig, 2006; Hirst and D'Souza, 2006). The enzymatic 'A' (but not 'B') components synthesized by each of these Gram-negative bacilli are also processed by serine-type proteases like furin. However, after proteolysis, these 'A' components form A_1 and A_2 subunits linked by a disulfide bond that is subsequently reduced by a disulfide isomerase located within the endoplasmic reticulum of targeted eukaryotic cells. Table 4.3 contains a list of rather diverse bacterial protein toxins that require extracellular (prokaryotic or eukaryotic) activation by proteolysis.

C. botulinum C2II

There is trypsin activation, between residues K_{181} and A_{182}, of the 81-kDa C2II precursor into an activated C2II molecule (60 kDa) (Blöcker *et al.*, 2000). Unlike the Ibp to Ib conversion

Table 4.3 Examples of bacterial toxins activated by extracellular proteases

Toxin site	Protease	Activation
Aeromonas hydrophila aerolysin	Trypsin or furin	K_{427} or R_{432}
Bacteroides fragilis toxin	Endogenous bacteroides	R_{211}
C. botulinum neurotoxin A	Endogenous clostridial	Between 'A' and 'B' subunits at K_{437} –K_{447}
C. tetani neurotoxin	Trypsin or clostripain	Between 'A' and 'B' subunits at R_{448}–R_{454}
C. difficile toxin B	Autocatalysis by toxin B	L_{543}
C. perfringens epsilon toxin	Trypsin	K_{13}
C. perfringens enterotoxin	Trypsin or chymotrypsin	K_{25} or N_{36}
C. septicum alpha toxin	Furin or trypsin	R_{398}
C. diphtheriae diphtheria toxin	Furin	Between 'A' and 'B' subunits at R_{193}
Enterococcus faecalis cytolysin	Subtilisin-like serine protease	E_6 of both subunits
E. coli heat-labile enterotoxin	Trypsin	Within 'A' subunit at R_{192}
E. coli heat-stable enterotoxin A	Endogenous	N_{54}
Helicobacter pylori vacuolating toxin A	Trypsin	Between p37 and p58 domains
P. aeruginosa exotoxin A	Furin	R_{279} located near translocation domain
S. dysenteriae shiga toxin	Furin	Within 'A' subunit at R_{251}
V. cholerae cholera toxin	Trypsin	Within 'A' subunit at R_{192}

by proteolysis, formation of activated C2II from precursor generates relatively stable C2II homoheptamers in solution (Barth *et al.*, 2000). With an ADP-ribosyltransferase designated as C2I, the activated C2II complex readily mediates biological effects on many different cell types. This specifically involves the formation of ion-permeable channels in lipid membranes (Schmid *et al.*, 1994). Earlier electron microscopy of C2II oligomers on lipid bilayers reveals annular heptameric structures with an inner diameter of 20–40 Å and an outer diameter of 110–130 Å (Barth *et al.*, 2000).

Although the C2II precursor binds to cells, it is not readily activated by cell-surface proteases like PA83 and it will neither dock with C2I nor facilitate cytotoxicity (Ohishi, 1987; Ohishi and Yanagimoto, 1992). This is very much like toxins found within the iota family. X-ray crystallography of C2II has been recently reported (Schleberger *et al.*, 2006), but at relatively low resolution. In this same study, the crystal structures for C2I were much better resolved (1.75 Å) and the findings suggest little conformational distortion of the molecule between pH 6.1 and pH 3.0. This latter point is particularly pertinent as per the role that an acidic environment (endosome or extracellular fluid) plays in C2II-mediated translocation of C2I via a C2II-formed channel.

B. anthracis PA83

Unlike the other 'B' components of clostridial and bacillus binary toxins, the PA precursor (PA83) is proteolytically activated by several routes into PA63 (Ezzell and Abshire, 1992; Petosa *et al.*, 1997; Brossier *et al.*, 2000; Liu and Leppla, 2003). This activation event can occur via exogenously added trypsin, an unidentified protease(s) in serum (Ezzell and Abshire, 1992; Brossier *et al.*, 2000), or rather ubiquitous cell-surface proteases (furin or furin-like) that recognize a unique consensus sequence $_{164}$RKKR$_{167}$ (Klimpel *et al.*, 1992; Gordon *et al.*, 1995; Leppla, 2006). *C. spiroforme* Sb does have a RSAR site located 208 residues from the N-terminus that may be an evolutionary remnant of a furin cleavage site (Gibert *et al.*, 1997). Furin activation of protein toxins is rather commonly exploited by prokaryotes (Gordon *et al.*, 1995), as per single-chain molecules possessing ADP-ribosyltransferase activity like *Corynebacterium diphtheriae* diphtheria toxin and *Pseudomonas aeruginosa* exotoxin A. Both of these toxins enter cells by receptor-mediated endocytosis and subsequently inhibit protein synthesis.

Proteolysis of PA83 on a cell surface still remarkably occurs at 4°C or on chemically-fixed cells (Leppla, 2006). A furin-resistant variant of PA83 is not readily internalized into cells (Beauregard *et al.*, 2000). Such results suggest that surface-associated PA83 is not 'wasted' (i.e. internalized) without proteolytic activation into PA63 and subsequent transport of its protein cargo (EF and/or LF) into the cytosol. Another governing event includes the 20-kDa peptide (designated as PA20) generated after proteolysis of PA83, which slows PA63 clearance from the cell surface (Beauregard *et al.*, 2000). The non-covalent association of PA63 with PA20 may prevent a series of events such as heptamer formation, localization of PA63 into lipid rafts (Abrami *et al.*, 2003), and ultimately endocytosis. Altogether, this further optimizes EF plus LF docking opportunities to PA63. To our knowledge, similar work with the 20-kDa precursor peptides from other *Clostridium* binary toxins has not been reported in the literature. Clearly, such efforts might yield similar results with the 'B' components of clostridial binary toxins.

The PA63 molecule readily assembles into SDS-resistant, hydrophobic homoheptamers that form pH-dependent (pH $\leq$ 7), ion-permeable channels in membranes. These channels are composed of extended β-barrels and readily obstructed by low-molecular-weight molecules (Blaustein *et al.*, 1989; Milne and Collier, 1993; Batra *et al.*, 2001; Nassi *et al.*, 2002; Liu and Leppla, 2003; Liu *et al.*, 2003b). The PA63 heptamer forms a ringed structure of 160 Å (diameter) by 85 Å (height) containing a central lumen diameter of 35 Å (Petosa *et al.*, 1997).

Structure and function

C. perfringens Ib and Ia

The importance of various domains on Ib activity has been determined by deletion mutagenesis and antibody studies (Fig. 4.1). Similar to the effects upon PA and C2II, deletion of just 10

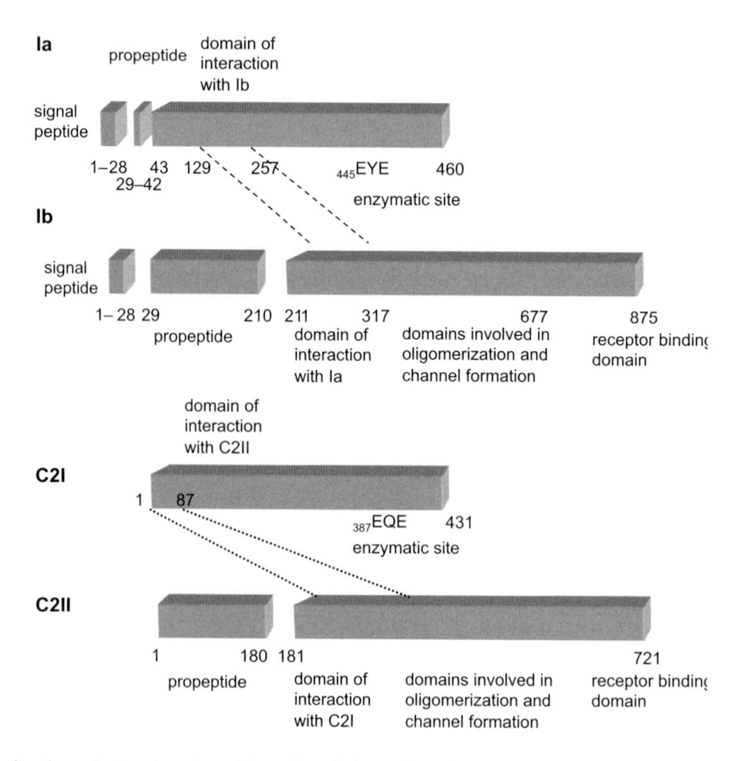

Figure 4.1 Organization of structural and functional domains within the *C. perfringens* iota and *C. botulinum* C2 toxins.

amino acids from the C-terminus (domain 4) effectively prevents Ib binding to Vero cells. C-terminal peptides of Ib containing more than 200 amino acids represent competitive inhibitors of iota cytotoxicity *in vitro* (Marvaud *et al.*, 2001). Deletion of just 27 N-terminal Ib residues within domain 1′ prevents Ia docking and intoxication. Such a truncation has little effect upon Ib binding to the cell surface, as this protein variant is a rather effective competitor of native ι toxin. Three monoclonal antibodies against a common N-terminal epitope within residues 28–66 also have no effect upon Ib binding or cytotoxicity (Marvaud *et al.*, 2001). It is possible that these immunoreagents do not occupy the Ib site necessary for Ia docking. Perhaps Ib oligomerization and/or docking of Ia readily displace these antibodies. There are to date no data linked to how avidly Ia docks with Ib oligomers, which is also the case for any of the other clostridial binary toxins and their 'A-B' molecular interactions.

In contrast to the N-terminal binding antibodies toward Ib, there are two distinct monoclonal antibodies recognizing unique Ib epitopes within the C-terminus (residues 632–655). These reagents afford protection against iota cytotoxicity via two different mechanisms. One antibody prevents Ib binding to cells, as determined by flow cytometry (Marvaud *et al.*, 2001) and Western blot analysis (Stiles *et al.*, 2002b). The other antibody has very little effect upon Ib binding yet it efficiently prevents Ib oligomerization on the cell surface. In perhaps similar fashion, two human monoclonal antibodies against *B. anthracis* PA83 prevent PA63 oligomerization and lethal toxin activity (Wang *et al.*, 2004). Results for the latter C-terminal binding antibody against Ib further demonstrate the importance of Ib oligomerization on biological activity of ι toxin, which is a common theme among all of the clostridial and bacillus binary toxins. Akin to polyclonal antibody studies with domain 4 of C2II (Blöcker *et al.*, 2000), monoclonal antibodies that recognize the same domain of Ib do not bind and thus do not afford any *in vitro* protection towards cell-bound Ib with Ia. All of the aforementioned monoclonal antibodies generated towards Ib recognize Ibp or *C. spiroforme* Sb in an ELISA and Western blot, but none cross-react with *B. anthracis* PA. It is quite surprising that C2II is also recognized in an ELISA by one of the Ib monoclonal antibodies that prevent Ib interactions with cells; however, this immunoreagent does not neutralize C2 cytotoxicity. Such a finding is quite intriguing, as C2II and Ib bind unique receptors via their C-terminus and share little sequence homology within this epitope (Fritz *et al.*, 1995; Blöcker *et al.*, 2000). Perhaps there are previously unrecognized conformational similarities that exist between the C-terminal regions of Ib and C2? To further enhance our understanding of ι toxin, future investigations focused upon domains 2 and 3 of Ib should be done to more clearly delineate residues required for oligomerization and channel formation. This type of detailed work has been done with *B. anthracis* PA, as described below.

No one has published the crystal structure for Ib to date, but that for Ia has been reported by Tsuge *et al.* (2003). Their extensive analysis of Ia shows two domains sharing conformational, but little sequence, similarity (Fig. 4.2a). A similar duplicate domain structure is evident for a related *B. cereus* ADP-ribosyltransferase called vegetative insecticidal protein 2 (VIP2) (Han *et al.*, 1999) and *C. botulinum* C2I (Schleberger *et al.*, 2006; Fig. 4.2b). VIP2 is transported into cells via a 'B' component called VIP1, much like the other clostridial/bacillus binary toxins described in detail throughout this chapter. The C-domains of Ia and VIP2 share 40% sequence identity as well as similar surface charges. The first glutamic acid found in the catalytic $_{378}$EXE$_{380}$ motif of Ia promotes ADP-ribosyltransferase, but not nicotinamide adenine dinucleotide (NAD)-glycohydrolase, activity (Sakurai *et al.*, 2003). Site-directed mutagenesis or mass spectrometry of cyanogen bromide and trypsin-generated peptides of Ia reveals that the C-terminal residues R$_{295}$ and E$_{380}$ are also important for Ia catalysis (Perelle *et al.*, 1993; van Damme *et al.*, 1996). These residues are conserved amongst various ADP-ribosyltransferases (Carroll *et al.*, 1984; Jung *et al.*, 1993; Han *et al.*, 1999; Sakurai *et al.*, 2003) and an additional triad, $_{338}$STS$_{340}$, is also commonly located near the active site. The same amino acids are also necessary for the enzymatic activity of *C. botulinum* (C2I) and *C. difficile*

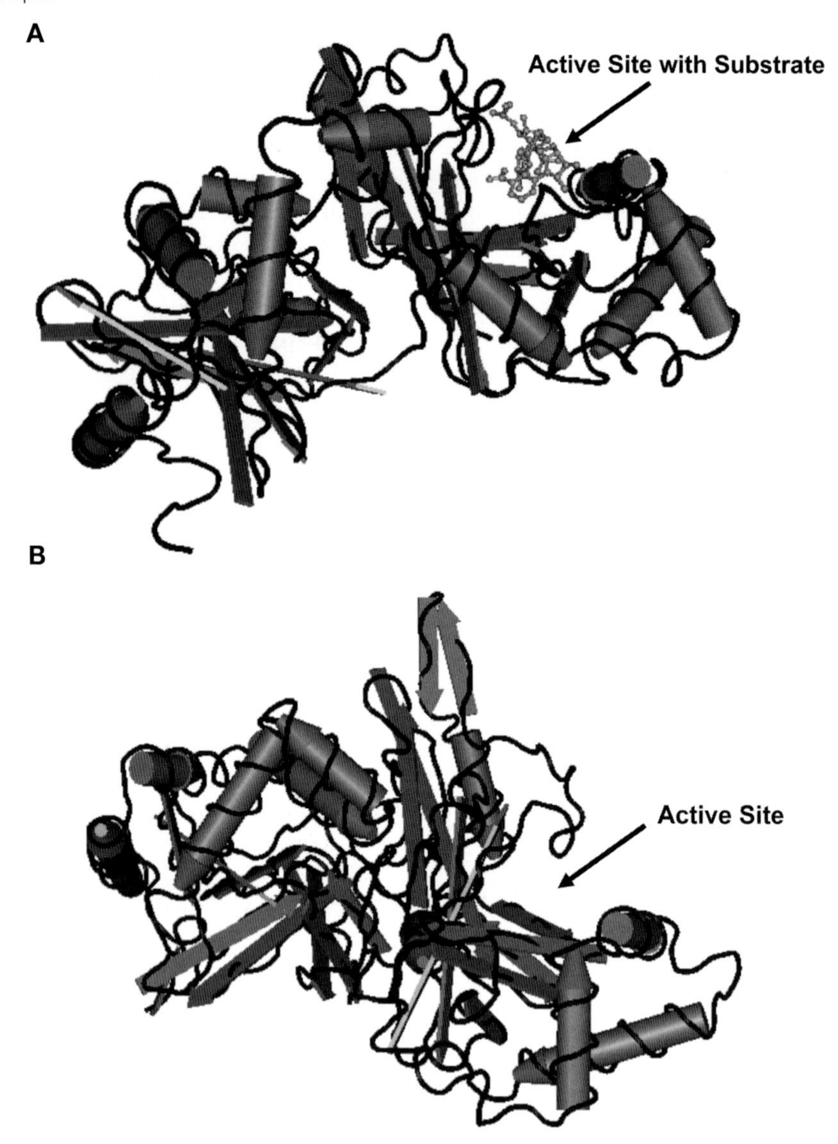

Figure 4.2 (a) Crystal structure of *C. perfringens* Ia with NAD substrate (Tsuge *et al.*, 2003). (b) Crystal structure of *C. botulinum* C2I at pH 6.1 (Schleberger *et al.*, 2006). These figures were generated by using Entrez's 3-D database and software for molecular modelling (Chen *et al.*, 2003).

(CDTa) binary toxins (Barth *et al.*, 1998b; Gülke *et al.*, 2001).

Extensive mutagenesis studies of Ia focused upon the NAD binding cavity reveal that Y_{246} and N_{255} are both important for ADP-ribosyltransferase, but not NAD-glycohydrolase, activity (Sakurai *et al.*, 2003). A Y_{251} residue is involved in both activities. The enzymatic activity of ι toxin is inhibited by EDTA chelation of divalent cations associated with actin, but low temperature (0°C) remarkably decreases activity by only 50% versus that evidenced at 37°C (Just *et al.*, 1990). All ADP-ribosyltransferases within the binary toxin group (C2I, CDTa, Ia, Sa, and VIP2) target G actin. This substrate is a commonly conserved protein found throughout Nature that plays a pivotal role in both the cytoskeleton, as well as intracellular trafficking, of all eukaryotic cells (Wertman and Drubin, 1992; Chowdhury *et al.*, 2002). In fact, the inter-relatedness of eukaryotic and prokaryotic cells is further evidenced by the latter containing an ancient homologue of actin within their cytoskeletal matrix (Lowe *et al.*, 2004).

In comparison to 'A' components found in other clostridial and bacillus binary toxins, the Ib docking region on Ia is more centrally located within residues 129–257 (Marvaud *et al.*, 2002). This is in contrast to C2I residues 1–87 needed for binding to C2II (Barth *et al.*, 2002ab), LF residues 1–254 interactions with PA (Arora and Leppla, 1993), or CDTa residues 1–240 docking to CDTb (Gülke *et al.*, 2001). Overall, these data probably reflect evolutionary variation among the 'A' and 'B' proteins comprising these related bacterial binary toxins.

The basic mechanism of ADP-ribosylation employed by C2, CDT, CST, ι, and VIP is evident among diverse bacteria (Masignani *et al.*, 2006). All ADP-ribosylating toxins use host derived NAD as a source of ADP-ribose. There are at least four bacterial groups of ADP-ribosylating toxins based upon their modified intracellular targets, which include: (1) elongation factor two (EF2) targeted by *C. diphtheriae* diphtheria toxin and *P. aeruginosa* exotoxin A via an N- and C-terminal active site, respectively; (2) heterotrimeric G-proteins targeted by *B. pertussis* pertussis toxin, *E. coli* heat-labile enterotoxin, and *V. cholerae* cholera toxin by way of N-terminal active sites; (3) Rho and Ras GTPases modified by *C. botulinum* C3 exoenzyme and *P. aeruginosa* exoenzyme S through C-terminal active sites; and (4) G actin. This last group is targeted by the toxins highlighted in this chapter and include *C. botulinum* C2 toxin (Aktories *et al.*, 1986b) and the ι-toxin family consisting of *C. difficile* CDT (Popoff *et al.*, 1988), *C. perfringens* ι toxin (Perelle *et al.*, 1995), as well as *C. spiroforme* CST (Popoff and Boquet, 1988; Stiles and Wilkins, 1986). All actin-modifying toxins have a C-terminally located active site on their 'A' components.

The actin-ADP-ribosylating toxins of clostridial and bacillus origin form two obvious groups based upon enzymatic activity. The *C. botulinum* C2 toxin can mono-ADP-ribosylate R_{177} of β/γ-non-muscle, as well as γ-smooth muscle, types of G actin (Aktories *et al.*, 1986; Ohishi and Tsuyama, 1986; Vandekerckhove *et al.*, 1988; Just *et al.*, 1993). Recent findings from Uematsu *et al.* (2007) show that disassembly of actin by C2 toxin subsequently induces microtubule assembly and polarization among human leukaemic cell lines. The ι-toxin family is less discriminating as it mono-ADP-ribosylates R_{177} of all known G-actin isoforms (Mauss *et al.*, 1990). To further reveal the ubiquitous nature of these enzymes, a type-III secretion protein (SpvB) of *Salmonella enterica* is an ADP-ribosyltransferase with enzymatic and structural properties very similar to Ia of ι toxin (Hochmann *et al.*, 2006). 'A' components like CDTa, Ia, and Sa each possess a LKDKE sequence within N-terminal residues 10–19 (Popoff, 2000). This five-residue region is evidently important for binding G actin. The C2I molecule has a unique actin-binding sequence and location, $_{44}$LKTKE$_{48}$, which likely explains its distinct substrate specificity. In a nicotinamide excess environment, the enzymatic components of both C2 and ι toxins remove an ADP-ribosyl moiety from modified actin. However, the C2I molecule does not displace ADP-ribose from Ia-modified actin of skeletal muscle origins (Just *et al.*, 1990).

Filamentous (F) actin that is commonly incorporated into the cytoskeleton does not represent a substrate target for any of the clostridial and bacillus binary toxins. Mono-ADP-ribosylation of G-actin inhibits monomer assembly into F-actin strands (Aktories *et al.*, 1986b; Aktories *et al.*, 1986a), thus decreasing F-actin concentrations while concomitantly increasing G-actin concentrations inside a cell (Aktories *et al.*, 1989; Aktories and Wegner, 1989). Within an actin–gelsolin complex, in which the latter molecule facilitates actin nucleation and polymerization, there is ADP-ribosylation of an actin molecule by both ι as well as C2 toxins (Wille *et al.*, 1992; Just *et al.*, 1993; Ballweber *et al.*, 2001). This activity blocks additional nucleation activity and subsequent formation of F actin. In addition to the actin-gelsolin complex, both ι and C2 toxins can also mono-ADP-ribosylate a G-actin molecule complexed with ATPase. A modified actin molecule then results in an increased exchange, but decreased hydrolysis, of ATP (Geipel *et al.*, 1989).

From a pathogen's perspective, disruption of a eukaryote's cytoskeleton and diminishing ATP hydrolysis can prevent phagocytosis or extend survival within a cell after internalization (Al-Mohanna *et al.*, 1987). Intracellular trafficking of vesicles necessary for cell homeostasis is also effectively terminated when the cytoskeleton becomes dysfunctional. Ultimately, these cumulative events induce cell death with subsequent release of valuable, intracellular nutrients for the pathogen and other microbes within that microenvironment. From a scientist's perspective, toxins that modify actin have become invaluable tools for studying the cytoskeleton and numerous cell processes.

C. botulinum C2II and C2I

As described by different groups for both *C. perfringens* Ib and *B. anthracis* PA, studies also reveal that the C-terminus of C2II facilitates binding to cell-surface receptor (Fig. 4.1). Deletion of only seven C-terminal residues effectively prevents C2II interactions with cells (Blöcker *et al.*, 2000). Antiserum specific for the C-terminus (domain 4; residues 592–721), but not that directed towards domains 1 (residues 1–264) or 3 (residues 490–592), blocks C2II binding to cells. The domain 4 antiserum neutralizes C2 cytotoxicity *in vitro* when preincubated with C2II, but this is not the case after C2II has bound to the cell surface. As previously described for Ib monoclonal antibodies (Marvaud *et al.*, 2001; Stiles *et al.*, 2002b), neutralizing epitopes within the C2II molecule are perhaps sterically hindered after C2II–cell receptor interactions. Deletion studies targeting the N-terminus of C2II precursor show that this region (residues 1–181), deleted after proteolytic activation of the C2II precursor, may have an impact upon proper folding of the molecule (Blöcker *et al.*, 2000). Mutagenesis of the C2II gene within a conserved region encompassing amino acids 303–331 (domain 2) show that voltage-gating, but not chloroquine binding or translocation of C2I into the cytosol, is effectively lost after a $E_{307}K$ mutation (Blöcker *et al.*, 2003). Domain 2 is involved in membrane insertion and channel formation.

Residues 1–87 of C2I mediate both binding to C2II heptamers and subsequent translocation across the cytoplasmic membrane into the cytosol (Barth *et al.*, 2002ab). Alignment of C2I with *B. cereus* VIP2 shows that amino acids 1–225 of C2I are akin to residues 60–275 of VIP2. Each of these regions contains four exposed α-helices within residues 1–87 in C2I and 60–133 in VIP2 (Han *et al.*, 1999). Analysis of X-ray crystallography data for other bacterial ADP-ribosyltransferases like *B. pertussis* pertussis toxin (Stein *et al.*, 1994), *C. diphtheriae* diphtheria toxin (Choe *et al.*, 1992), *E. coli* heat-labile enterotoxin (Sixma *et al.*, 1991), and *P. aeruginosa* exotoxin A (Li *et al.*, 1996) shows that the C-terminus of each contains two antiparallel β-sheets flanked by a pair of α-helices. There is also a highly conserved catalytic domain containing a $_{387}EXE_{389}$ motif found in both prokaryotic, as well as eukaryotic, ADP-ribosyltransferases (van Damme *et al.*, 1996). Among these enzymes of different origins, the overall sequence homology within the C-terminus is rather low. Studies targeting the EXE motif of *C. botulinum* C2I show that an $E_{387}Q$ mutation prevents ADP-ribosyltransferase, but not NAD-glycohydrolase, activity while the same alteration of E_{389} inhibits both (Barth *et al.*, 1998b).

B. anthracis PA, LF, and EF

X-ray structures of PA (Petosa *et al.*, 1997), LF (Pannifer *et al.*, 2001), and EF (Drum *et al.*, 2001) have provided invaluable structure/function data that now serve as 'templates' for the protein components of other *Clostridium* and *Bacillus* binary toxins not crystallized to date. There is much more literature available on modes of action for the *B. anthracis* oedema and lethal toxins versus all of that combined for the clostridial binary toxins (Barth *et al.*, 2004; Young and Collier, 2007). It is important to 'briefly' describe these bacillus toxins such that the reader has a better global perspective upon very similar proteins of clostridial origins.

The PA molecule consists of four distinct domains, of which the latter are applicable for the 'B' components (C2II, CDTb, Ib, Sb, and VIP1) of other binary toxins described in this chapter. Domains 1 (amino acids 1–249), 2 (amino acids 250–487), 3 (amino acids 488–596), and 4 (amino acids 597–735) respectively dock to an enzyme component(s), cause channel formation in lipid membranes, promote oligomerization, and bind to a specific cell-surface receptor(s). Very little amino acid homology exists between PA domains 1 and 4 with their equivalents found on other clostridial or bacillus binary toxins. This really is not so surprising, as these regions respectively afford a unique docking site for distinct 'A' components and receptor-binding specificity (Perelle *et al.*, 1997b; Barth *et al.*, 1998ab; Blöcker *et al.*, 2000; Marvaud *et al.*, 2001; Stiles *et al.*, 2002a).

Domain 1 contains a proteolytic cleavage site (R_{167}) that releases a 20-kDa peptide (PA20) from PA83 (Klimpel *et al.*, 1992). This event subsequently generates PA63 heptamers. The remaining segment of domain 1 (designated as 1′) faces into the channel lumen, unlike the peripherally located domain 4 (Petosa *et al.*, 1997). Domain 1′ binds two calcium molecules (coordinated via residues D_{177} or D_{235} and D_{179}, D_{181}, E_{188}) that preserve a proper PA63 structure which (1) resists further proteolytic degradation, (2) promotes heptamer formation, and (3) facilitates PA63 docking with EF and/or LF (Gupta *et al.*, 2003ab). No one has yet reported calcium associated with 'B' components of other binary toxins. Although their role is still unclear, various divalent cations (i.e. calcium, cobalt, and/or zinc) are required for production of *C. spiroforme* CST *in vitro* (Carman *et al.*, 1987).

Although various residues (R_{178}, K_{197}, R_{200}, F_{202}, L_{203}, P_{205}, I_{207}, I_{210}, and K_{214}) within domain 1′ have been defined as critical for the docking of EF and LF (Chauhan and Bhatnagar, 2002; Cunningham *et al.*, 2002), the maximum number of EF and/or LF molecules interacting with a PA63 heptamer has been rather controversial. Molar estimates range from seven EF/LF molecules to one PA63 heptamer (Singh *et al.*, 1999), via analysis by non-denaturing gel electrophoresis, to a more plausible three EF/LF molecules per PA63 heptamer (Cunningham *et al.*, 2002; Mogridge *et al.*, 2002ab). The latter ratio (considered by most now to be more accurate) was determined by multiple methods that include gel filtration chromatography, multiangle laser light scattering, and oligomer-deficient variants of PA63. Similar A:B stoichiometries are not yet available for the other binary toxins, but initial studies involving native gels and C2 toxin have been reported by Kaiser *et al.* (2006). Additionally, both EF and LF can bind to the same PA63 heptamer, thus forming a toxic protein package possessing multiple (oedema and lethal) activities (Pimental *et al.*, 2004).

On the other end of the PA molecule and distal to domain 1′, the importance of domain 4 in PA binding to cells was first determined through epitope mapping with monoclonal antibodies (Little *et al.*, 1988, 1996; Little and Lowe, 1991). Many other studies have since confirmed these early immunoreagent-based results, such as mutagenesis-based efforts revealing that residues Y_{681}, N_{682}, D_{683}, and P_{686} critically affect PA binding to receptor (Varughese *et al.*, 1999; Bradley *et al.*, 2003; Rosovitz *et al.*, 2003). An exposed loop (residues 703–722) also affects PA binding to cells as per deletions of nine or sixteen amino acids from this region (Brossier *et al.*, 1999). Furthermore, the far C-terminus of PA critically facilitates binding to cells (Singh *et al.*, 1991; Varughese *et al.*, 1999), as deletions of just 5–12 amino acids abrogate binding perhaps through conformational changes within the PA molecule (Leppla, 2006). Similar work with C2II and Ib reveals the same exquisite sensitivity regarding deletions within their C-terminus and subsequent effects upon biological activity (Blöcker *et al.*, 2000; Marvaud *et al.*, 2001). Evidently this represents a conserved structural trait found amongst all 'B' components within the clostridial and bacillus group of binary toxins.

The sequence similarities that exist between PA, C2II, and Ib are primarily localized within central domains 2 and 3 (amino acids 488–596). These domains respectively participate in channel formation/enzyme translocation and oligomerization of PA63 (Blaustein *et al.*, 1989; Schmid *et al.*, 1994; Petosa *et al.*, 1997; Benson *et al.*, 1998; Kimura *et al.*, 1998; Mogridge *et al.*, 2001; Sellman *et al.*, 2001; Knapp *et al.*, 2002; Blöcker *et al.*, 2003). As per co-crystal work with receptor, domain 2 of PA63 heptamers also contacts multiple residues on the receptor molecule (Lacy *et al.*, 2004, Scobie *et al.*, 2007). This interaction likely regulates pH-dependent formation of the PA channel necessary for subsequent translocation of 'A' components into the cytosol. Domain 2 contains a common 'Greek Key' motif (residues 262–368) that forms a β-hairpin, amphipathic loop which inserts into the membrane. Again, by playing upon protein toxin analogies, PA63 shares little sequence homology with the α haemolysin of *Staphylococcus aureus* but there are striking similarities in how these heptameric, pore-forming proteins produced by quite different bacteria insert into membranes (Song *et al.*, 1996). Residues D_{425} plus F_{427} of PA, which are conserved among the C2II, CDTb, Ib, and Sb molecules (Perelle *et al.*, 1993, 1997a; Gibert *et al.*, 1997; Kimura *et al.*, 1998), play a role in both channel formation as well as translocation of 'A' molecules into the cytosol (Sellman *et al.*, 2001). Alanine mutations of W_{346}, M_{350}, and L_{352} found within domain 2 of PA yield PA63 heptamers that do not facilitate LF-induced cytotoxicity. This result is likely due to dysfunctional membrane insertion and/or enzyme translocation (Batra *et al.*, 2001). Analysis of PA63 crystals generated at pH 6 and 7.5 shows that residues 342–355 become externalized at a lower pH (Petosa *et al.*, 1997). This type of molecular movement promotes a more hydrophobic state leading to oligomerization (Koehler *et al.*, 1991; Batra *et al.*, 2001). Additional mutagenesis studies of a surface-exposed, hydrophobic region within domain 3 reveal that specific residues F_{552}, F_{554}, I_{562}, L_{566}, and I_{574} may be involved in oligomer formation (Ahuja *et al.*, 2001; Mogridge *et al.*, 2001). These same residues exist in the 'B' components of other *Clostridium* and *Bacillus* binary toxins. Except for a study with *C. botulinum* C2II (Blöcker *et al.*, 2003), very little structure–function analysis has occurred within domains 2 and 3 of 'B' components from the other binary toxins.

The 'A' component of *B. anthracis* lethal toxin is LF, a zinc metalloprotease specific for a family of eukaryotic proteins (MAPKK) that transduce cell signals (Klimpel *et al.*, 1994; Duesbery *et al.*, 1998; Vitale *et al.*, 2000). A crystal structure for LF complexed with the N-terminus of MAPKK has been reported (Pannifer *et al.*, 2001). After entering a cell's cytosol via PA63 heptamers, LF binds to various types of MAPKK via the latter's C-terminal regulatory region to then cleave a proline-rich, N-terminal site (Duesbery *et al.*, 1998; Vitale *et al.*, 2000; Chopra *et al.*, 2003). Macrophages have been considered the primary target for lethal toxin, but additional work suggests that other cell types are also adversely affected by this toxin (Guidi-Rontani, 2002; Agrawal *et al.*, 2003; Moayeri *et al.*, 2003; Tournier *et al.*, 2005). It is still uncertain what role truncated MAPKK actually plays in toxicity as evidenced by their cleavage in macrophages that are otherwise resistant to lethal toxin-induced cell death (Watters *et al.*, 2001; Salles *et al.*, 2003). A recent microarray-based study by Ha *et al.* (2007), using lethal toxin-resistant murine macrophages, further suggests that mitochondrial proteins (Bnip3 and Bnip3L) play a role in LF-induced death. Resistant cells evidently downregulate the synthesis of Bnip3 and Bnip3L, which are involved in cell necrosis/apoptosis. Overall physiological effects of lethal toxin at the cellular level include (1) increased permeability of macrophage membranes, (2) depletion of intracellular ATP, and (3) ultimately cell lysis (Hanna *et al.*, 1992; Popov *et al.*, 2002; Kim *et al.*, 2003).

Proinflammatory cytokines may also play a role in *B. anthracis* toxin activity, but this concept still remains rather controversial (Erwin *et al.*, 2001; Popov *et al.*, 2002; Kim *et al.*, 2003; Gold *et al.*, 2004). A similar role of host-induced cytokines in conjunction with any clostridial binary toxin has not been described in the literature. A cytokine-independent hypoxia induced by *B. anthracis* lethal toxin can cause terminal necrosis of both the liver and metaphyseal bone marrow (Moayeri *et al.*, 2003). Overall, the lethal toxin evidently provides a defence mechanism for the bacterium that weakens a host's immune system by compromising the normal functions of macrophages and dendritic cells (Fukao, 2004; Ribot *et al.*, 2006).

In a similar structural fashion as PA, each LF molecule contains four distinct domains. The N-terminal domain 1 of LF (amino acids 1–262) docks to PA (Quinn *et al.*, 1991) and shares a multi-α helix bundle plus β-sheet structure within LF domain 4. Although there are structural similarities between domains 1 and 4 of LF, there is little sequence homology and no common function. Within domain 1 of both LF and EF, there is a VYYEIGK sequence that facilitates docking to PA63 heptamers (Gupta *et al.*, 2001; Kumar *et al.*, 2001). For LF, more recent studies have shown that two additional residues (E_{135} and D_{182}) play a key role in nanomolar affinity with PA heptamer by binding to the latter's K_{197} residues on adjacent PA molecules (Melnyk *et al.*, 2006). Decreased pH (as that found in acidified endosomes) would disrupt these charge-based interactions and thus cause LF release from the PA heptamer. The overall surface of an LF molecule that makes contact on PA heptamer encompasses ~1200 Å. The VYYEIGK sequence used by EF and LF does not appear in other binary toxin 'A' components, thus PA does not bind or internalize heterologous 'A' molecules (Perelle *et al.*, 1997b). The only known complementation of heterologous components within the *Clostridium* and *Bacillus* binary toxin group involves the enteric-acting *C. difficile* CDT, *C. perfringens* ι toxin, and *C. spiroforme* CST of the iota family (Simpson *et al.*, 1987; Perelle *et al.*, 1997b; Stiles *et al.*, 2002a).

Domain 2 (residues 263–297 and 385–550) of LF conformationally mimics the C-terminal catalytic domain of VIP2 from *B. cereus*. However, a glutamic acid needed for the ADP-ribosyltransferase activity of VIP2 is replaced by a lysine in LF. The latter molecule is devoid of such enzymatic activity. Overall, there is very little amino acid sequence identity that exists between VIP2 and LF, which suggests an evolutionary divergence (maybe convergence?) of similar 'A' genes within bacterial binary toxins. Specific domain 2-localized residues (L_{293}, K_{294}, R_{491}, L_{514}, N_{516}) not located within the active site, yet seemingly critical for LF interaction with MAPKK substrate, are important for the proteolytic activity of LF (Liang *et al.*, 2004).

Domain 3 (amino acids 303–382) of LF forms an α-helical bundle imbedded between the second and third helices of domain 2, shares a hydrophobic surface with domain 4, and provides substrate specificity (Pannifer *et al.*, 2001). Domain 4 (amino acids 552–809) contains the active site consisting of a HEXXH motif ($_{686}$HEFGH$_{690}$). An HEXXH sequence is quite common among various zinc metalloproteases (Quinn *et al.*, 1991; Klimpel *et al.*, 1994), including the lambda toxin of *C. perfringens* (Jin *et al.*, 1996) that may activate Ia and/or Ib of ι toxin *in vitro* and *in vivo* (Gibert *et al.*, 2000).

In addition to LF, *B. anthracis* produces another 'A' component designated as EF. The EF molecule is a calcium/calmodulin-dependent adenylate cyclase (Leppla, 1982; Drum *et al.*, 2001, 2002) that shares structural homology and epitopes with an adenylate cyclase from *Bordetella pertussis* (Masin *et al.*, 2006). Calmodulin interactions with EF are dependent upon calcium levels, that when elevated, inhibit catalysis by EF. Hydrophobic and hydrophilic residues (501–540 and 616–798) of EF induce a conformational shift after the binding of calmodulin. EF–calmodulin interactions subsequently expose the catalytic core of EF that is located within three globular domains spanning residues 294–622 (Munier *et al.*, 1993; Drum *et al.*, 2002; Shen *et al.*, 2002). The binding of calmodulin to EF also stabilizes residues 579–591 of the latter, which may affect ATP binding and catalysis.

The N-terminus of EF (or LF as stated before) docks with PA via a VYYEIGK sequence (Little *et al.*, 1994; Kumar *et al.*, 2001), while residues 291–776 are sufficient for catalysis (Drum *et al.*, 2000). Residues 342–358 of EF contain a motif (GXXXXGKT) commonly found in other ATP-binding proteins (Fry *et al.*, 1986). The *C. difficile* CDTb, *C. perfringens* Ib, and *C. spiroforme* Sb molecules also have a curious ATP-binding motif within their N-terminus; however, this region appears dysfunctional and unnecessary for the biological activity of these common 'B' components lacking any known enzymatic properties (Gibert *et al.*, 1997). To date, no one has ever described the binding of ATP to any 'B' component among the clostridial or bacillus binary toxins.

Once EF enters a cell, it can trigger a profound calcium influx that is followed by greatly inflated levels of intracellular cyclic-AMP (Kumar *et al.*, 2002). This effect occurs in various cell types. Elevated levels of intracellular cyclic-AMP (cAMP) induced by EF, which can be a thousand-fold higher than normal physiological concentrations (Leppla, 2006), are non-lethal *in vitro*. However, recent *in vivo* studies suggest that oedema toxin can be lethal in mice (Firoved *et al.*, 2005). The intracellular half-life of EF is approximately two hours (Leppla, 1982), but similar studies have not been reported for other clostridial and bacillus binary 'A' components. Increased cAMP levels among leucocytes can profoundly decrease a host's immune response (Confer and Eaton, 1982); thus, EF seemingly enables *B. anthracis* to flourish in a host (Kumar *et al.*, 2002). It is possible that judicious use of EF (with PA) could therapeutically 'attenuate' the immune system when activated by other bacterial toxins, such as those possessing superantigenic properties (Krakauer *et al.*, 2005). Recent *in vivo* studies by Tessier *et al.* (2007) show that the vascular leakage caused by *B. anthracis* oedema toxin involves histamine, prostanoids, and neurokinins. The effects of EF are no doubt in concert with LF during disease, with the latter molecule affecting MAPKK signalling pathways important in macrophage activation, nitric oxide production, and cytokine release (Popov *et al.*, 2002; Mock and Mignot 2003; Tournier *et al.*, 2005; Cui *et al.*, 2007; Firoved *et al.*, 2007).

The intoxication process

'B' binding to cell surfaces

To come within range of any intracellular substrate, and like other bacterial protein toxins, the clostridial and bacillus binary toxins must first bind to the exterior of a targeted cell. This initial intoxication step for clostridial and bacillus binary toxins occurs by 'B' interactions with a cell-surface receptor(s). Depending upon the *Clostridium* or *Bacillus* binary toxin, a 'B' molecule is evident in either a preformed solution-based, or cell-surface generated, homoheptameric complex that acts as a docking platform for 'A' molecules.

Recent work regarding the discovery of cell-surface receptors for binary toxins has been quite fruitful for various research groups. This is especially the case with *B. anthracis* PA, in which the receptor is a ubiquitous protein (Escuyer and Collier, 1991) first identified as tumour endothelium marker (TEM) 8 (variant two). TEM 8 consists of 368 amino acids and a von Willebrand factor A domain (Bradley *et al.*, 2001) that is commonly found on many integrins (Whittaker and Hynes, 2002; Bradley *et al.*, 2003). Although the role that TEM 8 naturally plays in healthy cells is unknown, its expression in tumour cells can be quite elevated and may aid in tubule formation/angiogenesis (St. Croix *et al.*, 2000). Variant one of TEM 8 (Liu and Leppla, 2003), and a ubiquitous protein expressed by the human capillary morphogenesis gene two (Scobie *et al.*, 2003), also function as PA receptors. These molecules all contain an imbedded metal ion-dependent adhesin site (MIDAS) important for PA, and probably natural, but as yet unknown ligand, interactions. The high-affinity binding of PA to receptor is ionically mediated by magnesium or manganese, but not calcium, via a carboxylate moiety provided by D_{683} of PA (Bradley *et al.*, 2003; Rosovitz *et al.*, 2003). *B. anthracis* PA clearly binds to various proteins possessing evolutionarily conserved, metal ion-containing domains on the cell surface (Whittaker and Hynes, 2002). Perhaps further work done with the 'B' components of clostridial binary toxins will also yield similar, and exciting, results. With time, such findings may include a common (or perhaps not so common?) family of proteins exploited as cell-surface receptors.

To further understand the binding and oligomerization properties of clostridial and bacillus 'B' components on cell surfaces, the potential role played by lipid rafts has been explored by different research groups. Lipid rafts are cholesterol-rich, detergent-insoluble (at 4°C) regions on cell membranes that inadvertently serve as entry/exit sites for various bacteria, viruses, and toxins (Miyata *et al.*, 2002; Simons and Ehehalt, 2002). It has been discovered that lipid rafts and a clathrin-dependent process do indeed promote oligomerization of PA, as well as internalization, of the *B. anthracis* toxins. However, the PA receptor is not initially localized within rafts (Abrami *et al.*, 2003). Through an ill-defined process mimicking ligand-dependent clustering of B-cell receptors into rafts (Cheng *et al.*, 1999), activated PA63 also causes receptor localization into lipid rafts. The close proximity of PA63–receptor complexes evidently facilitates PA63 heptamerization into a docking platform most favourable for EF and LF interactions. The latter molecules ultimately enter the cytosol by way of acidified endosomes and cation-selective channels generated by heptameric PA63 (Gordon *et al.*, 1988; Blaustein *et al.*, 1989; Milne *et al.*, 1993). The PA-generated channels use a lumen-facing F_{427} from each PA63 molecule, and proton gradient, to translocate 'A' molecules into the cytosol via a Brownian-based ratchet mechanism (Krantz *et al.*, 2006). There is evidence that translocated EF probably remains attached to the endosomal membrane versus floating throughout the cytosol (Guidi-Rontani *et al.*, 2000).

There has been relatively little work done with lipid rafts and the other binary bacterial toxins, although *C. perfringens* Ib localizes into these membrane microdomains on Vero cells that are sensitive to ι toxin (Hale *et al.*, 2004; Nagahama *et al.*, 2004). The Ibp molecule, which binds to cells but does not promote iota toxicity, is not associated with lipid rafts on the cell surface.

In addition to PA and Ib, receptor-binding studies have also been reported for C2II. Like the 'B' molecules of other binary toxins, the C2II precursor and proteolytically activated molecule can both bind to cells (Ohishi and Miyake, 1985). Activated C2II has rather unique haemagglutinating properties with erythrocytes from different species. This binding is competitively inhibited by various sugars such as *N*-acetylgalactosamine, *N*-acetylglucosamine, L-fucose, galactose, or mannose (Sugii and Kozaki, 1990). Trypsin or pronase pretreatment of human erythrocytes prevents C2II-induced haemagglutination, thus suggesting that the receptor is a glycoprotein not yet definitively identified to date. Work reported by Fritz *et al.* (1995) from Freiburg reveals that chemical mutagenesis of CHO cells generates mutants that subsequently do not bind C2II. A lack of binding occurs because these mutants are devoid of N-acetylglucosaminyltransferase I activity that facilitates the formation of asparagine-linked carbohydrates (Eckhardt *et al.*, 2000). It can be concluded that the receptor for C2II contains a complex or hybrid carbohydrate structure. In contrast to C2 toxin, the mutant CHO cells are still susceptible to ι toxin; therefore, this finding further demonstrates that C2II and Ib recognize unique receptors. The C2II molecule, like PA63 and Ib, also forms voltage-dependent channels in lipid membranes (Schmid *et al.*, 1994; Bachmeyer *et al.*, 2001; Blöcker *et al.*, 2003). A cluster of hydrophobic and hydrophilic amino acids within C2II residues 303–331 may play a critical role in C2II insertion into the membrane (Blöcker *et al.*, 2003). This same region is also found within PA residues 325–356 located in domain 2.

Although the C2II molecule binds and facilitates C2I-mediated cytotoxicity in all tested vertebrate cells (Ohishi *et al.*, 1984; Sugii *et al.*, 1990; Eckhardt *et al.*, 2000), the receptor for Ib is not as readily distributed throughout Nature (Stiles *et al.*, 2000). The Ib receptor is a protein(s) resistant to various proteases, but not pronase. Rather extensive pretreatment of cells with lectins or glycosidases does not affect Ib binding, thus suggesting that the receptor is not composed of carbohydrate. Experiments with polarized CaCo-2 (human colon) cells show that the Ib receptor is namely localized upon the basolateral membrane (Blöcker *et al.*, 2001). These findings are similar to those for PA with another human colon cell line (Beauregard *et al.*, 1999). Additionally, at 37°C (but not 4°C) the Ib molecule can traverse a CaCo-2 cell monolayer from either the apical or basolateral surface independent of Ia (Richard *et al.*, 2002). Ib that has travelled from either side of the cell to the other can internalize Ia found on the distal surface, even when the latter component of ι toxin is added three hours after Ib.

Western blot experiments reveal that Ib rapidly binds to cells at 37°C and then forms a large (greater than 200 kDa) complex within 1 min (Nagahama *et al.*, 2002; Stiles *et al.*, 2002b). This complex is evident on/in cells for at least six hours, which optimizes the chances for Ia-Ib interactions needed to create biologically active holotoxin. To date, the composition of this complex which likely contains receptor has not been reported in the literature. The rapid formation of long-lived, Ib-based complexes on cells is intriguing in light of earlier work reported by Sakurai and Kobayashi (1995). They showed that Ia injected intravenously into mice two hours after Ib causes death. These results suggest that Ib 'hangs around' in the body and is available for Ia docking over an extended period. When neutralizing Ib sera is given to mice only five minutes after an Ib injection, there is no protection afforded against Ia-induced lethality. Overall, these data reveal that Ib likely binds to cells very rapidly *in vivo* but then is not readily removed from the cell's surface. This makes sense for any clostridial or bacillus binary toxin, as they are not toxic unless the 'A' component docks and becomes internalized into a targeted cell. Among these binary toxins it does seem in a very odd and imperfect way that 'A' interactions with 'B' components are mediated by random chance.

During other *in vivo* studies, Sakurai and Kobayashi (1995) discovered that when Ib is injected intradermally into guinea pigs, and Ia is given intraperitoneally, a dermonecrotic lesion forms at the Ib injection site. Ia is able to 'locate' distantly placed Ib bound to the surface of cells, thus suggesting that the Ia molecule is not readily taken up by the body in a non-productive (i.e. non-toxic) sense. Perhaps this 'homing' characteristic can be exploited in future experiments involving ι toxin as a protein shuttle for medicinal compounds. Similar *in vivo* discoveries have also been reported for the C2 toxin in both mice and rats (Simpson, 1982). An important question still surrounding this particular issue involves the number of 'A' molecules needed to trigger cellular internalization by a 'B' heptamer. To date, receptor-binding studies for *C. difficile* CDTb, *C. spiroforme* Sb, and *B. cereus* VIP1 are lacking in the literature although CDTb and Sb likely bind to the same, or a very similar, receptor molecule(s) as *C. perfringens* Ib.

'A' docking and internalization

There are two major pathways for the uptake of bacterial protein toxins that ultimately act intracellularly. One route involves retrograde trafficking through the Golgi apparatus and endoplasmic reticulum, as is used by the *S. dysenteriae* shiga (Sandvig and Van Deurs 1996) and *V. cholerae* cholera (Orlandi *et al.*, 1993) toxins. This pathway is inhibited by brefeldin A, a small molecular-weight macrolide (280 Da) produced by fungi and which causes protein accumulation within the endoplasmic reticulum. A second intracellular route exploited by bacterial toxins involves translocation from acidified early endosomes into the cytosol. This intoxication avenue is employed by diphtheria toxin as well as the clostridial and bacillus binary toxins described in this chapter. Subsequent transport of vesicles from early to late endosomes involves microtubules that are readily depolymerized by nocodazole, a chemically-synthesized molecule of 300 Da (Gruenberg and Howell 1989; Sakai *et al.*, 1991). Neither brefeldin A nor nocodazole influences the biological activity of C2 or ι toxins; however, nocodazole can delay the effects of *B. anthracis* lethal toxin in various cell types (Abrami *et al.*, 2004). Translocation of the C2, ι, or oedema/lethal toxins across the endosomal membrane is blocked by a macrolide antibiotic, bafilomycin A (622 Da). This antibiotic inhibits vacuolar-type ATPases that internally acidify the endosome. Diminishment of endosomal pH naturally induces conformational changes within the heptameric 'B' complex and translocation of the 'A' component(s) then ensues across the endosomal membrane, most likely through a 'B'-based channel. This process can be artificially induced *in vitro* on the cell surface by simply lowering pH of the surrounding medium (Gordon *et al.*, 1988; Menard *et al.*, 1996; Barth *et al.*, 2000; Blöcker *et al.*, 2001). However, there are some unique pH requirements for artificially getting C2 and ι toxin into cells, as the latter requires a lower pH (≤ 5.0) versus the former (pH ≤ 5.5) for direct cytosolic entry through the cell membrane.

Entry of ι toxin from the endosome into the cytosol differs from the C2 and lethal/oedema toxins of *B. anthracis* as per inhibitory effects of chloroquine, monensin, nigericin, and ammonium chloride (Gibert *et al.*, 2007) (Fig. 4.3). Besides preventing endosomal acidification, chloroquine can physically block the C2II-induced pore and prevent C2I translocation (Schmid *et al.*, 1994; Blöcker *et al.*, 2003). Monensin, like nigericin, is an electroneutral exchanger of monovalent cations for protons that subsequently abolishes the pH gradient through endosomal membranes. By virtue of its alkalinity, ammonium chloride increases pH within the endosomes. Although the biological activity of ι toxin is not inhibited by monensin alone, a combination of monensin and valinomycin proves partially inhibitory, and there is a distinct synergistic decrement of ι toxin activity with a monensin plus bafilomycin A combination. In fact, the entry requirements of Ia from the endosome mimic those previously described for FGF-1, a fibroblast growth factor (Wesche *et al.*, 2006). Altogether, a pH and/or membrane potential gradient between the endosome and cytosol are evidently required for efficient translocation of Ia, C2I, EF, and LF (Gibert *et al.*, 2007). For now, it is unclear whether the 'B' heptamers of the clostridial and bacillus binary toxins enter the cytosol with the 'A' components or perhaps the 'B' components remain attached to the endosomal membrane (Ohishi and Yanagimoto, 1992; Richard *et al.*, 2002). It is possible that 'B' heptamers which release their 'A' cargo into the cytosol are then recycled back onto the cell surface to pick up more 'A' molecules, but this has not been reported in the literature.

Studies with the C2, CDT and ι toxins show that the 'A' components use heat-shock protein 90 (Hsp90) to cross the endosomal membrane (Haug *et al.*, 2003). How this exactly occurs is

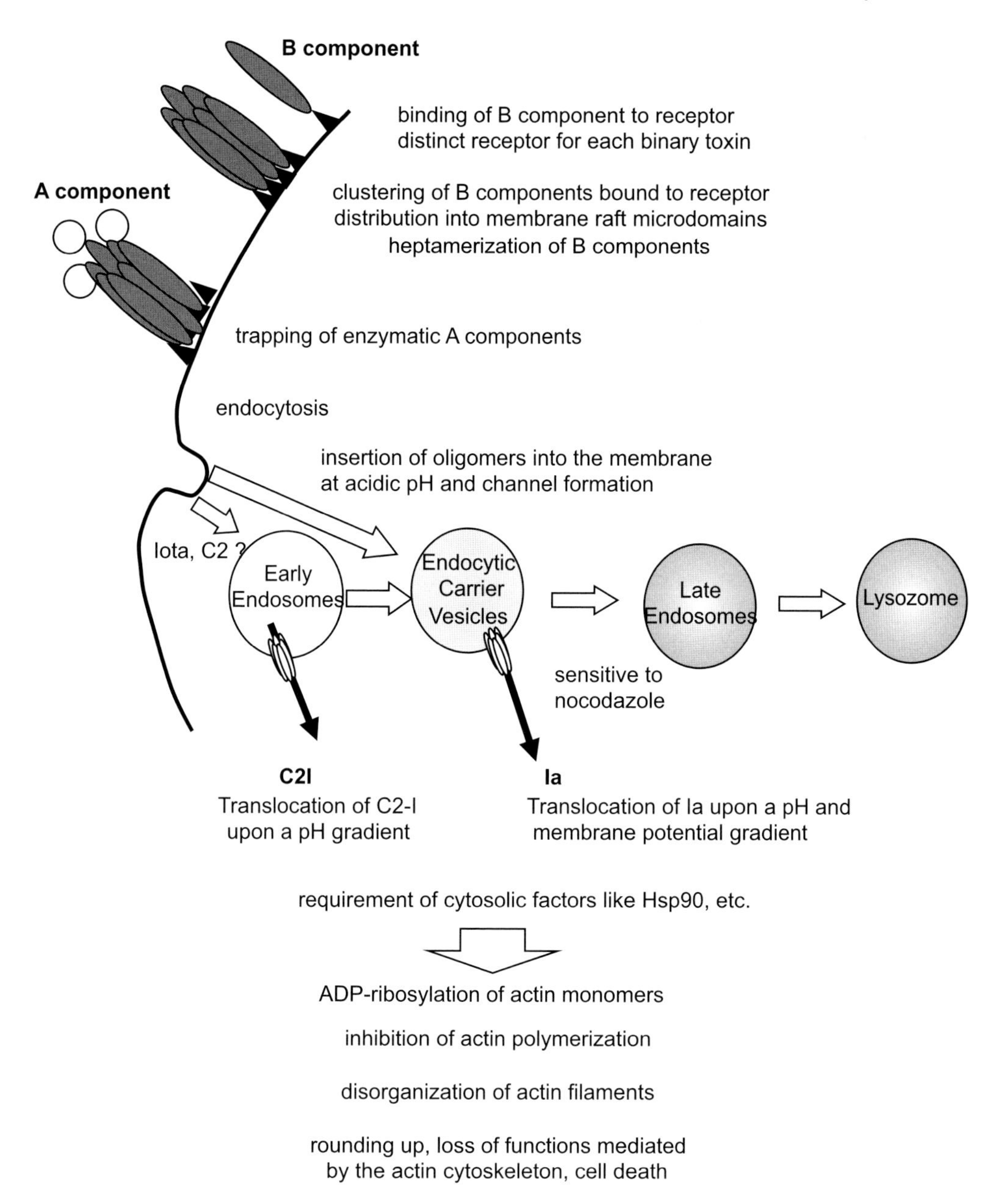

Figure 4.3 Intoxication process employed by *Clostridium* binary toxins.

not well defined, but the Hsp90 molecule is a highly conserved ATPase readily evident in all eukaryotic cells. In conjunction with other heat-shock proteins, Hsp90 provides an essential housekeeping role by regulating proteins involved in cell signalling (Pratt and Toft, 2003). Specific inhibitors of Hsp90 include antifungal compounds like geldanamycin (560 Da; competitive inhibitor for ATP) and radicicol (365 Da). These low-molecular-weight molecules effectively delay C2-, CDT-, or iota-induced cytotoxicity in various cell types. In contrast, the cytosolic entry of *B. anthracis* lethal toxins is not affected by Hsp90 inhibitors (Haug *et al.*, 2003). This latter result, like that recently reported by Gibert *et al.* (2007), further suggests distinct translocation mechanisms for 'A' components found among the *Clostridium* and *Bacillus* binary toxins.

It is quite possible that the 'A' components from clostridial and bacillus binary toxins unfold into a 'molten globule' and then thread through 'B'-generated channels in the membrane. Perhaps this occurs in a ratchet-type mechanism similar to that recently described for the *B. anthracis* toxins (Krantz *et al.*, 2006). With *B. anthracis* LF, and perhaps the 'A' components of other *Clostridium* and *Bacillus* binary toxins, the N-terminus first enters the 'B'-induced channel to begin an N- to C-terminal translocation into the cytosol (Zhang *et al.*, 2004). This process is akin to cell entry of the ADP-ribosyltransferase from *C. diphtheriae* diphtheria toxin. Ratts *et al.* (2003) report that Hsp90 and thioredoxin reductase, evident in the cell as a cytosolic complex, are both required for transporting diphtheria toxin from the endosome. Geldanamycin and radicicol are both needed to inhibit diphtheria cytotoxicity, which is a result in contrast with those from CDT, C2, or ι toxin studies. The thioredoxin reductase may cleave the disulfide bond existing between the 'A' and 'B' chains of diphtheria toxin. Such a cystine bond does not exist for clostridial and bacillus binary toxins; however, there is a precedent for such activation with other single-chain toxins like *C. tetani* tetanus toxin and *C. botulinum* neurotoxin A (Kistner *et al.*, 1992).

To summarize the intoxication process for *Clostridium* and *Bacillus* binary toxins, it is quite likely that other undefined factors are involved in the 'natural' cytosolic entry from endosomes. Clearly, a further understanding of toxin entry into a cell can lead to more efficient use of these proteins as molecular shuttles. Such knowledge can also provide new targets for potential therapeutics against bacterial toxins that necessarily gain entry into the cytosol for their biological effects upon a targeted cell.

Binary toxins as protein shuttles

Throughout time, various researchers from around the world have become intrigued with the possibility of employing bacterial toxins as protein-based vehicles for transporting other proteins, or biologically active entities, into targeted cells. As just two common examples, the single-chain exotoxin A of *P. aeruginosa* and *C. diphtheriae* diphtheria toxin have each been repeatedly modified for trafficking heterologous proteins (Pastan *et al.*, 1992; Francis *et al.*, 2000). More recently, the *Clostridium* and *Bacillus* binary toxins have also become tools for shuttling heterologous proteins, and even DNA, into cells. Again, as with many other aspects of clostridial and bacillus binary toxins, the knowledge first gleaned from *B. anthracis* toxins have advanced the overall field. For example, there have been many studies employing an LF fragment consisting of amino acids 1–254 (LF^{1-254}). Fusion of different proteins through either the N- or C-terminus leads to an LF chimera that becomes subsequently internalized by a targeted cell via heptameric PA63 (Milne *et al.*, 1995; Varughese *et al.*, 1998; Leppla *et al.*, 1999; Mehra *et al.*, 2001; Leppla, 2006). Fusion molecules consisting of LF residues 1–198, or deletion of just 40 N-terminal residues from wild-type LF without a fusion protein, are not cytotoxic. Such data further reveal an important role played by the N-terminus of LF in docking with PA and/or translocation into the cytosol.

One noticeable target for chimeras generated from *B. anthracis* lethal toxin has included the immune system. Certain LF-based chimeras trigger cytotoxic T-cell responses *in vivo* and/or *in vitro* by delivering major histocompatibility complex (MHC) class I epitopes from viruses and bacteria (Goletz *et al.*, 1997; Ballard *et al.*, 1998; Cao *et al.*, 2002). A variation upon this theme involves the anti-cancer capabilities of lethal toxin (Frankel *et al.*, 2002). *B. anthracis* lethal toxin has been used experimentally as an anti-tumour compound by exploiting the specific proteolytic properties of urokinase plasminogen activator, which is a surface protein commonly found on different types of malignant cells (Liu *et al.*, 2003a,b; Liu and Leppla, 2003). These particular constructs of lethal toxin replace the furin cleavage site (RKKR) on PA83 with SGRSA. The LF^{1-254} molecule is fused to an enzymatic domain of *P. aeruginosa* exotoxin A, which targets the protein synthesis machinery of eukaryotic cells.

It is not only proteins, but the DNA encoding for luciferase or green fluorescent protein has also been successfully transported into cells using chimeras of *B. anthracis* lethal toxin (Gaur *et al.*,

2002). In the future, this approach may represent upon further refinement an attractive alternative to viruses that are more commonly used today for gene therapy studies (Dobbelstein, 2003). Overall, the numerous shuttle experiments that have been reported for *B. anthracis* lethal toxin represent 'trend setting' uses that foretell a bright future for this, and other, binary toxins as research/biomedical tools.

In a way that is quite similar to *B. anthracis* LF, the N-terminal domain (residues 1–225) of *C. botulinum* C2I (designated as $C2I^{1-225}$) has also been engineered as a fusion protein with the catalytic domain of *S. enterica* SpvB or C3/C3-like ADP-ribosyltransferases (Barth *et al.*, 1998a, 2002ab; Pust *et al.*, 2007). A C3 molecule can be fused to either the N- or C-terminus of $C2I^{1-225}$ and then efficiently delivered into cells, via C2II, like wild-type C2I. Delivery into a targeted cell can occur even if $C2I^{1-225}$ is in the midst of a rather large, 75-kDa fusion composed of three different proteins (glutathione transferase–$C2I^{1-225}$–C3).

The C3 family of enzymes is produced by *B. cereus*, *C. botulinum*, *C. limosum*, and *S. aureus* (Just *et al.*, 1992; Wilde and Aktories, 2001; Wilde *et al.*, 2001). These ADP-ribosyltransferases are each approximately 23 kDa and, upon entry into the cytosol, inactivate small GTPases like Rho A, B, and C (Aktories *et al.*, 1989; Aktories *et al.*, 1992a; Sekine *et al.*, 1989). The crystal structure of *C. botulinum* C3 exoenzyme shares conformation with the C-terminal domain of C2I containing the active site (Han *et al.*, 2001). On their own, the C3/C3-like proteins penetrate cells rather poorly as greater than 10 μg/ml concentrations must be applied for an extended period (24h) to elicit any cytotoxic effects. In contrast though, a $C2I^{1-225}$-C3 fusion added to various cell types at much lower concentrations (100 ng/ml) with activated C2II elicits a typical 'C3 morphology'. These overt cellular changes occur within three hours after addition of the chimeric toxin, and are evidenced by cell contraction plus neurite-like protrusions (Barth *et al.*, 1998a). Potency of the $C2I^{1-225}$-C3 fusion toxin (with activated C2II) is approximately three hundred-fold higher than the native C3 toxin plus activated C2II combination. Removal of fusion toxin from the medium readily reverses the cytopathic effects within hours as per reorganized stress fibres and actin cytoskeleton (Barth *et al.*, 1999b).

Further refinement of the C2 shuttle system shows that a C2I fragment (amino acids 1–87; $C2I^{1-87}$) is just as effective as the larger $C2I^{1-225}$ construct for C3 transport into cells (Barth *et al.*, 2002b). The $C2I^{1-87}$ fragment represents a minimal part of C2I needed for not only translocation, but also competitive binding with native C2I for activated C2II. Clearly, the C2 shuttle system is quite versatile for translocating heterologous proteins into cells and subsequently studying various cell functions.

In addition to the *C. botulinum* C2 or *B. anthracis* lethal toxins, studies with *C. perfringens* ι toxin reveal that it too acts as an effective shuttle for heterologous proteins (Marvaud *et al.*, 2002). The minimal Ia fragment necessary for efficient transport of *C. botulinum* C3 into Vero cells, via Ib, consists of the catalytically inactive residues 129–157 ($Ia^{129-257}$). Use of this region for generating 'A'-based chimeras is rather unique, versus that previously described for the more N-terminal located C2I or LF molecules. Linkage of C3 to the C-terminus of various-sized Ia peptides proved most effective upon cultured cells in this study. However, every N-terminally linked conjugate tested was not cytotoxic when incubated with or without Ib. Quality control studies showed that each N- or C-linked conjugate of C3 to Ia is enzymatically active *in vitro*. Most of the N-linked conjugates also effectively dock to Ib on the cell surface, thus suggesting that C3-Ia conjugates may be inefficiently translocated. It is also possible that after entry of C3-Ia into the cytosol (assuming entry occurs), the enzyme may not adopt a biologically active conformation. Clearly, results from the iota studies differ from those for the *B. anthracis* lethal or *C. botulinum* C2 toxin systems. Such findings further underscore the inherent differences among these yet very similar *Clostridium* and *Bacillus* binary toxins. Unlike that described in the literature for *B. anthracis* lethal toxin, *in vivo* results are clearly lacking for both the C2- and iota-based shuttle systems.

Summary

From a historical perspective, the discovery of *C. perfringens* ι toxin in 1940 was the first for any *Clostridium* or *Bacillus* binary toxin (Bosworth, 1940). At that time, investigators had no idea that the ι toxin was composed of multiple proteins. It was not until 1956 that the multicomponent structure of *B. anthracis* toxins was initially reported, thus representing the first binary description for any of these (or other) bacterial toxins (Smith *et al.*, 1956; Smith, 2002). The passing of two to four more decades eventually revealed the multi-component nature of various *Clostridium* and *Bacillus* binary toxins discussed in detail throughout this chapter. These include the *C. botulinum* C2 toxin, *C. perfringens* ι toxin, *C. spiroforme* CST, *C. difficile* CDT, and *B. cereus*/*B. thuringiensis* VIP. Specific years for each of these reported discoveries are respectively 1980 (Ohishi *et al.*, 1980), 1986 (Stiles and Wilkins, 1986), 1989 (Popoff *et al.*, 1989), 1997 (Perelle *et al.*, 1997a) and 1999 (Han *et al.*, 1999).

It is, in our opinion, virtually impossible to present a comprehensive view of clostridial binary toxins without delving into literature describing the bacillus binary toxins. Hopefully, that type of 'global perspective' has been achieved in this chapter to give the reader a more 'balanced' view upon this fascinating, protein-based topic. Clearly, the oedema and lethal toxins of *B. anthracis* are the 'standard' for this group of binary toxins as evidenced by the sheer volume of publications. As just one unique example, specific identification of cell-surface receptors for PA represents relatively recent and exciting findings that are currently lacking for the other binary toxins (Bradley *et al.*, 2001, 2003; Liu and Leppla, 2003; Scobie *et al.*, 2003). Such discoveries can further spawn more fruitful discovery of therapeutics against the *B. anthracis* toxins and also spark the creative use of these toxins for various biomedical endeavours.

As mentioned throughout the text, there are both intriguing similarities but also remarkable differences found within the *Clostridium* and *Bacillus* binary toxins. Each of the 'B' precursors for bacterial binary toxins described in this chapter is proteolytically activated by serine-type proteases in solution; however, the PA83 molecule from *B. anthracis* is also uniquely processed on a targeted cell by surface-associated furin or furin-like proteins. Further evidence of versatility by *B. anthracis* toxins involves the PA63 heptamers that transport two different enzymatic molecules (EF and LF) into a cell. The 'B' heptamers from other clostridial binary toxins naturally shuttle only one type of enzyme, a mono-ADP-ribosyltransferase specific for G actin. Within the iota family, 'A' molecules can be transported by heterologous 'B' components from other clostridial species. The *C. botulinum* C2 toxin though seems rather distinct in that activated C2II exclusively transports C2I. From another perspective, and in contrast to the uniform requirement for proteolytic activation of 'B' components, the Ia molecule of ι toxin uniquely represents the only 'A' component in this binary group activated by proteases.

By 'looking into the future', it is quite plausible that further exploration with gene probes and specific antibodies will unveil new binary toxins produced by other bacteria. In fact a recent PCR-based study by Heffron and Poxton (2007) shows that quite a few *C. novyi* type A strains contain the *C. botulinum* C2I and/or C2II genes. Additionally, some strains of *B. cereus* associated with lethal pneumonia in humans possess the anthrax toxin genes (Hoffmaster *et al.*, 2004). Perhaps other bacteria from different genera may carry similar binary toxin genes? Clearly, such discoveries would reveal throughout Nature an even more diverse dispersion of a common binary toxin template. As evidenced throughout our biosphere, success breeds success at so many levels of life and from one generation to the next.

In addition to the extensive pathogenesis work described for the *B. anthracis* toxins, it still remains a mystery whether the other *Clostridium* and *Bacillus* binary toxins also afford such a unique, definitive advantage(s) for their bacterial hosts. For *C. spiroforme* it seems rather apparent that CST is intimately involved in pathogenesis, as this is the only protein toxin produced by the bacterium. However, for bacteria that produce other toxic proteins in addition to a binary toxin (i.e. *C. perfringens* and *C. difficile*), the story becomes rather complex. Such questions have most recently been at the clinical forefront with various epidemic strains of CDT-producing *C. difficile*. These pathogens continue to cause a most unpleasant, as well as sometimes lethal, form of enterocolitis found throughout the world and particularly within hospitals. This uncertainty of a toxin's role in pathogenesis is again a reflection of the literature not yet available for the clostridial, and some bacillus, binary toxins.

In conclusion, it is evident that the bacterial binary toxins presented throughout this chapter possess many more mysteries for future investigators to methodically unravel in the laboratory. A broad understanding of the past will help generate genuinely creative efforts that answer additional questions revolving around the *Clostridium*, and closely related *Bacillus*, binary toxins.

References

Abrami, L., Liu, S., Cosson, P., Leppla, S.H., and van der Goot, F.G. (2003). Anthrax toxin triggers endocytosis of its receptor via a lipid raft-mediated clathrin-dependent process. J. Cell Biol. *160*, 3321–3328.

Abrami, L., Lindsay, M., Parton, R.G., Leppla, S.H., and van der Goot, F.G. (2004). Membrane insertion of anthrax protective antigen and cytoplasmic delivery of lethal factor occur at different stages of the endocytic pathway. J. Cell Biol. *166*, 645–651.

Agrawal, A., Lingappa, J., Leppla, S.H., Agrawalo, S., Jabbar, A., Quinn, C., and Pulendran, B. (2003). Impairment of dendritic cells and adaptive immunity by anthrax lethal toxin. Nature *424*, 329–334.

Ahuja, N., Kumar, P., and Bhatnagar, R. (2001). Hydrophobic residues Phe552, Phe554, Ile562, Leu566, and Ile574 are required for oligomerization of anthrax protective antigen. Biochem. Biophys. Res. Commun. *287*, 542–549.

Aktories, K., Bärmann, M., Ohishi, I., Tsuyama, S., Jakobs, K.H., and Habermann, E. (1986). Botulinum C2 toxin ADP-ribosylates actin. Nature *322*, 390–392.

Aktories, K., and Koch, G. (1995). Modification of actin and of Rho proteins by clostridial ADP-ribosylating toxins. In Microbial Toxins and Virulence Factors in Disease, B. Iglewski, J. Moss, A.T. Tu, and M. Vaughan, eds., (New York, USA: Marcel Dekker), pp. 491–520.

Aktories, K., Reuner, K.-H., Presek, P., and Barmann, M. (1989). Botulinum C2 toxin treatment increases the G-actin pool in intact chicken cells: a model for the cytopathic action of actin-ADP-ribosylating toxins. Toxicon *27*, 989–993.

Aktories, K., and Wegner, A. (1989). ADP-ribosylation of actin by clostridial toxins. J. Cell Biol. *109*, 1385–1387.

Al-Mohanna, F. A., Ohishi, I., and Hallett, M.B. (1987). Botulinum C2 toxin potentiates activation of the neutrophil oxidase. Further evidence of a role for actin polymerization. FEBS Lett. *219*, 40–44.

Arora, N., and Leppla, S.H. (1993). Residues 1–254 of anthrax toxin lethal factor are sufficient to cause cellular uptake of fused polypeptides. J. Biol. Chem. *268*, 3334–3341.

Babudieri, S., Borriello, S.P., Pantosti, A., Luzzi, I., Testore, G.P., and Panichi, G. (1986). Diarrhoea associated with toxigenic *Clostridium spiroforme*. J. Infect. *12*, 278–279.

Bachmeyer, C., Benz, R., Barth, H., Aktories, K., Gibert, M., and Popoff, M. (2001). Interaction of *Clostridium botulinum* C2-toxin with lipid bilayer membranes and Vero cells: inhibition of channel function by chloroquine and related compounds *in vitro* and intoxication *in vivo*. FASEB J. *15*, 1658–1660.

Ballard, J.D., Collier, R.J., and Starnbach, M.N. (1998). Anthrax toxin as a molecular tool for stimulation of cytotoxic T lymphocytes: disulfide-linked epitopes, multiple injections, and role of CD4(+) cells. Infect. Immun. *66*, 4696–4699.

Ballweber, E., Galla, M., Aktories, K., Yeoh, S., Weeds, A.G. and Mannherz, H.G. (2001). Interaction of ADP-ribosylated actin with actin binding proteins. FEBS Lett. *508*, 131–135.

Barbut, F., Gariazzo, B., Bonne, L., Lalande, V., Burghoffer, B., Luiuz, R., and Petit, J.C. (2007). Clinical feature of *Clostridium difficile*-associated infections and molecular characterization of strains: results of a retrospective study, 2000–2004. Infect. Cont. Hosp. Epidemiol. *28*, 131–139.

Barth, H., Aktories, K., Popoff, M.R., and Stiles B.G. (2004). Binary bacterial toxins: biochemistry, biology, and applications of common *Clostridium* and *Bacillus* proteins. Microbiol. Mol. Biol. Rev. *68*, 373–402.

Barth, H., Blöcker, D., and Aktories, K. (2002a). The uptake machinery of clostridial actin ADP-ribosylating toxins – a cell delivery system for fusion proteins and polypeptide drugs. Naunyn-Schmiedeberg's Arch. Pharmacol. *366*, 501–512.

Barth, H., Blöcker, D., Behlke, J., Bergsma-Schutter, W., Brisson, A., Benz, R., and Aktories, K. (2000). Cellular uptake of *Clostridium botulinum* C2 toxin requires oligomerization and acidification. J. Biol. Chem. *275*, 18704–18711.

Barth, H., Olenik, C., Sehr, P., Schmidt, G., Aktories, K., and Meyer, D.K. (1999). Neosynthesis and activation of Rho by *Escherichia coli* cytotoxic necrotizing factor (CNF1) reverse cytopathic effects of ADP-ribosylated Rho. J. Biol. Chem. *274*, 27407–27414.

Barth, H., Hofmann, F., Olenik, C., Just, I., and Aktories, K. (1998a). The N-terminal part of the enzyme component (C2I) of the binary *Clostridium botulinum* C2 toxin interacts with the binding component C2II and functions as a carrier system for a Rho ADP-ribosylating C3-like fusion toxin. Infect. Immun. *66*, 1364–1369.

Barth, H., Preiss, J.C., Hofmann, F., and Aktories, K. (1998b). Characterization of the catalytic site of the ADP-ribosyltransferase *Clostridium botulinum* C2 toxin by site-directed mutagenesis. J. Biol. Chem. *273*, 29506–29511.

Barth, H., Roebling, R., Fritz, M., and Aktories, K. (2002b). The binary *Clostridium botulinum* C2 toxin as a protein delivery system: identification of the minimal protein region necessary for interaction of toxin components. J. Biol. Chem. *277*, 5074–5081.

Bartlett, J.G. (1994). *Clostridium difficile*: history of its role as an enteric pathogen and the current state of knowledge about the organism. Clin. Infect. Dis. *18*, S265–S272.

Batra, S., Gupta, P., Chauhan, V., Singh, A., and Bhatnagar, R. (2001). Trp 346 and Leu 352 residues in protective antigen are required for the expression of anthrax lethal toxin activity. Biochem. Biophys. Res. Commun. *281*, 186–192.

Beauregard, K.E., Collier, R.J., and Swanson, J.A. (2000). Proteolytic activation of receptor-bound anthrax protective antigen on macrophages promotes its internalization. Cell. Microbiol. *2*, 251–258.

Beauregard, K.E., Wimer-Mackin, S., Collier, R.J., and Lencer, W.I. (1999). Anthrax toxin entry into polarized epithelial cells. Infect. Immun. *67*, 3026–3030.

Benson, E.L., Huynh, P. D., Finkelstein, A., and Collier, R.J. (1998). Identification of residues lining the anthrax protective antigen channel. Biochemistry *37*, 3941–3948.

Billington, S.J., Wieckowski, E.U., Sarker, M.R., Bueschel, D., Songer, J.G., and McClane, B.A. (1998). *Clostridium perfringens* type E animal enteritis isolates with highly conserved, silent enterotoxin gene sequences. Infect. Immun. *66*, 4531–4536.

Blaustein, R.O., Koehler, T.M., Collier, R.J., and Finkelstein, A. (1989). Anthrax toxin: channel-forming activity of protective antigen in planar phospholipid bilayers. Proc. Natl. Acad. Sci. USA *86*, 2209–2213.

Blöcker, D., Bachmeyer, C., Benz, R., Aktories, K., and Barth, H. (2003). Channel formation by the binding component of *Clostridium botulinum* C2 toxin: glutamate 307 of C2II affects channel properties *in vitro* and pH-dependent C2I translocation *in vivo*. Biochemistry *42*, 5368–5377.

Blöcker, D., Barth, H., Maier, E., Benz, R., Barbieri, J.T., and Aktories, K. (2000). The C-terminus of component C2II of *Clostridium botulinum* C2 toxin is essential for receptor binding. Infect. Immun. 68, 4566–4573.

Blöcker, D., Behlke, J., Aktories, K., and Barth, H. (2001). Cellular uptake of the binary *Clostridium perfringens* iota-toxin. Infect. Immun. *69*, 2980–2987.

Blossom, D.B. and McDonald, L.C. (2007). The challenges posed by reemerging *Clostridium difficile* infection. Clin. Infect. Dis. *45*, 222–227.

Borriello, S., and Carman, R. (1983). Association of iota-like toxin and *Clostridium spiroforme* with both spontaneous and antibiotic-associated diarrhea and colitis in rabbits. J. Clin. Microbiol. *17*, 414–418.

Borriello, S., Davies, H.A., and Carman, R.J. (1986). Cellular morphology of *Clostridium spiroforme*. Vet. Microbiol. *11*, 191–195.

Bosworth, T. (1940). On a new type of toxin produced by *Clostridium welchii*. J. Comp. Path. *53*, 245–255.

Bourgogne, A., Drysdale, M., Hilsenbeck, S.G., Peterson, S.N., and Koehler, T.M. (2003). Global effects of virulence gene regulators in a *Bacillus anthracis* strain with both virulence plasmids. Infect. Immun. *71*, 2736–2743.

Bradley, K.A., Mogridge, J., Mourez, M., Collier, R.J., and Young, J.A. (2001). Identification of the cellular receptor for anthrax toxin. Nature *414*, 225–229.

Bradley, K.A., Mogridge, J., Rainey, G.J., Batty, S., and Young, J.A. (2003). Binding of anthrax toxin to its receptor is similar to alpha integrin-ligand interactions. J. Biol. Chem. *278*, 49342–49347.

Braun, M., Herholz, C., Straub, R., Choisat, B., Frey, J., Nicolet, J., and Kuhnert, P. (2000). Detection of the ADP-ribosyltransferase toxin gene (*cdt*A) and its activity in *Clostridium difficile* isolates from equidae. FEMS Microbiol. Lett. *184*, 29–33.

Brossier, F., Sirard, J.C., Guidi-Rontani, C., Duflot, E., and Mock, M. (1999). Functional analysis of the carboxy-terminal domain of *Bacillus anthracis* protective antigen. Infect. Immun. *67*, 964–967.

Brossier, F., Weber-Levy, M., Mock, M., and Sirard, J.-C. (2000). Role of toxin functional domains in anthrax pathogenesis. Infect. Immun. 68, 1781–1786.

Cao, H., Agrawal, D., Kushner, N., Touzjian, N., Essex, M., and Lu, Y. (2002). Delivery of exogenous protein antigens to major histocompatibility complex class I pathway in cytosol. J. Infect. Dis. *185*, 244–251.

Carman, R.J., and Borriello, S.P. (1982). Observations on an association between *Clostridium spiroforme* and *Clostridium perfringens* type E iota enterotoxaemia in rabbits. Eur. J. Chemother. Antibiot. *2*, 143–144.

Carman, R.J., and Borriello, S.P. (1984). Infectious nature of *Clostridium spiroforme*-mediated rabbit enterotoxaemia. Vet. Microbiol. *9*, 497–502.

Carman, R.J., and Evans, R.H. (1984). Experimental and spontaneous clostridial enteropathies of laboratory and free living lagomorphs. Lab. Anim. Sci. *34*, 443–452.

Carman, R.J., van Tassell, R.L., and Wilkins, T.D. (1987). Production of iota toxin by *Clostridium spiroforme*: a requirement for divalent cations. Vet. Microbiol. *15*, 115–120.

Carroll, S.F., and Collier, R.J. (1984). NAD binding site of diphtheria toxin: identification of a residue within the nicotinamide subsite by photochemical modification with NAD. Proc. Natl. Acad. Sci. USA *81*, 3307–3311.

Cataldi, A., Labruyere, E., and Mock, M. (1990). Construction and characterization of a protective antigen-deficient *Bacillus anthracis* strain. Mol. Microbiol. *4*, 1111–1117.

Chauhan, V., and Bhatnagar, R. (2002). Identification of amino acid residues of anthrax protective antigen involved in binding with lethal factor. Infect. Immun. *70*, 4477–4484.

Chen, J. *et al.* (2003). MMDB: Entrez's 3D-structure database. Nucleic Acids Res. *31*, 474–477.

Cheng, P.C., Dykstra, M.L., Mitchell, R.N., and Pierce, S.K. (1999). A role for lipid rafts in B cell antigen receptor signaling and antigen targeting. J. Exp. Med. *190*, 1549–1560.

Choe, S., Bennett, M.J., Fujii, G., Curmi, P.M., Kantardjieff, K.A., Collier, R.J., and Eisenberg, D. (1992). The crystal structure of diphtheria toxin. Nature *357*, 216–222.

Chopra, A.P., Boone, S.A., Liang, X., and Duesbery, N.S. (2003). Anthrax lethal factor proteolysis and inactivation of MAPK kinase. J. Biol. Chem. *278*, 9402–9406.

Chowdhury, H.H., Kreft, M., and Zorec, R. (2002). Distinct effect of actin cytoskeleton disassembly on exo- and endo-cytic events in a membrane patch of rat melanotrophs. J. Physiol. *545*, 879–886.

Chvyrkova, I., Zhang, X.C., and Terzyan, S. (2007). Lethal factor of anthrax toxin binds monomeric form of protective antigen. Biochem. Biophys. Res. Commun. *360*, 690–695.

Confer, D.L., and Eaton, J.W. (1982). Phagocyte impotence caused by an invasive bacterial cyclase. Science *217*, 948–950.

Cui, X., Li, Y., Laird, M.W., Subramanian, M., Moayeri, M., Leppla, S.H., Fitz, Y., Su, J., Sherer, K., and Eichacker, P.Q. (2007). *Bacillus anthracis* oedama and lethal toxin have different hemodynamic effects but function together to worsen shock and outcome in a rat model. J. Infect. Dis. *195*, 572–580.

Cunningham, K., Lacy, D.B., Mogridge, J., and Collier, R.J. (2002). Mapping the lethal factor and oedama factor binding sites on oligomeric anthrax protective antigen. Proc. Natl. Acad. Sci. USA *99*, 7049–7053.

Dai, Z., Sirard, J.C., Mock, M., and Koehler, T.M. (1995). The atxA gene product activates transcription of the anthrax toxin genes and is essential for virulence. Mol. Microbiol. *16*, 1171–1181.

Devriese, P.P. (1999). On the discovery of *Clostridium botulinum*. J. Hist. Neurosci. *8*, 43–50.

Dobbelstein, M. (2003). Viruses in therapy – royal road or dead end? Virus Res. 92, 219–221.

Drum, C.L., Shen, Y., Rice, P.A., Bohm, A., and Tang, W.J. (2001). Crystallization and preliminary X-ray study of the oedama factor exotoxin adenylyl cyclase domain from *Bacillus anthracis* in the presence of its activator, calmodulin. Acta. Crystallogr. D. Biol. Crystallogr. *57*, 1881–1884.

Drum, C.L., Yan, S.Z., Bard, J., Shen, Y.Q., Lu, D., Soelaiman, S., Grabarek, Z., Bohm, A., and Tang, W.J. (2002). Structural basis for the activation of anthrax adenylyl cyclase exotoxin by calmodulin. Nature *415*, 396–402.

Drum, C.L., Yan, S.Z., Sarac, R., Mabuchi, Y., Beckingham, K., Bohm, A., Grabarek, Z., and Tang, W.J. (2000). An extended conformation of calmodulin induces interactions between the structural domains of adenylyl cyclase from *Bacillus anthracis* to promote catalysis. J. Biol. Chem. *275*, 36334–36340.

Duesbery, N.S., Webb, C.P., Leppla, S.H., Gordon, V.M., Klimpel, K.R., Copeland, T.D., Ahn, N.G., Oskarsson, M.K., Fukasawa, K., Paull, K.D., and Vande Woude, G.F. (1998). Proteolytic inactivation of MAP-kinase-kinase by anthrax lethal factor. Science *280*, 734–737.

Eckhardt, M., Barth, H., Blöcker, D., and Aktories, K. (2000). Binding of *Clostridium botulinum* C2 toxin to asparagine-linked complex and hybrid carbohydrates. J. Biol. Chem. *275*, 2328–2334.

Erwin, J.L., DaSilva, L.M., Bavari, S., Little, S.F., Friedlander, A.M., and Chanh, T.C. (2001). Macrophage-derived cell lines do not express proinflammatory cytokines after exposure to *Bacillus anthracis* lethal toxin. Infect. Immun. 69, 1175–1177.

Escuyer, V., and Collier, R.J. (1991). Anthrax protective antigen interacts with a specific receptor on the surface of CHO-K1 cells. Infect. Immun. *59*, 3381–3386.

Ezzell, J.W., and Abshire, T.G. (1992). Serum protease cleavage of *Bacillus anthracis* protective antigen. J. Gen. Microbiol. *138*, 543–549.

Fernie, D.S., Knights, J.M., Thomson, R.O., and Carman, R.J. (1984). Rabbit enterotoxaemia: purification and preliminary characterization of a toxin produced by *Clostridium spiroforme*. FEMS Microbiol. Lett. *21*, 207–211.

Firoved, A.M., Miller, G.F., Moayeri, M., Kakkar, R., Shen, Y., Wiggins, J.F., McNally, E.M., Tang, W.J., and Leppla, S.H. (2005). *Bacillus anthracis* oedama toxin causes extensive tissue lesions and rapid lethality in mice. Am. J. Pathol. *167*, 1309–1320.

Firoved, A.M., Moayeri, M., Wiggins, J.F., Shen, Y., Tang, W.J., and Leppla, S.H. (2007). Anthrax oedama toxin sensitizes DBA/2J mice to lethal toxin. Infect. Immun. *75*, 2120–2125.

Fox, G. E., *et al.*, 1980. The phylogeny of prokaryotes. Science *209*, 457–463.

Francis, J.W., Brown Jr., R.H., Figueiredo, D., Remington, M.P., Castillo, O., Schwarzschild, M.A., Fishman, P.S., Murphy, J.R., and vanderSpek, J.C. (2000). Enhancement of diphtheria toxin potency by replacement of the receptor binding domain with tetanus toxin C-fragment: a potential vector for delivering heterologous proteins to neurons. J. Neurochem. *74*, 2528–2536.

Frankel, A.E., Powell, B.L., Duesbery, N.S., Vande Woude, G.F., and Leppla, S.H. (2002). Anthrax fusion protein therapy of cancer. Curr. Protein Pept. Sci. *3*, 399–407.

Fritz, G., Schroeder, P., and Aktories, K. (1995). Isolation and characterization of a *Clostridium botulinum* C2 toxin-resistant cell line: evidence for possible involvement of the cellular C2II receptor in growth regulation. Infect. Immun. 63, 2334–2340.

Fry, D.C., Kuby, S.A., and Mildvan, A.S. (1986). ATP-binding site of adenylate kinase: mechanistic implications of its homology with *ras*-encoded p21, F_1-ATPase, and other nucleotide-binding proteins. Proc. Natl. Acad. Sci. USA 83, 907–911.

Fujii, N., Kubota, T., Shirakawa, S., Kimura, K., Ohishi, I., Moriishi, K., Isogai, E., and Isogai, H. (1996). Characterization of component-I gene of botulinum C2 toxin and PCR detection of its gene in clostridial species. Biochem. Biophys. Res. Commun. 220, 353–359.

Fukao, T. (2004). Immune system paralysis by anthrax lethal toxin: the roles of innate and adaptive immunity. Lancet Infect. Dis. *4*, 166–170.

Gaur, R., Gupta, P.K., Goyal, A., Wels, W., and Singh, Y. (2002). Delivery of nucleic acid into mammalian cells by anthrax toxin. Biochem. Biophys. Res. Commun. *297*, 1121–1127.

Geipel, U., Just, I., Schering, B., Haas, D., and Aktories, K. (1989). ADP-ribosylation of actin causes increase in the rate of ATP exchange and inhibition of ATP hydrolysis. Eur. J. Biochem. *179*, 229–232.

Geric, B., Carman, R.J., Rupnik, M., Genheimer, C.W., Sambol, S.P., Lyerly, D.M., Gerding, D.N., and Johnson, S. (2006). Binary toxin-producing, large clostridial toxin-negative *Clostridium difficile* strains are enterotoxic but do not cause disease in hamsters. J. Infect. Dis. *193*, 1143–1150.

Geric, B., Johnson, S., Gerding, D.D., Grabnar, M., and Rupnik, M. (2003). Frequency of binary toxin genes among *Clostridium difficile* strains that do not produce large clostridial toxins. J. Clin. Microbiol. *41*, 5227–5232.

Gibert, M., Marvaud, J-C., Periera, Y., Hale, M.L., Stiles, B.G., Boquet, P., Lamaze, C. and Popoff, M.R. (2007). Differential requirement for the transloca- tion of clostridial binary toxins: Iota toxin requires a membrane potential gradient. FEBS Lett. *581*, 1287–1296.

Gibert, M., Perelle, S., Daube, G., and Popoff, M.R. (1997). *Clostridium spiroforme* toxin genes are related to *C. perfringens* iota toxin genes but have a different genomic localization. Syst. Appl. Microbiol. *20*, 337–347.

Gibert, M., Petit, L., Raffestin, S., Okabe, A., and Popoff, M.R. (2000). *Clostridium perfringens* iota-toxin requires activation of both binding and enzymatic components for cytopathic activity. Infect. Immun. *68*, 3848–3853.

Gkiourtzidis, K., Frey, J., Bourtzi-Hatzopoulou, E., Iliadis, N., and Sarris, K. (2001). PCR detection and prevalence of alpha-, beta-, beta 2-, epsilon-, iota- and enterotoxin genes in *Clostridium perfringens* isolated from lambs with clostridial dysentery. Vet. Microbiol. *82*, 39–43.

Gold, J.A., Hoshino, Y., Hoshino, S., Jones, M.B., Nolan, A., and Weiden, M.D. (2004). Exogenous gamma and alpha/beta interferon rescues human macrophages from cell death induced by *Bacillus anthracis*. Infect. Immun. *72*, 1291–1297.

Goletz, T.J., Klimpel, K.R., Arora, N., Leppla, S.H., Keith, J.M., and Berzofsky, J.A. (1997). Targeting HIV proteins to the major histocompatibility com- plex class I processing pathway with a novel gp120- anthrax toxin fusion protein. Proc. Natl. Acad. Sci. USA *94*, 12059–12064.

Goncalves, C., Decre, D., Barbut, F., Burghoffer, B., and Petit, J.-C. (2004). Prevalence and characterization of a binary toxin (actin-specific ADP-ribosyltransferase) from *Clostridium difficile*. J. Clin. Microbiol. *42*, 1933–1939.

Gordon, V.M., Leppla, S.H., and Hewlett, E.L. (1988). Inhibitors of receptor-mediated endocytosis block the entry of *Bacillus anthracis* adenylate cyclase toxin but not that of *Bordetella pertussis* adenylate cyclase toxin. Infect. Immun. *56*, 1066–1069.

Gordon, V.M., Klimpel, K.R., Arora, N., Henderson, M.A., and Leppla, S.H. (1995). Proteolytic activation of bacterial toxins by eukaryotic cells is performed by furin and by additional cellular proteases. Infect. Immun. *63*, 82–87.

Guidi-Rontani, C. (2002). The alveolar macrophage: the Trojan horse of *Bacillus anthracis*. Trends. Microbiol. *10*, 405–409.

Guidi-Rontani, C., Weber-Levy, M., Mock, M., and Cabiaux, V. (2000). Translocation of *Bacillus anthracis* lethal and oedema factors across endosome membranes. Cell. Microbiol. *2*, 259–264.

Guignot, J., Mock, V. and Fouet, A. (1997). AtxA activates the transcription of genes harbored by both *Bacillus anthracis* virulence plasmids. FEMS Microbiol. Lett. *147*, 203–207.

Gülke, I., Pfeifer, G., Liese, J., Fritz, M., Hofmann, F., Aktories, K., and Barth, H. (2001). Characterization of the enzymatic component of the ADP-ribosyltransferase toxin CDTa from *Clostridium difficile*. Infect. Immun. *69*, 6004–6011.

Gupta, P.K., Chandra, H., Gaur, R., Kurupati, R.K., Chowdhury, S., Tandon, V., Singh, Y., and Maithal, K. (2003a). Conformational fluctuations in anthrax protective antigen: a possible role of calcium in the folding pathway of the protein. FEBS Lett. *554*, 505–510.

Gupta, P.K., Kurupati, R.K., Chandra, H., Gaur, R., Tandon, V., Singh, Y., and Maithal, K. (2003b). Acid induced unfolding of anthrax protective antigen. Biochem. Biophys. Res. Commun. *311*, 229–232.

Gupta, P., Singh, A., Chauhan, V., and Bhatnagar, R. (2001). Involvement of residues 147VYYEIGK153 in binding of lethal factor to protective antigen of *Bacillus anthracis*. Biochem. Biophys. Res. Commun. *280*, 158–163.

Ha, S-D., Ng, D., Lamothe, J., Valvano, M.A., Han, J., and Kim, S.O. (2007). Mitochondrial proteins Bnip3 and Bnip3L are involved in anthrax lethal toxin-induced macrophage cell death. J. Biol. Chem. *282*, 26275–26283.

Hale, M.L., Marvaud, J-C., Popoff, M.R., and Stiles, B.G. (2004). Detergent-resistant membrane microdomains facilitate Ib oligomer formation and biological activity of *Clostridium perfringens* iota-toxin. Infect. Immun. *72*, 2186–2193.

Hall, I.C. and O'Toole, E. (1935). Intestinal flora in new-born infants with a description of a new patho- genic anaerobe, *Bacillus difficilis*. Am. J. Dis. Child. *49*, 390–402.

Han, S., Arvai, A.S., Clancy, S.B., and Tainer, J.A. (2001). Crystal structure and novel recognition motif of Rho ADP-ribosylating C3 exoenzyme from *Clostridium botulinum*: structural insights for recognition speci- ficity and catalysis. J. Mol. Biol. *305*, 95–107.

Han, S., Craig, J.A., Putnam, C. D., Carozzi, N.B., and Tainer, J.A. (1999). Evolution and mechanism from structures of an ADP-ribosylating toxin and NAD complex. Nature Struct. Biol. *6*, 932–936.

Hanna, P.C., Kochi, S., and Collier, R.J. (1992). Biochemical and physiological changes induced by anthrax lethal toxin in J774 macrophage-like cells. Mol. Biol. Cell. *3*, 1269–1277.

Haug, G., Leemhuis, J., Tiemann, D., Meyer, D.K., Aktories, K., and Barth, H. (2003). The host cell chaperone Hsp90 is essential for translocation of the binary *Clostridium botulinum* C2 toxin into the cytosol. J. Biol. Chem. *278*, 32266–32274.

Heffron, A., and Poxton, I.R. (2007). A PCR approach to determine the distribution of toxin genes in closely related *Clostridium* species: *Clostridium botulinum* type C and D neurotoxins and C2 toxin, and *Clostridium novyi* alpha toxin. J. Med. Microbiol. *56*, 196–201.

Helgason, E., Okstad, O.A., Caugant, D.A., Johansen, H.A., Fouet, A., Mock, M., Hegna, I., and Kolsto, A-B. (2000). *Bacillus anthracis*, *Bacillus cereus*, and *Bacillus thuringiensis* – one species on the basis of genetic evidence. Appl. Environ. Microbiol. 66, 2627–2630.

Hirst, T.R., and D'Souza, J.M. (2006). *Vibrio cholerae* and *Escherichia coli* thermolabile enterotoxin. In The Comprehensive Sourcebook of Bacterial Protein Toxins, 3rd edn, J.E. Alouf and M.R. Popoff, eds. (Paris: Academic Press), pp. 270–290.

Hochmann, H., Pust, S., von Figura, G., Aktories, K., and Barth, H. (2006). *Salmonella enterica* SpvB ADP-ribosylates actin at position arginine-177 characterization of the catalytic domain within the SpvB protein and a comparison to binary clostridial actin-ADP-ribosylating toxins. Biochemistry *45*, 1271–1277.

Hoffmaster, A.R., and Koehler, T.M. (1997). The anthrax toxin activator gene atxA is associated with CO_2-enhanced non-toxin gene expression in *Bacillus anthracis*. Infect. Immun. *65*, 3091–3099.

Hoffmaster, A.R., Ravel, J., Rasko, D.A., Chapman, G.D., Chute, M.D., Marston, C.K., De, B.K., Sacchi, C.T., Fitzgerald, C., Mayer, L.W., Maiden, M.C., Priest, F.G., Barker, M., Jiang, L., Cer, R.Z., Rilstone, J., Peterson, S.N., Weyant, R.S., Galloway, D.R., Read, T.D., Popovic, T., and Fraser, C.M. (2004). Identification of anthrax toxin genes in a *Bacillus cereus* associated with an illness resembling inhalation anthrax. Proc. Natl. Acad. Sci. USA *101*, 8449–8454.

Ivanova, N., Sorokin, A., Anderson, I., Galleron, N., Candelon, B., Kapatral, V., Bhattacharyya, A., Reznik, G., Mikhailova, N., Lapidus, A., Chu, L., Mazur, M., Goltsman, E., Larsen, N., D'Souza, M., Walunas, T., Grechkin, Y., Pusch, G., Haselkorn, R., Fonstein, M., Ehrlich, S.D., Overbeek, R., and Kyrpides, N. (2003). Genome sequence of *Bacillus cereus* and comparative analysis with *Bacillus anthracis*. Nature *423*, 87–91.

Jin, F., Matsushita, O., Katayama, S., Jin, S., Matsushita, C., Minami, J., and Okabe, A. (1996). Purification, characterization, and primary structure of *Clostridium perfringens* lambda-toxin, a thermolysin-like metalloprotease. Infect. Immun. *64*, 230–237.

Jung, M., Just, I., van Damme, J., Vandekerckhove, J., and Aktories, K. (1993). NAD-binding site of the C3-like ADP-ribosyltransferase from *Clostridium limosum*. J. Biol. Chem. *268*, 23215–23218.

Just, I., Geipel, U., Wegner, A., and Aktories, K. (1990). De-ADP-ribosylation actin by *Clostridium perfringens* iota-toxin and *Clostridium botulinum* C2 toxin. Eur. J. Biochem. *192*, 723–727.

Just, I., Mohr, C., Schallehn, G., Menard, L., Didsbury, J.R., Vandekerckhove, J., van Damme, J., and Aktories, K. (1992). Purification and characterization of an ADP-ribosyltransferase produced by *Clostridium limosum*. J. Biol. Chem. *267*, 10274–10280.

Just, I., Wille, M., Chaponnier, C., and Aktories, K. (1993). Gelsolin-actin complex is target for ADP-ribosylation by *Clostridium botulinum* C2 toxin in intact human neutrophils. Eur. J. Pharmacol. *246*, 293–297.

Kaiser, E., Haug, G., Hliscs, M., Aktories, K., and Barth, H. (2006). Formation of a biologically active toxin complex of the binary *Clostridium botulinum* C2 toxin without cell membrane interaction. Biochemistry *45*, 13361–13368.

Kaneuchi, C., Miyazato, T., Shinjo, T., and Mitsuoka, T. (1979). Taxonomic study of helically coiled, sporeforming anaerobes isolated from the intestines of humans and other animals: *Clostridium cocleatum* sp. nov. and *Clostridium spiroforme* sp. nov. Int. J. Sys. Bacteriol. *29*, 1–12.

Katz, L., Lamont, J.T., Trier, J.S., Sonnenblick, E.B., Rothman, S.W., Broitman, S.A. and Rieth, S. (1978). Experimental clindamycin associated colitis in rabbits. Evidence for toxin-mediated mucosal damage. Gastroenterology *74*, 246–252.

Keim, P., and Smith, K.L. (2002). *Bacillus anthracis* evolution and epidemiology. Curr. Top. Microbiol. Immunol. *271*, 21–32.

Kim, S.O., Jing, Q., Hoebe, K., Beutler, B., Duesbery, N.S., and Han, J. (2003). Sensitizing anthrax lethal toxin-resistant macrophages to lethal toxin-induced killing by tumor necrosis factor-alpha. J. Biol. Chem. *278*, 7413–7421.

Kimura, K., Kubota, T., Ohishi, I., Isogai, E., Isogai, H., and Fujii, N. (1998). The gene for component-II of botulinum C2 toxin. Vet. Microbiol. *62*, 27–34.

Kistner, A., and Habermann, E. (1992). Reductive cleavage of tetanus toxin and botulinum neurotoxin A by the thioredoxin system from brain. Evidence for two redox isomers of tetanus toxin. Naunyn Schmiedeberg's Arch. Pharmacol. *345*, 227–234.

Klimpel, K.R., Molloy, S.S., Thomas, G., and Leppla, S.H. (1992). Anthrax toxin protective antigen is activated by a cell surface protease with the sequence specificity and catalytic properties of furin. Proc. Natl. Acad. Sci. USA *89*, 10277–10281.

Klimpel, K.R., Arora, N. and Leppla, S.H. (1994). Anthrax toxin lethal factor contains a zinc metalloprotease consensus sequence which is required for lethal toxin activity. Mol. Microbiol. *13*, 1093–1100.

Knapp, O., Benz, R., Gibert, M., Marvaud, J-C., and Popoff, M.R. (2002). Interaction of *Clostridium perfringens* iota-toxin with lipid bilayer membranes. J. Biol. Chem. 277, 6143–6152.

Koehler, T.M., and Collier, R.J. (1991). Anthrax toxin protective antigen: low-pH-induced hydrophobicity and channel formation in liposomes. Mol. Microbiol. 5, 1501–1506.

Krakauer, T., Little, S.F., and Stiles, B.G. (2005). *Bacillus anthracis* oedama toxin inhibits *Staphylococcus aureus* enterotoxin B effects *in vitro*: a potential protein therapeutic? Infect. Immun. *73*, 7069–7073.

Krantz, B.A., Finkelstein, A., and Collier, R.J. (2006). Protein translocation through the anthrax toxin transmembrane pore is driven by a proton gradient. J. Mol. Biol. 355, 968–979.

Kumar, P., Ahuja, N., and Bhatnagar, R. (2001). Purification of anthrax oedama factor from *Escherichia coli* and identification of residues required for binding to anthrax protective antigen. Infect. Immun. *69*, 6532–6536.

Kumar, P., Ahuja, N., and Bhatnagar, R. (2002). Anthrax oedama toxin requires influx of calcium for inducing cyclic AMP toxicity in target cells. Infect. Immun. *70*, 4997–5007.

Kurazono, H., Hosokawa, M., Matsuda, H., and Sakaguchi, G. (1987). Fluid accumulation in the ligated intestinal loop and histopathological changes of the intestinal mucosa caused by *Clostridium botulinum* C2 toxin in the pheasant and chicken. Res. Vet. Sci. *42*, 349–353.

Lacy, D.B., Wigelsworth, D.J., Melnyk, R.A., Harrison, S.C., and Collier, R.J. (2004). Structure of heptameric protective antigen bound to an anthrax toxin receptor: a role for receptor in pH-dependent pore formation. Proc. Natl. Acad. Sci. USA *101*, 13147–13151.

Leppla, S.H. (1982). Anthrax toxin oedama factor: a bacterial adenylate cyclase that increases cyclic AMP concentrations in eukaryotic cells. Proc. Natl. Acad. Sci. USA *79*, 3162–3166.

Leppla, S.H. (2006). *Bacillus anthracis* toxins. In The Comprehensive Sourcebook of Bacterial Protein Toxins, 3rd edn, J.E. Alouf and M.R. Popoff, eds. (Paris: Academic Press), pp. 323–347.

Leppla, S.H., Arora, N., and Varughese, M. (1999). Anthrax toxin fusion proteins for intracellular delivery of macromolecules. J. Appl. Microbiol. *87*, 284.

Leuber, M., Orlik, F., Schiffler, B., Sickmann, A., and Benz, R. (2006). Vegetative insecticidal protein (Vip1Ac) of *Bacillus thuringiensis* HD201: evidence for oligomer and channel formation. Biochemistry *45*, 283–288.

Li, J., Miyamoto, K., and McClane, B.A. (2007). Comparison of virulence plasmids among *Clostridium perfringens* type E isolates. Infect. Immun. *75*, 1811–1819.

Li, M., Dyda, F., Benhar, I., Pastan, I., and Davies, D.R. (1996). Crystal structure of the catalytic domain of *Pseudomonas* exotoxin A complexed with a nicotinamide adenine dinucleotide analog: implications for the activation process and for ADP ribosylation. Proc. Natl. Acad. Sci. USA *93*, 6902–6906.

Liang, X., Young, J.J., Boone, S.A., Waugh, D.S., and Duesbery, N.S. (2004). Involvement of domain II in toxicity of anthrax lethal factor. J. Biol. Chem. *279*, 52473–52478.

Little, S.F., Leppla, S.H., Burnett, J.W., and Friedlander, A.M. (1994). Structure–function analysis of *Bacillus anthracis* oedama factor by using monoclonal antibodies. Biochem. Biophys. Res. Commun. *199*, 676–682.

Little, S.F., Leppla, S.H., and Cora, E. (1988). Production and characterization of monoclonal antibodies to the protective antigen component of *Bacillus anthracis* toxin. Infect. Immun. *56*, 1807–1813.

Little, S.F, and Lowe, J.R. (1991). Location of receptor-binding region of protective antigen from *Bacillus anthracis*. Biochem. Biophys. Res. Commun. *180*, 531–537.

Little, S.F., Novak, J.M., Lowe, J.R., Leppla, S.H., Singh, Y., Klimpel, K.R., Lidgerding, B.C., and Friedlander, A.M. (1996). Characterization of lethal factor binding and cell receptor binding domains of protective antigen of *Bacillus anthracis* using monoclonal antibodies. Microbiology *142*, 707–715.

Liu, S., Schubert, R.L., Bugge, T.H., and Leppla, S.H. (2003b). Anthrax toxin: structures, functions and tumour targeting. Exp. Opin. Biol. Ther. *3*, 843–853.

Liu, S., and Leppla, S.H. (2003). Cell surface tumor endothelium marker 8 cytoplasmic tail-independent anthrax toxin binding, proteolytic processing, oligomer formation, and internalization. J. Biol. Chem. *278*, 5227–5234.

Liu, S., Aaronson, H., Mitola, D.J., Leppla, S.H., and Bugge, T.H. (2003a). Potent antitumor activity of a urokinase-activated engineered anthrax toxin. Proc. Natl. Acad. Sci. USA *100*, 657–662.

Loo, V.G., Poirier, L., Miller, M. A., Oughton, M., Libman, M. D., Michaud, S., Bourgault, A-M., Nguyen, T., Frenette, C., Kelly, M., Vibien, A., Brassard, P., Fenn, S., Dewar, K., Hudson, T.J., Horn, R., Rene, P., Monczak, Y., and Dascal, A. (2005). A predominantly clonal multi-institutional outbreak of *Clostridium difficile*-associated diarrhea with high morbidity and mortality. N. Engl. J. Med. *353*, 2442–2449.

Lowe, J., van der Ent, F., and Amos, L.A. (2004). Molecules of the bacterial cytoskeleton. Annu. Rev. Biophys. Biomol. Struct. 33, 177–198.

Lucey, B.P. (2004). William H. Welch, MD, and the discovery of *Bacillus welchii*. Arch. Pathol. Lab. Med. *128*, 1193–1195.

McDonald, L.C., Killgore, G.E., Thompson, A., Owens, R.C., Kazakova, S.V., Sambol, S.P., Johnson, S., and Gerding, D.N. (2005). An epidemic, toxin gene-variant strain of *Clostridium difficile*. N. Engl. J. Med. *353*, 2433–2441.

Margulis, L., Jorgensen, J.Z., Dolan, S., Kolchinsky, R., Rainey, F.A., and Lo, S.-C. (1998). The *Arthromitus* stage of *Bacillus cereus*: intestinal symbionts of animals. Proc. Natl. Acad. Sci. USA *95*, 1236–1241.

Marvaud, J-C., Smith, T., Hale, M.L., Popoff, M.R., Smith, L.A., and Stiles, B.G. (2001). *Clostridium perfringens* iota-toxin: mapping of receptor binding and Ia docking domains on Ib. Infect. Immun. *69*, 2435–2441.

Marvaud, J-C., Stiles, B.G., Chenal, A., Gillet, D., Gibert, M., Smith, L.A., and Popoff, M.R. (2002). *Clostridium perfringens* iota toxin. Mapping of the Ia domain involved in docking with Ib and cellular internalization. J. Biol. Chem. *277*, 43659–43666.

Masignani, V., Pizza, M. and Rappuoli, R. (2006). Molecular, functional and evolutionary aspects of ADP-ribosylating toxins. In The Comprehensive Sourcebook of Bacterial Protein Toxins, 3rd edn, J.E. Alouf and M.R. Popoff, eds. (Paris: Academic Press), pp. 213–244.

Masin, J., Sebo, P., and Locht, C. (2006). *Bordetella* protein toxins. In The Comprehensive Sourcebook of Bacterial Protein Toxins, 3rd edn, J.E. Alouf and M.R. Popoff, eds. (Paris: Academic Press), pp. 291–309.

Mauss, S., Chaponnier, C., Just, I., Aktories, K., and Gabbiani, G. (1990). ADP-ribosylation of actin isoforms by *Clostridium botulinum* C2 toxin and *Clostridium perfringens* iota toxin. Eur. J. Biochem. *194*, 237–241.

Mehra, V., Khanna, H., Chandra, R., and Singh, Y. (2001). Anthrax-toxin-mediated delivery of a 19 kDa antigen of *Mycobacterium tuberculosis* into the cytosol of mammalian cells. Biotechnol. Appl. Biochem. 33, 71–74.

Melnyk, R.A., Hewitt, K.M., Lacy, D.B., Lin, H.C., Gessner, C.R., Li, S., Woods, V.L., and Collier R.J. (2006). Structural determinants for the binding of anthrax lethal factor to oligomeric protective antigen. J. Biol. Chem. *281*, 1630–1635.

Menard, A., Altendorf, K.D., Breves, D.D., Mock, M., and Montecucco, C. (1996). The vacuolar ATPase proton pump is required for the cytotoxicity of *Bacillus anthracis* lethal toxin. FEBS Lett. *386*, 161–164.

Michelet, N., Granum, P.E., and Mahillon, J. (2006). *Bacillus cereus* enterotoxins, bi- and tri-component cytolysins, and other hemolysins. In The Comprehensive Sourcebook of Bacterial Protein Toxins, 3rd edn, J.E. Alouf and M.R. Popoff, eds. (Paris: Academic Press), pp. 779–790.

Mignot, T., Mock, M., and Fouet, A. (2003). A plasmid-encoded regulator couples the synthesis of toxins and surface structures in *Bacillus anthracis*. Mol. Microbiol. *47*, 917–927.

Milne, J.C., Blanke, S.R., Hanna, P.C., and Collier, R.J. (1995). Protective antigen-binding domain of anthrax lethal factor mediates translocation of a heterologous protein fused to its amino- or carboxy-terminus. Mol. Microbiol. *15*, 661–666.

Milne, J.C., and Collier, R.J. (1993). pH-dependent permeabilization of the plasma membrane of mammalian cells by anthrax protective antigen. Mol. Microbiol. *10*, 647–653.

Miyata, S., Minami, J., Tamai, E., Matsushita, O., Shimamota, S., and Okabe, A. (2002). *Clostridium perfringens* epsilon-toxin forms a heptameric pore within the detergent-insoluble microdomains of Madin–Darby canine kidney cells and rat synaptosomes. J. Biol. Chem. *277*, 39463–39468.

Moayeri, M., Haines, D., Young, H.A., and Leppla, S.H. (2003). *Bacillus anthracis* lethal toxin induces TNF-α-independent hypoxia-mediated toxicity in mice. J. Clin. Invest. *112*, 670–682.

Mock, M., and Mignot, T. (2003). Anthrax toxins and the host: a story of intimacy. Cell. Microbiol. *5*, 15–23.

Mogridge, J., Cunningham, K., and Collier, R.J. (2002a). Stoichiometry of anthrax toxin complexes. Biochemistry *41*, 1079–1082.

Mogridge, J., Cunningham, K., Lacy, D.B., Mourez, M., and Collier, R.J. (2002b). The lethal and oedama factors of anthrax toxin bind only to oligomeric forms of the protective antigen. Proc. Natl. Acad. Sci. USA 99, 7045–7048.

Mogridge, J., Mourez, M. and Collier, R.J. (2001). Involvement of domain 3 in oligomerization by the protective antigen moiety of anthrax toxin. J. Bacteriol. *183*, 2111–2116.

Munier, H., Bouhss, A., Gilles, A.M., Krin, E., Glaser, P., Danchin, A., and Barzu, O. (1993). Structural flexibility of the calmodulin-binding locus in *Bordetella pertussis* adenylate cyclase. Reconstitution of catalytically active species from fragments or inactive forms of the enzyme. Eur. J. Biochem. *217*, 581–586.

Myers, G.S. *et al.* (2006). Skewed genomic variability in strains of the toxigenic bacterial pathogen, *Clostridium perfringens*. Genome Res. *16*, 1031–1040.

Nagahama, M., Nagayasu, K., Kobayashi, K., and Sakurai, J. (2002). Binding component of *Clostridium perfringens* iota-toxin induces endocytosis in Vero cells. Infect. Immun. *70*, 1909–1914.

Nagahama, M., Yamaguchi, A., Hagiyama, T., Ohkubo, N., Kobayashi, K., and Sakurai, J. (2004). Binding and internalization of *Clostridium perfringens* iota-toxin in lipid rafts. Infect. Immun. *72*, 3267–3275.

Nakamura, S., Serikawa, T., Yamakawa, K., Nishida, S., Kozaki, S., and Sakaguchi, G. (1978). Sporulation and C2 toxin production by *Clostridium botulinum* type C strains producing no C1 toxin. Microbiol. Immunol. *22*, 591–596.

Nassi, S., Collier, R.J., and Finkelstein, A. (2002). PA_{63} channel of anthrax toxin: an extended β-barrel. Biochemistry *41*, 1445–1450.

Novichkov, P.S., Omelchenko, M.V., Gelfand, M.S., Mironov, A.A., Wolf, Y.I., and Koonin, E.V. (2004). Genome-wide molecular clock and horizontal gene transfer in bacterial evolution. J. Bacteriol. *186*, 6575–6585.

Ohishi, I. (1983a). Lethal and vascular permeability activities of botulinum C2 toxin induced by separate injections of the two toxin components. Infect. Immun. *40*, 336–339.

Ohishi, I. (1983b). Response of mouse intestinal loop to botulinum C2 toxin: enterotoxic activity induced by cooperation of nonlinked protein components. Infect. Immun. *40*, 691–695.

Ohishi, I. (1987). Activation of botulinum C2 toxin by trypsin. Infect. Immun. *55*, 1461–1465.

Ohishi, I., and Hama, Y. (1992). Purification and characterization of heterologous component IIs of botulinum C2 toxin. Microbiol. Immunol. *36*, 221–229.

Ohishi, I., Iwasaki, M., and Sakaguchi, G. (1980). Purification and characterization of two components of botulinum C2 toxin. Infect. Immun. *30*, 668–673.

Ohishi, I., and Miyake, M. (1985). Binding of the two components of C2 toxin to epithelial cells and brush borders of mouse intestine. Infect. Immun. *48*, 769–775.

Ohishi, I., Miyake, M., Ogura, H., and Nakamura, S. (1984). Cytopathic effect of botulinum C2 toxin on tissue-culture cells. FEMS Microbiol. Lett. *23*, 281–284.

Ohishi, I., and Okada, Y. (1986). Heterogeneities of two components of C2 toxin produced by *Clostridium botulinum* types C and D. J. Gen. Microbiol. *132*, 125–131.

Ohishi, I., and Tsuyama, S. (1986). ADP-ribosylation of nonmuscle actin with component I of C2 toxin. Biochem. Biophys. Res. Commun. *136*, 802–806.

Ohishi, I., and Yanagimoto, A. (1992). Visualizations of binding and internalization of two nonlinked protein components of botulinum C2 toxin in tissue culture cells. Infect. Immun. *60*, 4648–4655.

Okinaka, R.T., Cloud, K., Hampton, O., Hoffmaster, A.R., Hill, K.K., Keim, P., Koehler, T.M., Lamke, G., Kumano, S., Mahillon, J., Manter, D., Martinez, Y., Ricke, D., Svensson, R., and Jackson, P.J. (1999). Sequence and organization of pXO1, the large *Bacillus anthracis* plasmid harboring the anthrax toxin genes. J. Bacteriol. *181*, 6509–6515.

Orlandi, P.A., Curran, P.K., and Fishman, P.H. (1993). Brefeldin A blocks the response of cultured cells to cholera toxin. Implications for intracellular trafficking in toxin action. J. Biol. Chem. *268*, 12010–12016.

Orlik, F., Schiffler, B., and Benz, R. (2004). Anthrax toxin protective antigen: inhibition of channel function by chloroquine and related compounds and study of binding kinetics using the current noise analysis. Biophys. J. *88*, 1715–1724.

Pannifer, A.D., Wong, T.Y., Schwarzenbacher, R., Renatus, M., Petosa, C., Bienkowska, J., Lacy, D.B., Collier, R.J., Park, S., Leppla, S.H., Hanna, P., and Liddington, R.C. (2001). Crystal structure of the anthrax lethal factor. Nature *414*, 229–233.

Pannucci, J., Okinaka, R.T., Sabin, R., and Kuske, C.R. (2002). *Bacillus anthracis* pXO1 plasmid sequence conservation among closely related bacterial species. J. Bacteriol. *184*, 134–141.

Pastan, I., Chaudhary, V. and FitzGerald, D.J. (1992). Recombinant toxins as novel therapeutic agents. Ann. Rev. Biochem. 61, 331–354.

Perelle, S., Gibert, M., Boquet, P., and Popoff, M.R. (1993). Characterization of *Clostridium perfringens* iota-toxin genes and expression in *Escherichia coli*. Infect. Immun. *61*, 5147–5156.

Perelle, S., Domenighini, M., and Popoff, M.R. (1996). Evidence that Arg-295, Glu-378, and Glu-380 are active-site residues of the ADP-ribosyltransferase activity of iota toxin. FEBS Lett. *395*, 191–194.

Perelle, S., Gibert, M., Bourlioux, P., Corthier, G., and Popoff, M.R. (1997a). Production of a complete binary toxin (actin-specific ADP-ribosyltransferase) by *Clostridium difficile* CD196. Infect. Immun. *65*, 1402–1407.

Perelle, S., Scalzo, S., Kochi, S., Mock, M., and Popoff, M.R. (1997b). Immunological and functional comparison between *Clostridium perfringens* iota toxin, *C. spiroforme* toxin, and anthrax toxins. FEMS Microbiol. Lett. *146*, 117–121.

Petosa, C., Collier, R.J., Klimpel, K.R., Leppla, S.H., and Liddington, R.C. (1997). Crystal structure of the anthrax toxin protective antigen. Nature *385*, 833–838.

Pimental, R.A., Christensen, K.A., Krantz, B.A., and Collier, R.J. (2004). Anthrax toxin complexes: heptameric protective antigen can bind lethal factor and oedama factor simultaneously. Biochem. Biophys. Res. Commun. *322*, 258–262.

Popoff, M.R. 2000. Molecular biology of actin-ADP-ribosylating toxins. In Handbook of Experimental Pharmacology – Bacterial Protein Toxins (Vol. 145), K. Aktories and I. Just, eds. (Berlin, Germany: Springer-Verlag), pp. 275–306.

Popoff, M.R., Milward, F.W., Bancillon, B., and Boquet, P. (1989). Purification of the *Clostridium spiroforme* binary toxin and activity of the toxin on HEp-2 cells. Infect. Immun. *57*, 2462–2469.

Popoff, M.R., Rubin, E.J., Gill, D.M., and Boquet, P. (1988). Actin-specific ADP-ribosyltransferase produced by a *Clostridium difficile* strain. Infect. Immun. *56*, 2299–2306.

Popov, S.G., Villasmil, R., Bernardi, J., Grene, E., Cardwell, J., Popova, T., Wu, A., Alibek, D., Bailey, C., and Alibek, K. (2002). Effect of *Bacillus anthracis* lethal toxin on human peripheral blood mononuclear cells. FEBS Lett. *527*, 211–215.

Pratt, W.B., and Toft, D.O. (2003). Regulation of signaling protein function and trafficking by the hsp90/hsp70-based chaperone machinery. Exp. Biol. Med. *228*, 111–133.

Price, L.B., Hugh-Jones, M., Jackson, P.J., and Keim, P. (1999). Genetic diversity in the protective antigen gene of *Bacillus anthracis*. J. Bacteriol. *181*, 2358–2362.

Pust, S., Hochmann, H., Kaiser, E., von Figura, G., Heine, K., Aktories, K., and Barth, H. (2007). A cell-permeable fusion toxin as a tool to study the consequences of actin-ADP-ribosylation caused by the *Salmonella enterica* virulence factor SpvB in intact cells. J. Biol. Chem. *282*, 10272–10282.

Quinn, C.P., Singh, Y., Klimpel, K.R., and Leppla, S.H. (1991). Functional mapping of anthrax toxin lethal factor by in-frame insertion mutagenesis. J. Biol. Chem. *266*, 20124–20130.

Rasko, D.A., Ravel, J., Okstad, O.A., Helgason, E., Cer, R.Z., Jiang, L., Shores, K.A. Fouts, D.E., Tourasse, N.J., Angiuoli, S.V., Kolonay, J., Nelson, W.C., Kolsto, A.B., Fraser, C.M., and Read, T.D. (2004). The genome sequence of *Bacillus cereus* ATCC 10987 reveals metabolic adaptations and a large plasmid related to *Bacillus anthracis* pXO1. Nucleic Acids Res. *32*, 977–988.

Ratts, R., Zeng, H., Berg, E.A., Blue, C., McComb, M.E., Costello, C.E., vanderSpeck, J.C., and Murphy, J.R. (2003). The cytosolic entry of diphtheria toxin catalytic domain requires a host cell cytosolic translocation factor complex. J. Cell Biol. *160*, 1139–1150.

Read, T. D., et. al. (2003). The genome sequence of *Bacillus anthracis* Ames and comparison to closely related bacteria. Nature *423*, 81–86.

Ribot, W.J., Panchal, R.G., Brittingham, K.C., Ruthel, G., Kenny, T.A., Lane, D., Curry, B., Hoover, T.A., Friedlander, A.M., and Bavari. S. (2006). Anthrax lethal toxin impairs innate immune functions of alveolar macrophages and facilitates *Bacillus anthracis* survival. Infect. Immun. *74*, 5029–5034.

Richard, J.F., Mainguy, G., Gibert, M., Marvaud, J-C., Stiles, B.G., and Popoff, M.R. (2002). Transcytosis of iota-toxin across polarized CaCo-2 cells. Mol. Microbiol. *43*, 907–917.

Rosovitz, M.J., Schuck, P., Varughese, M., Chopra, A.P., Mehra, V., Singh, Y., McGinnis, L.M., and Leppla, S.H. (2003). Alanine scanning mutations in domain 4 of anthrax toxin protective antigen reveal residues important for binding to the cellular receptor and to a neutralizing monoclonal antibody. J. Biol. Chem. *278*, 30936–30944.

Ross, H.E., Warren, M.E., and Barnes, J.M. (1949). *Clostridium welchii* iota toxin: its activation by trypsin. J. Gen. Microbiol. 3, 148–152.

Rubin, E.J., Gill, D.M., Boquet, P., and Popoff, M.R. (1988). Functional modification of a 21-kilodalton G protein when ADP-ribosylated by exoenzyme C3 of *Clostridium botulinum*. Mol. Cell Biol. *8*, 418–426.

Sakurai, J., and Kobayashi, K. (1995). Lethal and dermonecrotic activities of *Clostridium perfringens* iota toxin: biological activities induced by cooperation of two nonlinked components. Microbiol. Immunol. 39, 249–253.

Sakurai, J., Nagahama, M., Hisatsune, J., Katunuma, N., and Tsuge, H. (2003). *Clostridium perfringens* iota-toxin, ADP-ribosyltransferase: structure and mechanism of action. Adv. Enzyme Regul. *43*, 361–377.

Salles, I.I., Tucker, A.E., Voth, D.E., and Ballard, J.D. (2003). Toxin-induced resistance in *Bacillus anthracis* lethal toxin-treated macrophages. Proc. Natl. Acad. Sci. USA *100*, 12426–12431.

Sandvig, K. (2006). The shiga toxins: properties and action on cells. In The Comprehensive Sourcebook of Bacterial Protein Toxins, 3rd edn, J.E. Alouf and M.R. Popoff, eds. (Paris: Academic Press), pp. 310–322.

Schering, B., Barmann, M., Chhatwal, G.S., Geipel, U., and Aktories, K. (1988). ADP-ribosylation of skeletal muscle and non-muscle actin by *Clostridium perfringens* iota toxin. Eur. J. Biochem. *171*, 225–229.

Schleberger, C., Hochmann, H., Barth, H., Aktories, K. and Schulz, G.E. (2006). Structure and action of the binary C2 toxin from *Clostridium botulinum*. J. Mol. Biol. *364*, 705–715.

Schmid, A., Benz, R., Just, I., and Aktories, K. (1994). Interaction of *Clostridium botulinum* C2 toxin with lipid bilayer membranes: formation of cation-selective channels and inhibition of channel function by chloroquine and peptides. J. Biol. Chem. *269*, 16706–16711.

Scobie, H.M., Rainey, G.J., Bradley, K.A., and Young, J.A. (2003). Human capillary morphogenesis protein 2 functions as an anthrax toxin receptor. Proc. Natl. Acad. Sci. USA *100*, 5170–5174.

Scobie, H.M., Marlett, J.M., Rainey, G.J.A., Lacy, D.B., Collier, R.J., and Young, J.A.T. (2007). Anthrax toxin receptor 2 determinants that dictate the pH threshold of toxin pore formation. PloS One 2(3), e329.

Sebaihia, M. *et al.* (2007). Genome sequence of a proteolytic (Group I) *Clostridium botulinum* strain Hall A and comparative analysis of the clostridial genomes. Genome Res. *17*, 1082–1092.

Sebaihia, M. *et al.* (2006). The multidrug-resistant human pathogen *Clostridium difficile* has a highly mobile mosaic genome. Nat. Genet. *38*, 779–786.

Sellman, B.R., Nassi, S., and Collier, R.J. (2001). Point mutations in anthrax protective antigen that block translocation. J. Biol. Chem. *276*, 8371–8376.

Shen, Y., Lee, Y.S., Soelaiman, S., Bergson, P., Lu, D., Chen, A., Beckingham, K., Grabarek, Z., Mrksich, M., and Tang, W.J. (2002). Physiological calcium concentrations regulate calmodulin binding and catalysis of adenylyl cyclase exotoxins. EMBO J. *21*, 6721–6732.

Shi, Y., Xu, W., Yuan, M. Tang, M., Chen, J., and Pang, Y. (2004). Expression of vip1/vip2 genes in *Escherichia coli* and *Bacillus thuringiensis* and the analysis of their signal peptides. J. Appl. Microbiol. *97*, 757–765.

Simons, K., and Ehehalt, R. (2002). Cholesterol, lipid rafts, and disease. J. Clin. Invest. *110*, 597–603.

Simpson, L.L. 1982. A comparison of the pharmacological properties of *Clostridium botulinum* type C1 and C2 toxins. J. Pharmacol. Exp. Ther. *223*, 695–701.

Simpson, L.L., Stiles, B.G., Zepeda, H.H., and Wilkins, T.D. (1987). Molecular basis for the pathological actions of *Clostridium perfringens* iota toxin. Infect. Immun. *55*, 118–122.

Simpson, L.L., Stiles, B.G., Zepeda, H., and Wilkins, T.D. (1989). Production by *Clostridium spiroforme* of an iotalike toxin that possesses mono(ADP-ribosyl) transferase activity: identification of a novel class of ADP-ribosyltransferases. Infect. Immun. *57*, 255–261.

Singh, Y., Klimpel, K.R., Quinn, C.P., Chaudhary, V.K., and Leppla, S.H. (1991). The carboxyl-terminal end of protective antigen is required for receptor binding and anthrax toxin activity. J. Biol. Chem. 266, 15493–15497.

Singh, Y., Klimpel, K.R., Goel, S., Swain, P.K., and Leppla, S.H. (1999). Oligomerization of anthrax toxin protective antigen and binding of lethal factor during endocytic uptake into mammalian cells. Infect. Immun. *67*, 1853–1859.

Sixma, T.K., Pronk, S.E., Kalk, K.H., Wartna, E.S., van Zanten, B.A.M., Witholt, B., and Hol, W.G. (1991). Crystal structure of a cholera toxin-related heat-labile enterotoxin from *E. coli*. Nature *351*, 371–377.

Smith, H. (2002). Discovery of the anthrax toxin: the beginning of studies of virulence determinants regulated *in vivo*. Int. J. Med. Microbiol. *291*, 411–417.

Smith, H., Williams, A.E., Pearce, J.H., Keppie, J., Harris-Smith, P.W., Fitsgeorge, R.B., and Witt, K. (1956). The chemical basis of the virulence of *Bacillus anthracis* VII. Two components of the anthrax toxin: their relationship to known immunizing aggressins. Br. J. Exp. Pathol. *37*, 263–271.

Song, L., Hobaugh, M.R., Shustak, C., Cheley, S., Bayley, H., and Gouaux, J.E. (1996). Structure of staphylococcal alpha-hemolysin, a heptameric transmembrane pore. Science *274*, 1859–1866.

Stanley, J.L., and Smith, H. (1961). Purification of factor I and recognition of a third factor of the anthrax toxin. J. Gen. Microbiol. *26*, 49–66.

Stare, B. G., Delmee, M., and Rupnik, M. (2007). Variant forms of the binary toxin CDT locus and tcdC gene in *Clostridium difficile* strains. J. Med. Microbiol. *56*, 329–335.

St. Croix, B., Rago, C., Velculescu, V., Traverso, G., Romans, K.E., Montgomery, E., Lal, A., Riggins, G.J., Lengauer, C., Vogelstein, B., and Kinzler. K.W. (2000). Genes expressed in human tumor endothelium. Science *289*, 1197–1202.

Stein, P.E., Boodhoo, A., Armstrong, G.D., Cockle, S.A., Klein, M.H., and Read, R.J. (1994). The crystal structure of pertussis toxin. Structure *2*, 45–57.

Stiles, B.G., Blöcker, D., Hale, M.L., Guetthoff, M.A., and Barth, H. (2002a). *Clostridium botulinum* C2 toxin: binding studies with fluorescence-activated cytometry. Toxicon *40*, 1135–1140.

Stiles, B.G., Hale, M.L., Marvaud, J.-C., and Popoff, M.R. (2000). *Clostridium perfringens* iota toxin: binding studies and characterization of cell surface receptor by fluorescence-activated cytometry. Infect. Immun. *68*, 3475–3484.

Stiles, B.G., Hale, M.L., Marvaud, J.-C., and Popoff, M.R. (2002b). *Clostridium perfringens* iota toxin: characterization of the cell-associated iota b complex. Biochem. J. *367*, 801–809.

Stiles, B.G., and Wilkins, T.D. (1986). Purification and characterization of *Clostridium perfringens* iota toxin:

dependence on two nonlinked proteins for biological activity. Infect. Immun. *54*, 683–688.

Sugii, S., and Kozaki, S. (1990). Hemagglutinating and binding properties of botulinum C2 toxin. Biochim. Biophys. Acta *1034*, 176–179.

Tessier, J., Green, C., Padgett, D., Zhao, W., Schwartz, L., Hughes, M., and Hewlett, E. (2007). Contributions of histamine, prostanoids, and neurokinins to oedama elicited by oedama toxin from *Bacillus anthracis*. Infect. Immun. *75*, 1895–1903.

Tournier, J.N., Quesnel-Hellmann, A., Mathieu, J., Montecucco, C., Tang, W.J., Mock, M., Vidal, D.R., and Goossens, P.L. (2005). Anthrax oedama toxin cooperates with lethal toxin to impair cytokine secretion during infection of dendritic cells. J. Immunol. *174*, 4934–4941.

Tsuge, H., Nagahama, M., Nishimura, H., Hisatsune, J., Sakaguchi, Y., Itogawa, Y., Katunuma, N., and Sakurai, J. (2003). Crystal structure and site-directed mutagenesis of enzymatic components from *Clostridium perfringens* iota-toxin. J. Mol. Biol. *325*, 471–483.

Turnbull, P.C. 2002. Introduction: anthrax history, disease and ecology. Curr. Top. Microbiol. Immunol. *271*, 1–19.

Uematsu, Y., Kogo, Y., and Ohishi, I. (2007). Disassembly of actin filaments by botulinum C2 toxin and actin-filament-disrupting agents induces assembly of microtubules in human leukaemia cell lines. Biol. Cell 99, 141–150.

van Damme, J., Jung, M., Hofmann, F., Just, I., Vandekerckhove, J., and Aktories, K. (1996). Analysis of the catalytic site of the actin ADP-ribosylating *Clostridium perfringens* iota toxin. FEBS Lett. *380*, 291–295.

Vandekerckhove, J., Schering, B., Bärmann, M., and Aktories, K. (1988). Botulinum C2 toxin ADP-ribosylates cytoplasmic β/γ-actin in arginine 177. J. Biol. Chem. *263*, 696–700.

Varughese, M., Chi, A., Teixeira, A.V., Nicholls, P.J., Keith, J.M., and Leppla, S.H. (1998). Internalization of a *Bacillus anthracis* protective antigen-c-Myc fusion protein mediated by cell surface anti-c-Myc antibodies. Mol. Med. *4*, 87–95.

Varughese, M., Teixeira, A.V., Liu, S., and Leppla, S.H. (1999). Identification of a receptor-binding region within domain 4 of the protective antigen component of anthrax toxin. Infect. Immun. *67*, 1860–1865.

Vitale, G., Bernardi, L., Napolitani, G., Mock, M., and Montecucco, C. (2000). Susceptibility of mitogen-activated protein kinase kinase family members to proteolysis by anthrax lethal factor. Biochem. J. *352*, 739–745.

Wang, F., Ruther, P., Jiang, I., Sawada-Hirai, R., Sun, S.M., Nedellec, R., Morrow, P.R., and Kang, A.S. (2004). Human monoclonal antibodies that neutralize anthrax toxin by inhibiting heptamer assembly. Human Antibodies *13*, 105–110.

Warren, G.W. 1997. Vegetative insecticidal proteins: novel proteins for control of corn pests. In Advances in Insect Control, the Role of Transgenic Plants, N.B. Carozzi and M. Koziel, eds. (London,: Taylor & Francis), pp. 109–121.

Watters, J.W., Dewar, K., Lehoczky, J., Boyartchuk, V., and Dietrich, W.F. (2001). Kif1C, a kinesin-like motor protein, mediates mouse macrophage resistance to anthrax lethal factor. Curr. Biol. *11*, 1503–1511.

Welch, W.H. and Flexner, S. (1896). Observations concerning *Bacillus aerogenes capsulatus*. J. Exp. Med. *1*, 5–45.

Wertman, K.F., and Drubin, D.G. (1992). Actin constitution: guaranteeing the right to assemble. Science *258*, 759–760.

Wesche, J., Malecki, J., Wiedlocha, A., Skjerpen, C.S., Claus, P., and Olsnes, S. (2006). FGF-1 and FGF-2 require the cytosolic chaperone Hsp90 for translocation into the cytosol and the cell nucleus. J. Biol. Chem. *281*, 11405–11412.

Whittaker, C.A., and Hynes, R.O. (2002). Distribution and evolution of von Willebrand/integrin A domains: widely dispersed domains with roles in cell adhesion and elsewhere. Mol. Biol. Cell 13, 3369–3387.

Wilde, C., and Aktories, K. (2001). The Rho-ADP-ribosylating C3 exoenzyme from *Clostridium botulinum* and related C3-like transferases. Toxicon 39, 1647–1660.

Wilde, C., Chhatwal, G.S., Schmalzing, G., Aktories, K., and Just, I. (2001). A novel C3-like ADP-ribosyltransferase from *Staphylococcus aureus* modifying RhoE and Rnd3. J. Biol. Chem. 276, 9537–9542.

Wille, M., Just, I., Wegner, A., and Aktories, K. (1992). ADP-ribosylation of the gelsolin-actin complex by clostridial toxins. J. Biol. Chem. 267, 50–55.

Williams R.C., Rees, M.L., Jacobs, M.F., Pragai, Z., Thwaite, J.E., Baillie, L.W., Emmerson, P.T., and Harwood, C.R. (2003). Production of *Bacillus anthracis* protective antigen is dependent on the extracellular chaperone, PrsA. J. Biol. Chem. *278*, 18056–18062.

Young, J.A. and Collier, R.J. (2007). Anthrax toxin: receptor binding, internalization, pore formation, and translocation. Ann. Rev. Biochem. *76*, 243–265.

Zhang, S., Finkelstein, A., and Collier, R.J. (2004). Evidence that translocation of anthrax toxin's lethal factor is initiated by entry of its N terminus into the protective antigen channel. Proc. Natl. Acad. Sci. USA *101*, 16756–16761.

Zhao, Y., and Melville, S.B. (1998). Identification and characterization of a sporulation-dependent promoter upstream of the enterotoxin gene (*cpe*) of *Clostridium perfringens*. J. Bacteriol. *180*, 136–142.

Molecular Epidemiology of Group I and II *Clostridium botulinum*

5

Miia Lindström, Maria Fredriksson-Ahomaa and Hannu Korkeala

Abstract

Clostridium botulinum, producing highly potent botulinum neurotoxin, is a diverse species consisting of four genetically and physiologically distinct groups (groups I–IV) of organisms. Groups I and II *C. botulinum* produce A, B, E and/or F toxins which cause human botulism. In addition, some strains of *Clostridium butyricum* and *Clostridium barati* produce type E and F toxins, respectively, and have thus been related to human illness. Human botulism appears in five different forms, such as the classical food botulism, infant botulism, wound botulism, adult infectious botulism, and iatrogenic botulism. Typical of all forms of human botulism is descending flaccid paralysis which may lead to death upon respiratory muscle failure. While the research and diagnostics of botulinum neurotoxigenic clostridia and botulism were based on toxin detection by the mouse bioassay until mid-1990s, the subsequent development of molecular detection and typing assays enabled rapid, sensitive, specific, and ethically acceptable molecular epidemiological detection, identification and strain characterization of these organisms, increasing our understanding of the epidemiology of botulinum neurotoxigenic clostridia and botulism.

Clostridium botulinum and botulism

Clostridium botulinum

Like other clostridia, *Clostridium botulinum* is an anaerobic spore-forming Gram-positive bacterium frequently found in soil and aquatic environments. Although otherwise harmless as a non-infective and non-invasive organism, *C. botulinum* is continuously the focus of large research interests due to its ability to produce the highly potent botulinum neurotoxin during its vegetative growth.

Based on the serological properties of the neurotoxins produced, *C. botulinum* strains are traditionally divided into seven toxinotypes, A–G. Phenotypically and genetically, *C. botulinum* strains form four very distinct groups of organisms, designated groups I–IV (Holdeman and Brooks, 1970; Smith and Hobbs, 1974; Hutson *et al.*, 1993). Generally, groups I and II cause botulism in humans and group III in animals. Apart from one report on human botulism related to group IV (Sonnabend *et al.*, 1981), this group has generally not been associated with illness. Owing to its distinct phenotypic and genetic properties, is has been proposed to be renamed *Clostridium argentinense* (Suen *et al.*, 1988). Of *C. botulinum* strains related to human illness, group I cultures produce toxin A, B, and/or F and group II cultures produce toxin B, E or F. Dual-toxic strains have been reported from early on (Giménez and Ciccarelli, 1970; Giménez, 1984; Hatheway and McCroskey, 1987; Franciosa *et al.*, 1997; Kirma *et al.*, 2004), as well as those producing one type of toxin but carrying a silent gene for another (Franciosa *et al.*, 1994, 2004; Kirma *et al.*, 2004). Approximately half of strains producing A toxin have been estimated to carry a silent B gene (Franciosa *et al.*, 1994). In addition to *C. botulinum*, some strains of *Clostridium butyricum* and *Clostridium baratii* also produce type E and F neurotoxins, respectively.

Physiologically, group I and II *C. botulinum* strains are significantly different. Group I strains are proteolytic, while group II strains are non-proteolytic and utilize mainly carbohydrates. The spores of group I have a high heat resistance (Schmidt 1964; Ito *et al.*, 1967; Stumbo 1973; Smelt 1980; Lynt *et al.*, 1981; Scott and Bernard, 1982) and their vegetative cells are mesophilic, being able to grow at a temperature range of 12–48°C (Smith and Sugiyama, 1988) and having an optimum growth temperature around or slightly above the body temperature of 37°C. Group II spores have a moderate heat resistance (Lynt *et al.*, 1983; Peck *et al.*, 1992a,b, 1993; Lindström *et al.*, 2003) and, as psychrotrophic organisms, group II vegetative cells are able to grow at refrigerated temperatures down to 3°C (Schmidt *et al.*, 1961; Eklund *et al.*, 1967a,b; Graham *et al.*, 1997). Their optimal growth temperature is 26–30°C. The physiological diversity is further multiplied by the fact that the neurotoxigenic *C. butyricum* and *C. barati* are distinct from group I and II *C. botulinum*. As a consequence of this miscellaneous selection of physiologies, botulinum neurotoxigenic *Clostridia* cause human health threats through a number of ways (Peck, 1997, 2006; Lindström *et al.*, 2006a).

Botulinum neurotoxin

Botulinum neurotoxins are 150-kDa zinc-dependent metalloproteases (Montecucco and Schiavo, 1993). The protein is consisted of a 100-kDa heavy chain and a 50-kDa light chain linked by a disulphide bond. The neurotoxin targets in humans and animals are motoric nerve ends where the toxin blocks neurotransmitter release causing flaccid paralysis. The heavy chain is responsible for specific binding to the presynaptic membrane and translocation of the light chain across the membrane into the cytosol. The light chain possesses catalytic activity cleaving proteins involved in acetylcholine vesicle docking and fusion to the presynaptic membrane (Montecucco and Schiavo, 1994), inhibiting neurotransmission and resulting in paralysis of the post-synaptic muscle.

C. botulinum secretes the neurotoxin during its vegetative growth, with the peak production occurring at late logarithmic phase (Lövenklev *et al.*, 2004). The toxin molecule is secreted as a progenitor toxin that, in addition to the neurotoxin, contains several non-toxic neurotoxin associated proteins (NAP) probably protecting the neurotoxin and assisting in absorption from the gastrointestinal tract into circulation (Fujinaga *et al.*, 1997). The maximal potency of the botulinum neurotoxins is achieved after enzymatic cleavage of the toxin molecule (Das Gupta and Sugiyama, 1972; Ohishi and Sakaguchi, 1977). In proteolytic group I *C. botulinum*, the cleavage of toxin is achieved by endogenous enzymes, but neurotoxins produced by non-proteolytic group II strains probably require external proteases, such as the pancreatic trypsin (Duff *et al.*, 1956; Das Gupta and Sugiyama, 1972), for their activation. Gene encoding a candidate endogenous protease, responsible for the cleavage in group I *C. botulinum*, has been identified in the genome of the group I strain ATCC 3502 (Sebaihia *et al.*, 2007).

Botulinum neurotoxin genes

The genes encoding botulinum neurotoxins may generally be located chromosomally, on a plasmid, or encoded by a bacteriophage. In group I and II *C. botulinum*, genes encoding types A, B, E, and F toxins have been considered to be located on bacterial chromosomes (Sebaihia *et al.*, 2007), while types C and D toxin genes in group III *C. botulinum* are mediated by bacteriophages (Eklund *et al.*, 1972; Eklund and Poysky, 1974), and type G toxin genes in group IV are carried by a plasmid (Eklund *et al.*, 1988, Zhou *et al.*, 1995). However, recent research has revealed that the A3 and A4 toxin subtypes are encoded by plasmids (Marshall *et al.*, 2007). Moreover, the presence of IS elements in the neurotoxin gene cluster of some type A1 and A2 strains (Dineen *et al.*, 2004) and the tremendous structural similarities between the neurotoxin gene clusters in type E and type A2–A4 strains (Chen *et al.*, 2007) suggest that there is far more to be learned on the genetic mobility and regulation of the botulinum neurotoxin encoding genes.

The *botA–botG* (also named *cntA/A–cntA/G*) genes encoding the different toxin serotypes A through G show a 50–70% amino acid identity (Whelan *et al.*, 1992; Campbell *et al.*, 1993a; Hutson *et al.*, 1994). The different toxin types are serologically different (Smith and Sugiyama, 1988) and polyclonal antisera can be used to distinguish between serotypes (Solomon and Lilly, 2001). Structural variation in the binding domain of the heavy chain results in receptor specificity and probably explains the sensitivity of different animal species to different toxin types. Neurotoxin gene sequencing and monoclonal antibodies binding specific epitopes at the neurotoxin molecule have been used to identify novel neurotoxin subtypes (Smith and Sugiyama, 1988; Smith *et al.*, 2005; Hill *et al.*, 2007b; Chen *et al.*, 2007). At least four type A neurotoxin subtypes, four type B, four type F, and six type E neurotoxin subtypes have been identified to date. The subtypes within a serotype show up to 32% amino acid differences, with the largest variation being within serotype A, and the lowest variation (up to 6%) within type E.

Neurotoxin gene cluster sequencing has suggested that different toxin serotypes and subtypes are associated with a distinct selection of NAPs (Kimura *et al.*, 1990; Thompson *et al.*, 1990; East *et al.*, 1992, 1994, 1996a,b; Poulet *et al.*, 1992; Fujii *et al.*, 1993a,b; Willems *et al.*, 1993; East and Collins, 1994; Fujinaga *et al.*, 1994; Hutson *et al.*, 1994; Ohyama *et al.*, 1995; Moriishi *et al.*, 1996; Kubota *et al.*, 1996, 1998; Bhandari *et al.*, 1997; Nakajima *et al.*, 1998; Dineen *et al.*, 2003, 2004; Zhang *et al.*, 2003; Sakaguchi *et al.*, 2005; Franciosa *et al.*, 2006; Chen *et al.*, 2007) (Fig. 5.1). All toxin clusters include *ntnh* (also named *cntB*) which is co-transcribed with the neurotoxin gene (Hauser *et al.*, 1994; Henderson *et al.*, 1996; Dineen *et al.*, 2004). The resulting non-toxic non-haemagglutinin (NTNH) component is responsible for protection of the neurotoxin from environmental factors such as a low pH in the stomach of humans and animals (Kitamura *et al.*, 1969). The neurotoxin gene cluster of many type A1, B, C, D and G strains also includes two or three genes encoding haemagglutinin (HA) components of 17–70 kDa (Fig. 5.1), suggested to assist in targeted absorption of the neurotoxin in the gastro-intestinal (GI) tract (Fujinaga *et al.*, 1997) through disruption of intestinal epithelial intercellular junctions (Matsumura *et al.*, 2007). The HA components are missing in all so far identified type E and F neurotoxin gene clusters. Other potentially related components are encoded by the *p47* and *orfX1–orfX3* genes residing the type A2–A4, E, and F neurotoxin gene clusters, but the function of these proteins is currently unknown. In types A, B, C, D, F and G toxin-producing strains, the neurotoxin gene cluster carries *botR* (also named *cntR*) encoding the alternative sigma factor BotR shown to positively regulate the neurotoxin synthesis (Marvaud *et al.*, 1998; Raffestin *et al.*, 2005). No *botR* homologues have been identified in type E neurotoxin-encoding gene clusters, and it has remained unclear which mechanism regulates type E toxin synthesis in type E neurotoxigenic *Clostridia*.

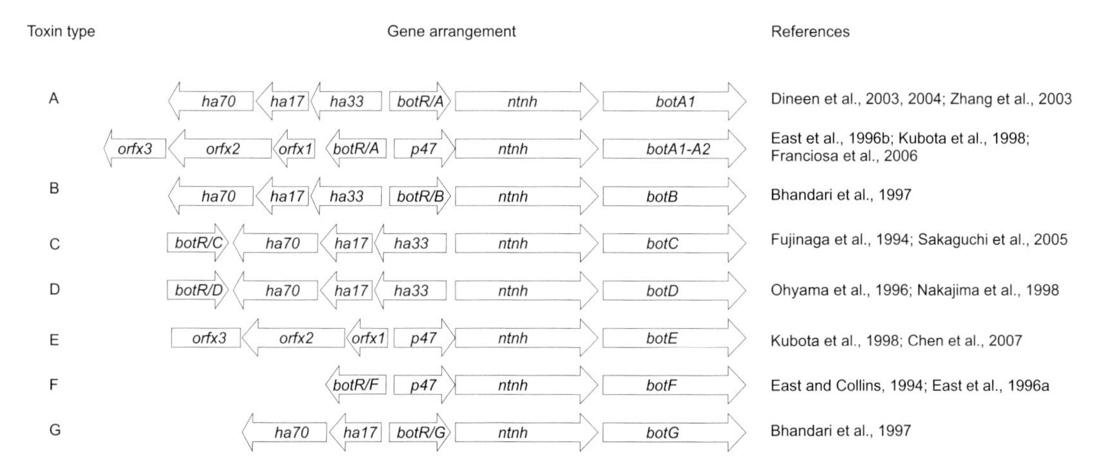

Figure 5.1 Structure of type A–G botulinum neurotoxin gene clusters

Botulism

Botulism is a consequence of botulinum neurotoxin entering the human or animal body and inhibiting acetylcholine release in motoric nerve endings. Typical of the disease is a descending flaccid paralysis which in humans starts in the densely innervated areas of the head and proceeds caudally. First symptoms typically include impaired swallowing and speech, drooling, double vision and facial paralysis. Later stages are characterized by paralysis of limbs and finally respiratory muscles, which may lead to death unless treated. While food-borne botulism is an intoxication which follows ingestion of toxin with food, the other forms of botulism are infections due to spore germination, growth, and toxin production *in vivo*.

Food-borne botulism

The best known form of human botulism is the classical food poisoning botulism, an intoxication that follows the consumption of food containing preformed neurotoxin. Depending on the amount of toxin ingested, the incubation period typically varies from 12 to 72 h, but has been reported to extend to several weeks (Gao *et al.*, 1990). In addition to the paralytic signs of botulism, the food-borne form is often manifested by gastrointestinal symptoms. Mild cases with gastrointestinal symptoms dominating are extremely rarely reported (Sobel *et al.*, 2007). However, in many botulism outbreaks there have been patients suffering from mild symptoms in addition to the severe cases hospitalized. The mild cases often remain unconfirmed (Korkeala *et al.*, 1998; Lindström *et al.*, 2006b) or undiagnosed. Considering the high risk of botulism related to modern processing and packaging of foods (Lindström *et al.*, 2006a), undiagnosed mild cases are probably much more common than is generally considered. To get a better understanding of this form of food botulism, health care professionals should be reminded of the potential diagnosis and symptoms of mild botulism, and easy-to-use rapid diagnostic tests should be developed to aid in diagnosis.

Both group I and II *C. botulinum* are involved with food-borne botulism. Regions with a particularly high incidence of food-borne botulism include the Republic of Georgia, Poland, China, Russia, Kyrgyzstan, and certain ethnic populations in the North, e.g. Alaskan Inuits (Hauschild and Dodds, 1993; Peck, 2006). Typically food-borne botulism is caused by home-prepared foods such as cured meats, canned vegetables, and fermented fish products, as a consequence of insufficient heating and/or chilling procedures. Canned foods typically involve group I *C. botulinum*, while group II organisms are related to marine foods or preserved meat with mild heat processing. Outbreaks are usually sporadic and restricted to a family or a small gathering. Exceptionally large was the Thailand outbreak in 2006 affecting 209 people, 42 of whom developed respiratory failure, who had consumed home-prepared bamboo shoots contaminated with botulinum neurotoxin type A (Kongsaengdao *et al.*, 2006). Commercial foods are less frequently involved in food-borne botulism, but such outbreaks may be large and cause significant economic losses to the food industry (Korkeala *et al.*, 1998, Lindström *et al.*, 2006a,b). An example of a large type E outbreak due to commercial product *faseikh* (uneviscerated, salted mullet fish) affected 91 people, with 18 dying, in Cairo in 1991 (Weber *et al.*, 1993). An international type A outbreak affecting patients in several states in the USA and in Canada was identified in 2006 and was traced to commercial carrot juice (http://www.cbc.ca/health/story/2006/10/23/botulism-warnings.html, accessed 09 Nov 2007).

The case-fatality rate of food-borne botulism in developed countries is typically less than 10% (Sobel *et al.*, 2004). Worldwide, 52% of all cases were due to type B toxin, and 34 and 12% due to types A and E toxins, respectively, in 1995 (Hatheway 1995). Type F cases are extremely rare: in 1981–2002, only 13 (1%) of all botulism cases in the USA were due to type F toxin, with one being traced to a food source (Harvey *et al.*, 2002; Gupta *et al.*, 2005).

Infant botulism

Infant botulism is a consequence of ingested spores of botulinum neurotoxigenic clostridia germinating and growing neurotoxic cultures in the infantile gastrointestinal tract with poorly developed intestinal normal bacterial population (Arnon, 1986). Owing to the requirement of growth from spores *in vivo*, infant botulism is related mainly to the mesophilic group I *C. botulinum*. In addition to group I *C. botulinum* types A and B (Hatheway *et al.*, 1981; Hatheway and McCroskey, 1987), Bf (Barash and Arnon, 2004), and F (Hoffman *et al.*, 1982), cases due to type E and F toxins produced by *C. butyricum* (Aureli *et al.*, 1986; McCroskey *et al.*, 1986; Hatheway and McCroskey, 1987; Fenicia *et al.*, 2007a) and *C. baratii* (Hall *et al.*, 1985; Barash *et al.*, 2005; Keet *et al.*, 2005), respectively, have been reported. The disease generally affects babies under one year of age, with the youngest reported patient being only 54 h old (Barash *et al.*, 2005; Keet *et al.*, 2005). Clinical appearance varies from subclinical to sudden death (Arnon and Chin, 1979). Symptoms may include prolonged constipation, impaired feeding and lethargy, facial muscle paralysis, ptosis, and general weakness (Arnon, 1989; Clemmens and Bell, 2007). Infant botulism has been suggested to be a causative agent of cot death (Anonymous 1976; Arnon *et al.*, 1978). Case-fatality rate is less than 2% (Centers for Disease Control and Prevention, 1998). The only foodstuffs associated with infant botulism are honey (Aureli *et al.*, 2002), which often carries high numbers of *C. botulinum* spores (Arnon, 1992; Hauschild and Dodds, 1993; Nevas *et al.*, 2002, 2005c, 2006), and contaminated infant milk powder (Brett *et al.*, 2005). Dust is an important environmental source of spores (Arnon, 1992; Nevas *et al.*, 2005c).

Wound botulism

Wound botulism is increasingly diagnosed among injecting drug abusers due to the use of contaminated needles or impure heroin (Passaro *et al.*, 1998; Werner *et al.*, 2000; Athwal *et al.*, 2000, 2001; Mulleague *et al.*, 2002; Akbulut *et al.*, 2005; Aguirre-Balsalobre *et al.*, 2007; Artin *et al.*, 2007). Wound botulism follows *in vivo* germination, growth and toxin production from *C. botulinum* spores in profound wounds or abscesses providing anaerobic conditions. The vast majority of cases are due to the mesophilic group I organisms, mainly of toxin types A and B. The first case of type E wound botulism was reported in Sweden in 2007 (Artin *et al.*, 2007). The authors reported the case to be caused by *C. botulinum* type E; however, species identification was not described and it remains a possibility that the case was actually due to a neurotoxic strain of the mesophilic *C. butyricum* (Anniballi *et al.*, 2002) rather than the psychrotrophic group II *C. botulinum* type E.

The clinical outcome of wound botulism is similar to other forms of botulism, but the wound form is often accompanied by a mixed infection and may be due to several botulinum toxin-producing strains (Akbulut *et al.*, 2005). The median incubation period is 7 days, and estimated case-fatality rate is 15% (Hatheway, 1995).

Infectious botulism in adults

The adult form of infectious botulism is rare and resembles infant botulism in its pathogenesis and clinical status. It results from colonization of the intestinal tract by toxin-producing clostridia (Chia *et al.*, 1986; McCroskey and Hatheway, 1988; McCroskey *et al.*, 1991; Fenicia *et al.*, 1999, 2007a). People with altered intestinal bacterial population due to abdominal surgery (Freedman *et al.*, 1986; Isachsohn *et al.*, 1985), prolonged antimicrobial treatment, or gastrointestinal wounds and abscesses are at risk (Chia *et al.*, 1986). The history of an infectious botulism case typically lacks association to foods with high risk of botulism (McCroskey and Hatheway, 1988). Like in other infectious forms of botulism, cases are typically due to group I organisms. Particularly many reported cases have been due to type E or F neurotoxin produced by *C. butyricum* (Fenicia *et al.*, 1999, 2007a) or *C. barati*, respectively (Sobel *et al.*, 2005). Interestingly, type F toxin-related cases have been shown to have a distinct clinical course, characterized by a fulminant onset (Sobel *et al.*, 2005). It is not clear whether this phenomenon is due to distinct characteristics of the type F neurotoxin or to the distinct epidemiology of *C. barati*.

Other forms of botulism

Inhalation botulism is very rare and may result from aerosolization of the neurotoxin. A few human cases have been reported (Holzer, 1962). Iatrogenic botulism with local or generalized weakness develops as a consequence of therapeutic injection of the neurotoxin (Mezaki *et* al., 1996, Bakheit et al., 1997). Until these days this form has been rare; however, increased therapeutic and cosmetic use of botulinum neurotoxin and, unfortunately, increased release of counterfeit products with uncontrolled toxin potency (Brin et al., 2007) will probably lead to a dramatic increase in the number of cases worldwide. Depending on the toxic dose, iatrogenic botulism may manifest with exceptionally severe paralysis and result in long-term neural damage (Chertow et al., 2006; Souayah et al., 2006).

Treatment of botulism

The most important treatment in all forms of botulism is intensive symptomatic care, with particular attention to maintenance of respiratory support (Arnon et al., 1977; Johnson et al., 1979; Bossi et al., 2004). Administration of trivalent (A, B, E) antitoxin may neutralize the circulating neurotoxin, but is successful only shortly after ingestion of the toxin and becomes irrelevant when the toxin has left blood circulation and entered nerve endings. The antiserum may also raise severe allergic reactions (Boyer et al., 2001).

As for infant botulism, administration of a human-derived immune globulin has been reported to significantly shorten the duration of hospital stay (Thompson et al., 2005; Arnon et al., 2006). Specific treatment of wound botulism includes thorough surgical debridement and antimicrobial agents.

Molecular epidemiological tools applied for *Clostridium botulinum*

In the early research on *C. botulinum*, detection and identification of *C. botulinum* was merely based on botulinum toxin detection in culture supernatants and pure cultures of bacterial isolates by the mouse lethality assay. Toxin typing was achieved with neutralization of toxic material with specific antibodies and detection of lethality in mice. Toxin type was the lowest level of strain characterization and thus accurate epidemiological analysis was not possible. The use of large numbers of mice required in toxin assays caused concern, leading to active search for other diagnostic tools. The molecular era starting in the 1990s revolutionized bacterial diagnostics and revealed virulence gene sequencing, PCR-based detection methods and later molecular typing methods. In the laboratory diagnostics of botulism and neurotoxigenic *Clostridia*, these methods have significantly diminished the use of experimental animals, shortened assay performance times, and allowed an accurate strain characterization, improving epidemiological understanding of botulism.

Detection methods

A number of molecular detection protocols based on conventional PCR (Fach et al., 1993, 1995; Franciosa et al., 1994; Szabo et al., 1994; Hielm et al., 1996; Sciacchitano and Hirshfield, 1996; Takeshi et al., 1996; Aranda et al., 1997; Kakinuma et al., 1997; Braconnier et al., 2001; Dahlenborg et al., 2001, 2003; Lindström et al., 2001; Craven et al., 2002; Gauthier et al., 2005), real-time PCR (Akbulut et al., 2004, 2005; Yoon et al., 2005; Artin et al., 2007; Fenicia et al., 2007b; Raphael and Andreadis, 2007), and Southern hybridization (Fach et al., 1993, 1995; Aranda et al., 1997; Braconnier et al., 2001) have been published for *C. botulinum*. This has enabled more effective research on the prevalence of the bacterium in different niches. A vast majority of PCR and Southern hybridization assays are based on simple or multiplex detection of the neurotoxin genes (*bot*). However, a promising real-time PCR-based screening method based on a fluorescently labelled probe targeted to the highly conserved *ntnh* gene, located immediately upstream of *bot*, across all neurotoxigenic clostridia was published recently (Raphael and Andreadis, 2007). In conventional PCR, simple or multiple amplification products can be detected using gel electrophoresis (Franciosa et al., 1994; Lindström et al., 2001) or by further hybridization onto a membrane coated with cDNA probes to the PCR products (Gauthier et al., 2005). Compared with culture methods, molecular detection techniques are sensitive, specific, and rapid to perform. Depending on the assay type and sample matrix, detection limit is 10 fg of pure DNA to 10^3 cells/g of sample (reviewed in Lindström and Korkeala, 2006). A disadvantage of molecular detection assays is that they do not detect biologically active neurotoxin or even activity of genes. Moreover, single-nucleotide polymorphism at primer or probe annealing sites may cause false negative results. This problem has become evident only recently as extensive sequencing efforts have revealed a number of novel subtypes of the neurotoxin genes (Franciosa *et al.*, 2006; Chen *et al.*, 2007; Hill *et al.*, 2007b).

As both the *bot*/B and *bot*/F genes of group I and II *C. botulinum* are similar, PCR and hybridization techniques based on the neurotoxin gene do not reveal the physiological group of *C. botulinum*. From epidemiological perspective it would be extremely important to distinguish between the physiological groups (Lindström and Korkeala, 2006). Conventionally this is based on detecting digestion of casein or other protein substrate by group I organisms, but this method is slow and sometimes ambiguous. Molecular approaches based on 23S *rrn* (Campbell *et al.*, 1993b) or a *Hin*dIII fragment-based DNA probe (McKinney *et al.*, 1993) specific for group II and group I *C. botulinum*, respectively, have been published. Also some molecular typing methods such as ribotyping and amplified fragment length polymorphism (AFLP) (see below) are also suitable for differentiation between *C. botulinum* groups I and II (Hielm *et al.*, 1999; Keto-Timonen *et al.*, 2005) but like the probe-based assays, they are laborious in routine use. Recently published rapid PCR assays based on the *fldB* gene associated with phenylalanine metabolism (Dahlsten *et al.*, 2008) and *flaA* gene encoding a flagellin (Paul *et al.*, 2007) are expected to be very useful approaches to discriminate between group I and II *C. botulinum*.

Genotyping

Pulsed-field gel electrophoresis (PFGE)

The most widely applied molecular epidemiological tool for *C. botulinum*, like many other food-borne pathogenic bacteria, is pulsed-field gel electrophoresis (PFGE). The method is based on digestion of chromosomal DNA with rarely cutting restriction enzymes and visualization of the macrorestriction patterns in an agarose gel. The restriction fragments are large and may be up to 1000 kb in size, hence the gels are run under a pulsating electric field to allow the fragments to migrate efficiently. PFGE is highly discriminative and reproducible, allowing powerful characterization to strain level. The most successful typing of group I *C. botulinum* has been reported with a combination of *Sac*II and *Xho*I (Nevas *et al.*, 2005b), and of group II with *Sma*I and *Xho*I (Hielm *et al.*, 1998a). Species identification or phylogenetic analysis is not possible with PFGE. The assay is also laborious to perform and, particularly with clostridia, may experience problems due to endonuclease activity and thus DNA degradation. This problem has been solved by using a formaldehyde fixation of cells (Hielm *et al.*, 1998a) or adding thiourea in the electrophoresis buffer (Leclair *et al.*, 2006), with the combination of both being the most effective option (Leclair *et al.*, 2006).

Ribotyping

Ribosomal genes (*rrn*) are widely used as phylogenetic markers of bacteria. Ribotyping (*rrn* restriction pattern analysis) is based on a genomic restriction pattern hybridized with labelled cDNA probes targeted to *Escherichia coli* 16S *rrn* genes (Grimont and Grimont, 1986). Being less discriminatory than PFGE in bacterial strain differentiation but sufficiently efficient to discriminate between bacterial species makes ribotyping suitable for phylogenetic analysis. Ribotyping is highly reproducible, but as with PFGE, degradation of DNA may hinder the typing of some strains (Hielm *et al.*, 1999).

Amplified rDNA restriction analysis (ARDRA)

Another *rrn*-gene based analysis applied for *C. botulinum* is ARDRA, in which the 16S and 23S *rrn* genes are amplified, digested with a restriction enzyme, and separated in an agarose gel (Pourshafie *et al.*, 2005). The method is simple and relatively rapid to perform, and provides a fair distinction between group I *C. botulinum* type A isolates from different outbreaks and *C. sporogenes*. The use of ARDRA has not been reported for other *C. botulinum* or clostridial strains and hence its usefulness for e.g. phylogenetic analysis is unclear.

Amplified fragment length polymorphism (AFLP)

AFLP is based on digestion of genomic DNA with two restriction enzymes, followed by ligation of restriction site-specific adapters, and amplification of a subset of fragments by PCR (Vos *et al.*, 1995). Successful typing of large numbers of isolates has been obtained by combining the rarely cutting *Eco*RI and frequently cutting *Hind*III restriction enzymes (Keto-Timonen *et al.*, 2005, 2006), or combining *Eco*RI to the frequently cutting *Mse*I (Hill *et al.*, 2007b). Also the use of a single restriction enzyme, *Hind*III, has been reported (Brett *et al.*, 2005). Unlike PFGE and ribotyping, AFLP is suited to both strain differentiation and phylogenetic analysis, and thus satisfactorily discriminates between group I and group II *C. botulinum* strains (Keto-Timonen *et al.*, 2005, 2006; Hill *et al.*, 2007b). Furthermore, as a PCR-based technique, AFLP is not disturbed by DNA degradation.

Random amplification of polymorphic DNA (RAPD)

RAPD is based on PCR with randomly annealing universal primers under low-stringency conditions (Williams *et al.*, 1990). It is fast and easy to perform, and may be even more discriminatory than PFGE (Hyytiä *et al.*, 1999). All strains are typeable with RAPD. However, due to the random annealing of unspecific primers and non-stringent conditions, the method lacks reproducibility and may occasionally cause misleading results in epidemiological studies.

Repetitive element sequence-based PCR (Rep-PCR)

Rep-PCR is based on amplification of genomic repetitive extragenic elements with single or multiple consensus primers (Versalovic *et al.*, 1991). Amplification products form a fingerprint and are visualized in gel electrophoresis. Differentiation to the strain level is limited, restricting the applicability of the method in epidemiological research of *C. botulinum* (Hyytiä *et al.*, 1999). Like RAPD, Rep-PCR is rapid to perform. Its advantage over RAPD is its higher reproducibility due to the use of specific primers.

Variable number tandem-repeat sequence analysis (VNTR)

A multiple-locus VNTR sequence-based typing analysis on group I *C. botulinum* enabled efficient strain discrimination between type A toxin producing strains, which in AFLP showed high similarity (Smith *et al.*, 2007). The repeat sequences were based on the ATCC 3502 strain genome sequence (Sanger Centre, Cambridge, UK; Sebaihia *et al.*, 2007). The method has not been validated with other than type A serotype strains.

DNA sequencing

Single-locus and multiple-locus sequencing are currently commonly applied typing tools with many bacteria, both of which having been applied for *C. botulinum*. The 16S *rrn* gene sequence has widely been used as a target for phylogenetic analyses, but it lacks discriminatory power to distinguish bacterial strains: the 16S *rrn* genes of *C. botulinum* strains belonging to the same physiological group are 99–100% similar, while only 90% identity is observed between the 16S *rrn* sequences of group I and II strains. Sequencing variable regions of the flagellin-encoding genes *flaA1* and *flaA2* in group I and II *C. botulinum* strains revealed efficient discrimination between groups I and II, and further between some group I but not group II strains (Paul *et al.*, 2007).

The neurotoxin gene sequencing has revealed a large nucleotide sequence variation allowing subtyping of the serotypes A, B (Hill *et al.*, 2007b), and E (Chen *et al.*, 2007). This subtyping is in accordance with immunological studies based on monoclonal antibodies (Smith *et al.*, 2005). Within group I *C. botulinum*, an AFLP analysis with *Eco*RI and *Mse*I was consistent with the neurotoxin subtypes (Hill *et al.*, 2007a,b).

Comparative genome sequencing has been applied in the modern bacterial typing. The novel sequencing platforms allow rapid production of genomic sequence data and assembly of closely related species or strains. So far the only published *C. botulinum* genome sequence is of group I type A toxin-producing strain ATCC 3502 (Hall A) (Sebaihia *et al.*, 2007), but genome sequencing projects on group I types A, B, Bf, and F, and group II types B and E strains are ongoing.

DNA microarrays

Genome-based DNA microarrays have introduced new aspects into epidemiological research on bacteria. Through specific hybridization of labelled nucleic acids onto probe spots representing every gene in a genome, DNA microarrays enable simultaneous analysis of thousands of genes and thus several genomes at a time. DNA microarrays have been applied in genetic typing of several food-borne pathogenic bacteria, including group I *C. botulinum* (Lindström *et al.*, 2007; Sebaihia *et al.*, 2007, Carter *et al.*, 2007). The method reveals plenty of information, but currently suffers from high cost and requirement for specialized expertise in statistics and data analysis. Moreover, the method is dependent on genomic sequence information and is thus limited to species or strains that are close relatives to the strain represented on the array.

Molecular epidemiology of groups I and II *Clostridium botulinum*

While *C. botulinum* in general is a ubiquitous environmental bacterium and forms resistant endospores which may survive in the environment for long periods of time, group I and II strains seem to have a different epidemiology (Fig. 5.2). The main reservoir of group I spores is soil from where spores are transmitted into the GI tract of animals. Cease of respiratory functions upon death of the animal creates anaerobic conditions, allowing for multiplication from spores. Rotting carcasses enable spread of spores to other animals, vegetation and the environment. As for group II, the reservoir of spores, type E in particular, is marine and freshwater sediments. Fish get contaminated through gills and GI tract, and the bacterium spreads to the aquatic environment and animals higher up in the food chain. Multiplication occurs probably in dead organic material at bottom sediments.

In the 1960–1980s, epidemiological research on *C. botulinum* was merely focused on the prevalence of the organism in various niches and outbreak investigations. Central diagnostic

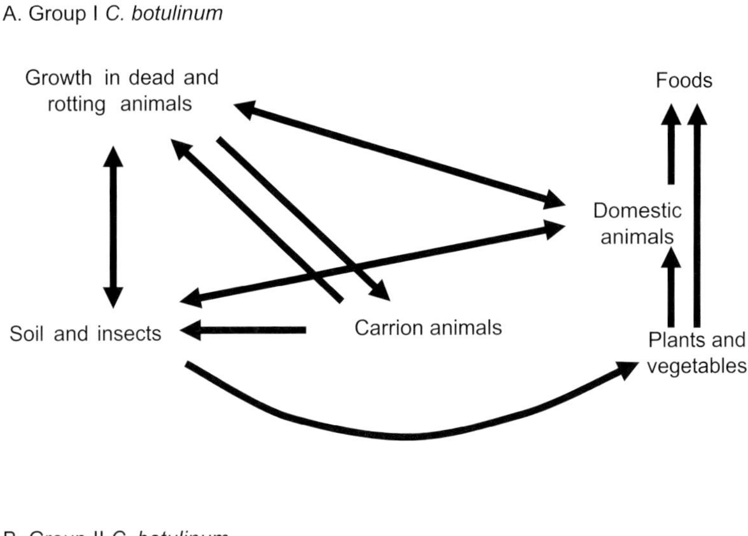

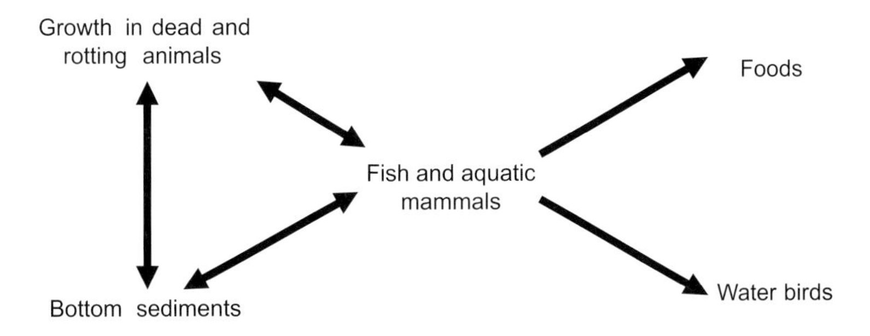

Figure 5.2 Epidemiology of group I and II *Clostridium botulinum*

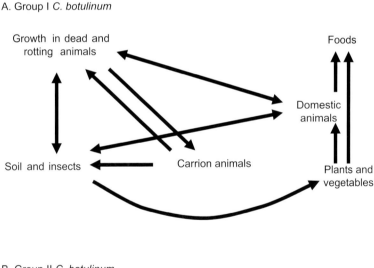

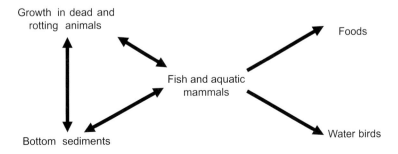

tools were culture methods combined with the mouse bioassay. Concerns over the high number of experimental animals required for screening studies and interference of environmental substances with the mouse assay (Szabo *et al.*, 1994) led to the implementation of rapid and sensitive PCR and Southern hybridization assays in screening studies (Tables 5.1–5.3). While the introduction of molecular techniques has not changed the estimated prevalence of *C. botulinum* in different niches, the possibility to investigate several multiple aliquots of a sample by PCR (Hielm *et al.*, 1996; Lindström and Korkeala, 2006) has allowed an accurate estimation of spore/cell counts of *C. botulinum* using the most probable number (MPN) technique (Tables 5.1–5.3). PCR-based surveys suggest that spore counts in the environment (up to 5×10^3 cells/kg) are generally some 10-fold higher than in unprocessed foods (up to 2×10^2 cells/kg), and 100-fold higher than in most processed foods (Tables 5.1–5.3). Development of genetic typing methods has provided a more profound insight into individual strains, and has thus increased epidemiological understanding of neurotoxigenic clostridia and facilitated outbreak investigations (Table 5.4).

Table 5.1 Prevalence of neurotoxigenic *Clostridia* in the environment as detected by molecular methods and further characterized by culture or molecular methods

Country	Sample material	Positive samples (%)	Mean spore count (spores/kg)[a]	Toxin type	Group or species of neurotoxigenic isolate[b]	Reference
Australia	Soil	94	400	B	NK	Szabo *et al.* (1994)
China	Soil	10–90	NE[c]	E	*C. butyricum*	Meng *et al.* (1999)
Finland	Fish farm sediment	68	2	E	*C. botulinum* II	Hielm *et al.*, (1998b)
Finland	Sea sediment	80	940	E	*C. botulinum* II	Hielm *et al.* (1996)
Finland	Sea and freshwater sediment	71	690	E	*C. botulinum* II	Hielm *et al.* (1998c)
Finland	Pig faeces	3	1400	B	*C. botulinum* I	Myllykoski *et al.* (2006)
Finland	Soil	31	50–1050	A, B	*C. botulinum* I	Nevas *et al.* (2006)
				E	*C. botulinum* II	Nevas *et al.* (2006)
Finland	Bees	10	110–510	B	*C. botulinum* I	Nevas *et al.* (2006)
				E	*C. botulinum* II	
				F	NK	
Finland	Pollen	7–15	60–690	B	*C. botulinum* I	Nevas *et al.* (2006)
				E	*C. botulinum* II	
France	Sediment	4	1–2	E	*C. botulinum* II	Fach *et al.* (2002)
Sweden	Pig faeces	62	99	B	NK	Dahlenborg *et al.* (2001)
Sweden	Cattle faeces	73	470	B	NK	Dahlenborg *et al.* (2003)
India	Soil	10	39	E	*C. botulinum* II	Dhaked *et al.* (2002)

[a]Reported by authors or extrapolated from reported data using the MPN technique for a single dilution (Thomas, 1942).
[b]All *C. botulinum* type A and E isolates are expected to be of group I and II, respectively. NK, the physiological group is not known.
[c] NE, not estimated.

Table 5.2 Prevalence of *Clostridium botulinum* in raw foods as detected by molecular methods and further characterized by culture or molecular methods

Country	Sample type	Positive samples (%)	Mean spore count (spores/kg)[a]	Toxin type	Group or species of neurotoxigenic isolate[b]	Reference
Denmark	Raw honey	28	NE[c]	A, B	*C. botulinum* I	Nevas *et al.* (2005a)
				E	*C. botulinum* II	Nevas *et al.* (2005a)
Finland	Different fish species	10–40	180	E	*C. botulinum* II	Hyytiä *et al.* (1998b)
Finland	River lamprey	2	100	E	*C. botulinum* II	Merivirta *et al.* (2006)
Finland	Fish skin and intestines	10	238	E	*C. botulinum* II	Hielm *et al.* (1998b)
Finland	Fish roe	8	58	E	*C. botulinum* II	Hyytiä *et al.* (1998b)
Finland	Raw honey	2–4	NE	B	*C. botulinum* I	Nevas *et al.* (2006)
France	Fish	17	4–12	A	*C. botulinum* I	Fach *et al.* (2002)
				E	*C. botulinum* II	Fach *et al.* (2002)
				B, F	NK	Fach *et al.* (2002)
France	Fish and shellfish	8	2–3	A	*C. botulinum* I	Carlin *et al.* (2004)
				B	NK	Carlin *et al.*, (2004)
France	Meat and poultry	8	2–4	A	*C. botulinum* I	Carlin *et al.* (2004)
				B	NK	Carlin *et al.* (2004)
Germany	Fish intestines gills skin	30	80	E	*C. botulinum* II	Hyytiä-Trees *et al.* (1999)
Norway	Raw honey	14	NE	B	*C. botulinum* I	Nevas *et al.* (2005a)
				F	NK	Nevas *et al.* (2005a)

[a]Reported by authors or extrapolated from reported data using the MPN technique for a single dilution (Thomas 1942).
[b] All type A and E isolates are expected to be of group I and II, respectively. NK, the physiological group is not known.
[c]NE, not estimated.

Prevalence of *Clostridium botulinum* group I and II strains

Environment

In general, *C. botulinum* is found ubiquitously in the environment in all parts of the world. Traditionally group I *C. botulinum* is considered to prevail in temperate regions of the world, while group II predominates in areas with a mild climate in the northern hemisphere. However, recent research with sensitive molecular tools has shown that the geographical distribution of these two physiological groups of *C. botulinum* is not unequivocal. For example, Nordic countries, known for their high environmental prevalence (70–80%) and spore counts (10^3 spores/kg) of group II *C. botulinum* type E (Johannsen, 1962, 1963; Cann *et al.*, 1965; Huss *et al.*, 1974; Huss, 1980; Ala-Huikku *et al.*, 1977; Hielm *et al.*, 1996, 1998b,c), have recently been shown with PCR to be widely contaminated also with group I type B spores (Nevas *et al.*, 2005a, 2006).

The prevalence group I *C. botulinum* types A and B varies geographically. Exceptionally high prevalences have been observed with conventional culture methods in soils, inland sediments, and former cattle grounds in the UK and Ireland (56–100%), Iceland (50–100%) (Hauschild and Dodds, 1993), Italy (86–100%) (Creti *et al.*, 1990), western USA (62%) (Smith, 1978),

Table 5.3 Prevalence of *Clostridium botulinum* in processed foods as detected by molecular methods and further characterized by culture or molecular methods

Country	Sample type and size	Positive samples (%)	Mean spore count (spores/kg)[a]	Toxin type	Group or species of neurotoxigenic isolate[b]	Reference
Argentina	Honey	75	25	A, B	*C. botulinum* I	Nevas *et al.*, 2002
Australia	Honey	29	36	A, B	*C. botulinum* I	Nevas *et al.*, 2002
Cuba	Honey	20	20	B	*C. botulinum* I	Nevas *et al.*, 2002
Denmark	Honey	24	NE[c]	B	*C. botulinum* I	Nevas *et al.*, 2005
Finland	Vacuum-packaged hot-smoked fish	7	35	E	*C. botulinum* II	Hyytiä *et al.*, 1998
Finland	Vacuum-packaged cold-smoked fish	3	160	E	*C. botulinum* II	Hyytiä *et al.*, 1998
Finland	Air-packaged hot-smoked fish	3	45	E	*C. botulinum* II	Hyytiä *et al.*, 1998
Finland	Vacuum-packaged vegetarian sausage	6	25	B	*C. botulinum* I	Lindström *et al.*, 2001
				E	*C. botulinum* II	
Finland	Canned deer meat	9	1	B	*C. botulinum* I	Lindström *et al.*, 2001
Finland	Honey	7	52	A, B	*C. botulinum* I	Nevas *et al.*, 2002
France	Honey	20	40	A, B	*C. botulinum* I	Nevas *et al.*, 2002
France	Aroma, sauce and gravy	2	0.3–0.6	ND[d]	ND	Carlin *et al.*, 2004
France	Thickening agents, starch	16	3–7	A	*C. botulinum* I	Carlin *et al.*, 2004
				B	ND	
France	Dehydrated dairy ingredients	12	2–5	ND	ND	Carlin *et al.*, 2004
Germany	Honey	2	0.8	A	*C. botulinum* I	Mäde *et al.*, 2000
Italy	Honey	50	20	A	*C. botulinum* I	Nevas *et al.*, 2002
Norway	Honey	9	NR[e]	B	*C. botulinum* I	Nevas *et al.*, 2005
				E	*C. botulinum* II	
Spain	Honey	25	60	A, B	*C. botulinum* I	Nevas *et al.*, 2002
Sweden	Honey	2	NR	E	*C. botulinum* II	Nevas *et al.*, 2005

[a]Reported by authors or extrapolated from reported data using the MPN technique for a single dilution (Thomas, 1942).
[b]All type A and E isolates are expected to be of group I and II, respectively.
[c]NE, not estimated.
[d]ND, not determined.
[e]NR, not reported.

Table 5.4 Human botulism outbreaks investigated with both molecular and conventional culture methods.

Country	Description of outbreak(s)	Sample(s)	Toxin type(s)	Organism	Genetic typing method and result	References
Canada	Food botulism, four outbreaks	Isolates from faecal and food samples, environ-mental isolates	E	*C. botulinum* II	PFGE, RAPD, and ribotyping, clonality of isolates representing each outbreak, clonality of temporally unrelated but geographically related environmental isolates and disease isolates	Leclair *et al.*,2006 Paul *et al.*, 2007
Finland	Infant botulism, one case	Intestinal content, vacuum cleaner dust	B	*C. botulinum* group I	PCR, detection of neurotoxin gene in patient and dust bag; PFGE, clonality of isolates from patient and dust bag	Nevas *et al.*, 2005c
Finland	Food botulism, one case	Gastric samples, suspected foods	E	*C. botulinum* group II	PCR, detection of neurotoxin gene from patient and fish eggs; PFGE, clonality of isolates from patient and fish eggs	Lindström *et al.*, 2004
Finland	Food botulism, one case	Gastric samples, suspected foods	B	*C. botulinum* group I	PCR, existence of non-disease-related strains in patient	Lindström *et al.*, 2004
Finland	Food botulism, two cases	Serum, suspected fish product	E	*C. botulinum* group II	PCR, detection of neurotoxin gene in patient and fish product; PFGE, clonality of fish isolate with a North American strain	Korkeala *et al.*, 1998
Germany	Wound botulism, six cases	Abscess	B	*C. botulinum*, group I	PCR, detection of neurotoxin gene; PFGE, clonality of isolates from three socially unrelated patients; 16S *rrn* sequencing, species identification of disease isolates	Kalka-Moll *et al.*, 2007; Kalka-Moll 2007
Italy	Food botulism, one case	Canned macrobiotic food	Ab	*C. botulinum* group I	PCR, identification of dual toxin production of the isolate	Franciosa *et al.*, 1997
Italy	Adult infectious botulism, two cases	Faecal samples	E	*C. butyricum*	PCR, identification of neurotoxigenic clostridia	Fenicia *et al.*, 1999
Japan	Infant botulism, one case	Faecal sample	B	*C. botulinum*	Nested-PCR, direct detection of neurotoxin gene from faeces without enrichment	Kakinuma *et al.*, 1997
Sweden	Wound botulism, one case	Pus	E	NR[a]	Real-time PCR, detection of neurotoxin gene	Artin *et al.*, 2007

Table 5.4 *continued*

Country	Description of outbreak(s)	Sample(s)	Toxin type(s)	Organism	Genetic typing method and result	References
UK	Infant botulism, one case	Intestinal and faecal samples, suspected foods, unopened foods of the same batch as the vehicle for botulism	A, B	*C. botulinum*	AFLP and PFGE, clonality of patient and suspected food isolates, identification of non-disease-related strains in foods, first case of infant botulism related to food other than honey	Brett *et al.*, 2005, Johnson *et al.*, 2005
UK	Wound botulism, 25 suspected cases	Tissue and abscess	A, B	*C. botulinum* group I [b]	Real-time PCR, detection of neurotoxin gene in 11 cases	Akbulut *et al.*, 2005

[a]NR, not reported.
[b]Physiological group assumed based on growth observed on CBI medium, which generally does not support the growth of group II *C. botulinum*.

China and Taiwan (32–47%) (Hauschild and Dodds, 1993), Brazil and Argentina (40–67%), Australia (100%), and Africa (86–100%) (Hauschild and Dodds, 1993). PCR studies are limited in number but have shown potential in revealing a high prevalence of group I type B strains in Finland (31%) (Nevas *et al.*, 2006). Group I type F strains are very rare in the environment; in fact, fewer than 20 strains have probably ever been isolated worldwide, with less than half of them from soil. Neurotoxigenic *C. butyricum* has been isolated in soils of China (Meng *et al.*, 1999), but they probably also exist in Italy where human botulism due to these strains have been reported (Fenicia *et al.*, 1999). Neurotoxigenic *C. barati* has not been reported in environmental samples.

Of group II strains, type E appears to be strongly associated with aquatic environments (reviewed in Hauschild and Dodds, 1993). Both molecular and conventional methods have revealed high prevalences in the Baltic Sea area (70–100%) and freshwaters in the Nordic countries (71–100%) (Hielm *et al.*, 19986, 1998c), Faroe Islands and Greenland (up to 100%) (Hauschild and Dodds 1993), Great Lakes District, Alaska, and Washington area in the USA (96–100%), Canada (up to 100%), Japan (25–100%), and Russia (52–100%) (Hauschild and Dodds, 1993).

Group II type B strains have been isolated in many countries, with Poland having particularly high incidence of food-borne botulism due to these strains (Hauschild and Dodds, 1993). A common vehicle is meat, which refers to other than aquatic source. United Kingdom (35–43%) and Ireland (18%) have reported relatively high prevalence of type B strains in coastal sediments, with at least part of these strains assumed to represent group II and to be of terrestrial origin (Cann *et al.*, 1968). Also a Japanese river area soil was contaminated with group II type B organisms (Yamakawa and Nakamura, 1992). In fact, Huss (1980) suggested that farming and forestation would expel type E and allure spread of type B *C. botulinum* in a dried seabed region. Although some internationally well-known group II type B and F strains isolated by Dr M. W. Eklund originate from Pacific Coast sea sediments (Leclair *et al.*, 2006), group II type B and F are probably less distinctively related to aquatic systems than type E. Their main reservoir has remained unclear, probably largely due to insufficient typing tools and efforts; a vast majority of prevalence studies on *C. botulinum* conducted to date do not report the physiological group of the organism. With type A and E toxic strains physiological grouping seems straightforward since all so far isolated wild-type strains of type A most probably represent group I and type

E strains represent group II. However, B and F-toxic strains exist in both groups, and it is not possible to assign these strains to group I or II based on results obtained in the conventional mouse bioassay, colony morphology, or growth at specified incubation temperatures routinely used in the diagnostics of *C. botulinum* (Lindström and Korkeala, 2006; Myllykoski *et al.*, 2006).

Animal GI tract

Apart from fish (see Raw foods, below), a limited number of reports on the prevalence of group I and II *C. botulinum* in animals' GI tract or faeces has been reported. Myllykoski *et al.* (2006) found that 3% of Finnish pigs carry group I *C. botulinum* type B, while prevalence in Swedish pigs was as high as 62% (Dahlenborg *et al.*, 2001). The difference is probably true and not due to methodological bias, since Myllykoski *et al.* (2006) showed that their PCR-based method was as sensitive as or even more sensitive than the assay used by Dahlenborg *et al.* (2001). In the Swedish study the authors did not isolate the type B organisms, but they state these to be of group II due to a relatively fast PCR-positive result after incubation at 30°C (Dahlenborg *et al.*, 2001). However, in the absence of data with true isolates and more specific tools, the possibility of these B-positive results due to group I cannot be excluded as group I type B spores are highly prevalent in the Nordic countries (Nevas *et al.*, 2005a, 2006). There are no reports on confirmed isolation of group II type B organisms in the Nordic countries. Johannsen (1963) and Huss (1980) found *C. botulinum* type B in sediment samples, but the physiological group of these organisms remained unclear.

In the authors' laboratory at the University of Helsinki, Finland, several unpublished studies on the occurrence of *C. botulinum* in animal GI tract have been conducted using PCR (Lindström *et al.*, 2001). The *botB* gene was detected in 2–18% of the faecal specimen of cattle and horses. The prevalence in both animals was generally higher during pasture than when fed inside (unpublished). Two percent of the Finnish mole population were positive for *botB* (unpublished). All type B isolates recovered from these studies were confirmed to represent group I with culture method or PCR (Dahlsten *et al.*, 2008).

Raw foods

As a consequence of the high environmental contamination, raw food and food raw materials will frequently be contaminated with *C. botulinum*. The most widely investigated raw foods include fish and other seafood. The prevalence of group II *C. botulinum* in raw fish is reported to be from 1% to 70% worldwide. A high prevalence in raw fish has been reported in Denmark (65%) (Huss *et al.*, 1974), Sweden (46%) (Johannsen, 1963), Finland (10–40%) (Hyytiä *et al.*, 1998b), Germany (30%) (Hyytiä *et al.*, 1999), Russia (35%) (Rouhbakhsh-Khaleghdoust, 1975), and in the USA (43–53%) (Eklund and Poysky, 1967; Baker *et al.*, 1990). PCR detection combined with most probable number based enumeration (MPN) yielded the prevalence of group II *C. botulinum* type E to be 10–40% in different raw fish species in Finland (Hielm *et al.*, 1996; Hyytiä *et al.*, 1998b). The highest prevalence of 40% was observed in Baltic herring, which is used to feed farmed fish cultivated for human consumption in the Nordic countries. Consequently, 15% of farmed rainbow trout carried type E spores (Hyytiä *et al.*, 1998b). A further study revealed that cultivation of fish in freshwater ponds and marine net cages possesses a significantly high risk for type E botulism (Hielm *et al.*, 1998b). Regular cleaning and removal of dead material from pool bottoms in self-cleaning ponds were shown to reduce the risk of fish contamination with type E spores (Hielm *et al.*, 1998b).

Apart from seafoods, only few screening studies on the contamination of food raw materials or unprocessed foodstuffs with *C. botulinum* have been reported, even less using molecular detection methods. Raw meat may occasionally be contaminated with type A and B spores (Roberts and Smart, 1976), again the physiological group of type B has remained unknown. Also one study reports 36% of raw meat to be rendered to carry type E (Klarmann, 1989). It is not clear whether meat becomes contaminated at slaughter or at later stages of processing. The fact that meat animals carry spores in their GI tract (Dahlenborg *et al.*, 2001, 2003; Myllykoski *et al.*, 2006) supports contamination of meat at an early stage.

Between 14% and 28% of raw honey samples taken from honeycombs in beehives have been shown to contain *C. botulinum* in Finland, Denmark and Norway (Nevas *et al.*, 2005a, 2006). While the majority of the positive samples carried *botB*, PCR revealed also amplification fragments for type A, E and F neurotoxin genes. A total of 226 isolates were recovered and were shown to represent type B, with the exception of one type A isolate (Nevas *et al.*, 2005a, 2006). All isolates have further been confirmed to represent group I with culture method, PCR (Dahlsten *et al.*, 2008), or AFLP (Keto-Timonen *et al.*, 2005).

Processed foods

The fact that processed foods contain *C. botulinum* spores indicates either a post-process contamination or, more likely, insufficiency of heat processes employed in the food industries to eliminate *C. botulinum* (Hyytiä *et al.*, 1999; Hyytiä-Trees *et al.*, 2000; Lindström *et al.*, 2003, 2006a). In Finland, time-temperature combinations of heat processes employed in the fish smoking industry were shown to commonly correspond to a less than a 1D process; thus eliminating less than one log unit of group II *C. botulinum* type E spores (Lindström *et al.*, 2003). Consequently, as many as 7% of vacuum-packaged hot-smoked and 3% of packaged cold-smoked fish products sold in Finland in the 1990s were shown to contain type E spores as investigated with PCR and further isolation of *C. botulinum* type E strains (Hyytiä *et al.*, 1999). In France, 1.6–16% of ingredients used to produce ready-to-eat foods, such as aroma, sauce and gravy, thickening agents, and dehydrated dairy products, were shown to be contaminated with *C. botulinum* types A and B by PCR (Carlin *et al.*, 2004).

Only older reports, merely based on conventional methods, are available for the USA. These reports show 1–5% of processed and packaged foods to contain *C. botulinum* spores (Pace *et al.*, 1967; Taclindo *et al.*, 1967; Insalata *et al.*, 1969; Hayes *et al.*, 1970; Abrahamsson and Riemann, 1971).

Genomic analysis of groups I and II *Clostridium botulinum*

The ecology and epidemiology of group I and II *C. botulinum* are apparently different (Fig. 5.2). Where the prevalence and spore counts of group I spores in soils in many places fall below 40% and 100/kg, respectively, strikingly high prevalence of 70–100% and spore counts of $>10^3$/kg of group II spores, type E in particular, have frequently been observed in aquatic environments. Furthermore, the genetic diversity within group I is generally smaller than within group II (Hyytiä *et al.*, 1998a, 1999; Hielm, *et al.*, 1998a,b; Keto-Timonen *et al.*, 2005; Nevas *et al.*, 2005b, Hill *et al.*, 2007a,b). These phenomena may be explained by the different reservoirs and niches for multiplication of group I and II. Aquatic environment probably provides conditions for more optimal multiplication of group II organisms than dry soil for group I. Group II organisms may thus undergo evolution more efficiently than group I.

Group I

The first report on fingerprinting group I *C. botulinum* strains with PFGE by Lin and Johnson (1995) revealed the relatively large genome size of 4 Mb for this species and served as the pioneering insight into the genome of group I *C. botulinum*. The PFGE protocol was further optimized by Nevas *et al.* (2005b) for genotyping of 55 group I strains. They showed that, despite satisfactory discrimination between type A strains, type B strains formed clusters of identical isolates. Two largest type B clusters were later identified to represent the majority of group I strains in Finland in a survey on *C. botulinum* in honey, honey production environment, and soil (Nevas *et al.*, 2006). Identical strains were also isolated in Norwegian and Danish honey (Nevas *et al.*, 2005a), indicating that the two main group I type B populations have been widely spread throughout the Nordic countries.

A more detailed analysis of the two Nordic group I type B clusters with DNA microarrays based on the genome of the group I type A strain ATCC 3502 (Sebaihia *et al.*, 2007) confirmed that strains within each cluster were identical (Lindström *et al.*, 2007), supporting the theory of a stable genome of group I *C. botulinum* (Sebaihia *et al.*, 2007). The two clusters differed from each other by 145 genes present in the ATCC 3502 strain. Genetic differences between the two clusters were confirmed by PCR and laboratory tests, and revealed different resistance to arsenic and cadmium, existing in high environmental concentrations in certain areas of the Nordic countries (Lindström *et al.*, 2007).

Large AFLP typing studies on clostridia support the physiological grouping (I-IV) of *C. botulinum* strains and the narrow biodiversity among group I strains as opposed to group II (Keto-Timonen *et al.*, 2005, 2006; Hill *et al.*, 2007a,b). Within group I, more variation among types A and B were observed than among type F strains which represent a minority among group I isolates (Keto-Timonen *et al.*, 2005; Hill *et al.*, 2007a,b). As for human botulism worldwide, type B neurotoxin causes approximately half of all cases (Hatheway, 1995), with the majority of cases being probably due to group I. In accordance, type B strains seem to possess a larger biodiversity than types A and F within group I worldwide, with several individual or clusters of type B or bivalent B strains being genetically distinct (Keto-Timonen *et al.*, 2005, 2006; Hill *et al.*, 2007a,b). A particularly diverse population of group I type B or bivalent B strains has recently been identified in California (Hill *et al.*, 2007a), while in the Nordic countries the population is relatively narrow and is dominated by two closely related clusters (Nevas *et al.*, 2006; Lindström *et al.*, 2007).

An interesting feature in the large AFLP-based typing studies is clustering of *C. botulinum* strains according to the neurotoxin subtype (Hill *et al.*, 2007a,b). Further analysis of the 40 DNA fragments included in the AFLP analysis suggested that the neurotoxin gene cluster was not involved in the fingerprints (Hill, 2008). Therefore it seems that strains representing different toxin subtypes are distinct in their genetic background. Moreover, the toxin subtypes seem to have evolved as a result of random mutagenesis affecting the entire bacterial genomes rather than as separate evolutionary units.

Insects seem to play an important role in the epidemiology of group I *C. botulinum*. The genome of the ATCC 3502 strain was shown to carry five genes putatively encoding chitinases responsible for degrading chitin, the major component of the insect exoskeleton (Sebaihia *et al.*, 2007). These five genes were shown to be present in all group I *C. botulinum* strains to date investigated with the ATCC 3502 microarrays (Lindström *et al.*, 2007; Sebaihia *et al.*, 2007), indicating chitinases are essential for group I *C. botulinum*. The central role of insects is also supported by the fact that group I *C. botulinum* is frequently found in beehives, bees, and honey (Nakano *et al.*, 1994; Nevas *et al.*, 2006).

Group II

As with group I, the genome size of group II *C. botulinum* was revealed with PFGE analysis to be approximately 4 Mb (Hielm *et al.*, 1998a). However, the genetic biodiversity of group II *C. botulinum*, type E in particular, was shown to be large (Hielm *et al.*, 1998a,b; Hyytiä *et al.*, 1998a; Leclair *et al.*, 2006) as opposed to group I. In a Finnish survey on type E contamination in marine and freshwater fish farms, type E strains isolated even from a narrow geographical location represented a large number of genotypes as investigated with PFGE and RAPD (Hielm *et al.*, 1998c; Hyytiä *et al.*, 1998a). The fact that 30% of the Finnish isolates carried one or two plasmids further increases the high biodiversity (Hielm *et al.*, 1998c). On the other hand, isolation of identical type E genotypes from various aquatic locations (Hielm *et al.*, 1998c) supports the theory of group II strains travelling long distances along with bottom water flows (Huss, 1980).

Unlike within group I, AFLP, PFGE and RAPD studies on group II *C. botulinum* show consistently three major clusters (Keto-Timonen *et al.*, 2005, 2006; Leclair *et al.*, 2006; Hill *et al.*, 2007b). Roughly, two of these clusters consist of type E isolates alone, and the third cluster contains group II type B and F strains. Comparison with other clostridial species in a cluster dendrogram based on AFLP typing with a species similarity level of 45% (Keto-Tiomonen *et al.*, 2006) suggests that the three group II clusters represent genetically different lineages and could even be considered as distinct species. A 16S *rrn* sequence analysis of group II strains supports the heterogeneity of group II *C. botulinum* (Hill *et al.*, 2007b).

The fact that group II type B and F strains cluster together and are distinct from type E may reflect differences in ecology and epidemiology. As indicated above, type E is strongly associated to water, while such a strong association cannot be established for the other group II strains. Further investigations with specific molecular typing tools are required to clarify the epidemiology of the group II *C. botulinum* type B and F lineage.

Contamination analyses

Molecular typing has been employed in industrial contamination analyses to show that foods may be contaminated with pathogens or spoilage organisms not only through raw materials, but also through processing and packaging equipment and production environment (Björkroth and Korkeala, 1997; Autio *et al.*, 1999; Lundén *et al.*, 2002). As has been discussed above, processed and packaged foods may carry group I and II *C. botulinum* and the general understanding supports raw materials to be the main source of spores. However, the possibility of post-process contamination has not been discussed and no contamination analyses on *C. botulinum* in food industries have been conducted. Such studies are needed to complement the current understanding of the epidemiology of food-borne botulism. The only reported contamination analysis on *C. botulinum* was conducted in honey production environment and showed that group I *C. botulinum* is a frequent contaminant in the entire honey production chain (Nevas *et al.*, 2006). Certain hygienic practices, including hand wash, change of footwear when entering production facilities, and use of new or heated wax frames in beehives, play a key role in preventing honey contamination with *C. botulinum* spores (Nevas *et al.*, 2006).

Molecular investigation of human botulism outbreaks

Long food shelf lives, increased travelling, and rapid adaptation of pathogens to new niches have contributed to infectious diseases and food epidemics becoming more widespread and difficult to solve. Molecular detection and typing have become an essential part of disease investigation (Miettinen *et al.*, 1999) and contamination analysis (Lundén *et al.*, 2002) making rapid and targeted diagnosis or problem-solving possible. As for botulism, outbreaks are typically solved by detection of botulinum neurotoxin in the patient and suspected source. Isolation of the neurotoxigenic organism and molecular typing of disease isolates is routinely neglected. However, with the current knowledge on the existence of the wide range of botulinum neurotoxigenic clostridial species causing botulism to humans through a wealth of sources, ignorance to isolate and thoroughly characterize the disease isolates is an imprudent disservice for the future clinicians, epidemiologists, and scientists working on neurotoxigenic clostridia. The few outbreaks with molecular epidemiological investigation are summarized in Table 5.4.

Food-borne botulism

Being an intoxication rather than an infection, food-borne botulism is rarely investigated using molecular detection and typing. The routine methodology involves toxin testing in the patient and in suspected food samples. The mouse bioassay is used to check disease isolates for neurotoxin production and the neurotoxin type, but only very rarely for species and genotype.

Although not able to replace the mouse bioassay, PCR-based detection methods have reduced the number of experimental animals required in outbreak investigations by allowing efficient screening of large number of samples. Ideally PCR is also used to determine the toxinotype of disease isolates shown to produce biologically active neurotoxin with the mouse test.

Three type E botulism outbreaks due to *C. botulinum* group II have been diagnosed in Finland in 1997, 1999 and 2006 (Korkeala *et al.*, 1998; Lindström *et al.*, 2004, 2006). In two cases PCR was employed in confirming toxin type in parallel with the mouse neutralization test, and correctly identified the outbreaks to be of type E (Korkeala *et al.*, 1998; Lindström *et al.*, 2004). Cultures on egg yolk-based medium showed lipase reaction and thus allowed distinction from neurotoxigenic *C. butyricum*. In the 1999 case, also a non-disease-related type B organism was detected with PCR and the isolate was further typed as group I *C. botulinum* (Lindström *et al.*, 2004). The 1997 case affected two German people who consumed vacuum-packaged fish manufactured in Finland. The fish was originally caught in Canada; indeed the related *C. botulinum* isolate resembled a North American isolate when genotyped with PFGE (Korkeala *et al.*, 1998).

An interesting molecular epidemiological study was conducted in the Canadian arctic where botulism due to consumption of aged marine mammals and fish is relatively common among the Inuit communities (Hauschild and Gauvreau, 1985; Austin *et al.*, 1999; Leclair *et al.*, 2006). PFGE typing of group II *C. botulinum* type E isolates from four unrelated botulism outbreaks occurring between 1995 and 2000 revealed distinct genotypes to characterize each outbreak and confirmed a link between patients and the suspected food source in at least three outbreaks. Interestingly, environmental type E strains isolated 5–7 years after the outbreaks in the same areas represented the same PFGE genotypes as the outbreak isolates, with an association being established between the isolation site of the environmental strain and the outbreak involving a similar genotype. Furthermore, similar PFGE genotypes were recovered from two botulism outbreaks occurring in subsequent years in the same geographical region (Leclair *et al.*, 2006). These findings and the high number of botulism outbreaks in the Canadian arctic suggest that this region has its own endemic *C. botulinum* population, which, in combination with insufficient processing and chilling procedures, causes a severe public health problem.

Infant botulism

The only case of infant botulism in Finland was investigated with PCR, and *C. botulinum* type B was observed in the infant's intestine (Nevas *et al.*, 2005c). Botulinum neurotoxin was not detected in the patient's serum, suggesting PCR on intestinal content combined with the clinical outcome to be a highly sensitive diagnostic tool for infant botulism. In a Japanese infant botulism case, confirmed toxin-positive by the mouse bioassay, nested-PCR enabled even a direct detection of type B neurotoxin gene in the patient's faeces without enrichment (Kakinuma *et al.*, 1997). However, the sensitivity of the nested-PCR method was not determined.

In the Finnish infant botulism case, PCR also showed a type B-positive result for one of 200 vacuum-cleaner dust bag samples and facilitated the identification and isolation of a type B isolate. The isolate was shown to represent group I by culture methods. The patient and dust isolates were shown to be linked by PFGE analysis (Nevas *et al.*, 2005c). Further PFGE and DNA microarray studies on group I *C. botulinum* type B strains in Finland have shown the infant botulism case isolates to represent the most common genetic lineage in the Nordic countries (Nevas *et al.*, 2005a, 2006; Lindström *et al.*, 2007). This shows that the high environmental occurrence of *C. botulinum* is a true public health threat, not only through foods.

A profound molecular epidemiological investigation was also carried out on an infant botulism case in the UK (Brett *et al.*, 2005; Johnson *et al.*, 2005). The primary diagnosis was based on clinical outcome and toxin detection in the patient's rectal washout and faecal specimens (Brett *et al.*, 2005). *C. botulinum* type B isolates were recovered from these specimens, and when typed with AFLP (Brett *et al.*, 2005) and PFGE (Johnson *et al.*, 2005), they showed several distinct genotypes. A subsequent analysis of foods recovered from the patient's home revealed an opened package of infant milk formula to contain two different genotypes of *C. botulinum* type B organisms with banding patterns identical to the rectal and faecal isolates, and two other genotypes probably unrelated to the case. Furthermore, a type A isolate with a unique genotype was recovered from a rice pudding from the baby's home, and an unopened milk formula from the same batch as the opened formula was shown to contain type B organisms. Although these isolates were of a unique genotype, the entire batch of 122,388 packages of the milk formula was recalled by the manufacturer. This case clearly demonstrates the necessity of a thorough molecular epidemiological investigation of all *C. botulinum* isolates recovered from patient and potential sources. Multiple genetically and physiologically distinct strains may be involved in a botulism outbreak and complicate the disease as well as its investigation. Non-disease-related isolates seem to frequently be involved in botulism outbreaks (Lindström *et al.*, 2004; Brett *et al.*, 2005).

Wound botulism

The emerging nature of wound botulism among intravenous drug abusers has raised a need for efficient testing. A recent report on a cluster of wound botulism cases in Germany introduces an example of an elegant outbreak investigation (Kalka-Moll *et al.*, 2007). Of six clinical cases, four were confirmed as botulism with laboratory diagnostics. Serum samples of two patients were positive as confirmed with polyvalent antiserum. To avoid redundant use of mouse tests, toxin typing was done with PCR from anaerobic cultures obtained from abscess specimens of three patients, and yielded type B. The culture isolates were confirmed to be *C. botulinum* by 16S *rrn* sequencing, Gram staining, colony morphology, and biochemical test series. The physiological group of the *C. botulinum* isolates was expectedly group I (Kalka-Moll, personal communication, 2007). PFGE typing of *C. botulinum* isolated from three patients revealed identical genotypes, indicating a common infectious source. All the patients were socially non-related, thus it was expected that the source of infection was contaminated heroin (Kalka-Moll *et al.*, 2007). This study shows that molecular typing of disease isolates may also have forensic value.

A real-time PCR assay for botulinum neurotoxins A, B, and E, and tetanus toxin proved to be a valuable tool in investigating suspected wound botulism cases in the UK (Akbulut *et al.*, 2005). Of 25 suspected cases, 10 were confirmed positive by the mouse bioassay. Nine of the bioassay-positive cases were positive in the PCR analysis, and two other cases were diagnosed based on a positive PCR result alone, suggesting the real-time PCR assay to be more sensitive than the mouse test. Positive PCR results were mainly obtained from enriched tissue or pus samples; however, in two cases direct PCR from a tissue sample yielded a positive result. On the other hand, PCR inhibition yielding no result was observed in five tissue samples when attempted for direct PCR without enrichment (Akbulut *et al.*, 2005). Hence the importance of proper sample pre-treatment procedures and the use of internal amplification controls in PCR assays needs to be emphasized.

Real-time PCR was applied in investigation of a wound botulism case in Sweden, revealing the first type E-related wound botulism case (Artin *et al.*, 2007). The PCR assay was optimized and involved an internal amplification control. The authors stated that the case was due to *C. botulinum* type E; however, they do not report the laboratory procedures for species identification (Artin *et al.*, 2007). Since group II *C. botulinum* type E is a psychrotrophic organism and, in fact, has never been related to infectious botulism before, it cannot be excluded that the causative agent was actually the mesophilic type E neurotoxigenic *C. butyricum*, frequently related to human infectious botulism (Aureli *et al.*, 1986; McCroskey *et al.*, 1986; Fenicia *et al.*, 1999, 2007a). This case demonstrates the complex taxonomy and epidemiology of botulinum neurotoxigenic clostridia, and underlines the necessity of a comprehensive analysis of the disease vehicle and the causative organism.

References

Abrahamsson, K., and Riemann, H. (1971). Prevalence of *Clostridium botulinum* in semipreserved meat products. Appl. Microbiol. *21*, 543–544.

Aguirre-Balsalobre, F., Gonzalez-Herrera, M., Mengual-Verdu, E., Garcia-Conca, V., and Hueso-Abancens, J.R. (2007). Simultaneous botulism in two brothers addicted to cocaine. Arch. Soc. Esp. Oftalmol. *82*, 583–6.

Akbulut, D., Grant, K.A., and McLauchlin, J. (2004). Development and application of real-time PCR assays to detect fragments of the *Clostridium botulinum* types A, B, and E neurotoxin genes for investigation of human food-borne and infant botulism. Foodborne Path. Dis. *1*, 247–257.

Akbulut, D., Grant, K.A., and McLauchlin, J. (2005). Improvement in laboratory diagnosis of wound botulism and tetanus among injecting illicit-drug users by use of real-time PCR assays for neurotoxin gene fragments. J. Clin. Microbiol. *43*, 4342–4348.

Ala-Huikku, K., Nurmi, E., Pajulahti, H., and Raevuori, M. (1977). The occurrence of *Clostridium botulinum* type E in Finnish trout farms and the prevention of toxin formation in fresh-salted vacuum-packed trout fillets. Nord. Vet. Med. *29*, 386–391.

Anniballi, F., Fenicia, L., Franciosa, G., and Aureli, P. (2002). Influence of pH and temperature on the growth of and toxin production by neurotoxigenic strains of *Clostridium butyricum* type E. J. Food Prot. *65*, 1267–1270.

Anonymous (1976). Botulism and sudden infant death. Lancet 2(8000), 1411–1412.

Aranda, E., Rodríguez, M.A., Asensio, M.A., and Córdoba, J.J. (1997). Detection of *Clostridium botulinum* types A, B, E, and F in foods by PCR and DNA probe. Lett. Appl. Microbiol. *25*, 186–190.

Arnon, S.S. (1986). Infant botulism: anticipating the second decade. J. Infect. Dis. *154*, 201–206.

Arnon, S.S. (1989). Infant botulism. In Anaerobic Infections in Humans, S.M. Finegold, and W.L. George, ed. (Toronto, Canada: Academic Press), p. 601.

Arnon, S.S. (1992). Infant botulism, In Textbook of Pediatric Infectious Diseases, R.D. Feigen, and J.D. Gerry, ed. (Philadelphia, USA: Saunders Press) pp. 1095–1102.

Arnon, S.S., and Chin, J. (1979). The clinical spectrum of infant botulism. Rev. Infect. Dis. *1*, 614–624.

Arnon, S.S., Midura, T.F., Clay, S.A., Wood, R.M., and Chin, J. (1977). Infant botulism: Epidemiological, clinical, and laboratory aspects. JAMA *237*, 1946.

Arnon, S.S., Midura, T.F., Damus, K., Wood, R.M., and Chin, J. (1978). Intestinal infection and toxin production by *Clostridium botulinum* as one cause of sudden infant death syndrome. Lancet *1*(8077), 1273–1277.

Arnon, S.S., Schechter, R., Maslanka, S.E., Jewell, N.P., and Hatheway, C.L. (2006). Human botulism immune globulin for the treatment of infant botulism. N. Engl. J. Med. *354*, 462–741.

Artin, I., Björkman, P., Cronqvist, J., Rådström, P., and Holst, E. (2007). The first case of type E wound botulism was diagnosed using real-time PCR. J. Clin. Microbiol. *45*, 3589–94.

Athwal, B.S., Gale, A.N., Brett, M.M., and Youl, B.D. (2000). Wound botulism in UK. Lancet *356*, 2011–2012.

Athwal, B.S., Gale, A.N., Brett, M.M., and Youl, B.D. (2001). Wound botulism in the UK. Lancet *357*, 234.

Aureli, P., Fenicia, L., Pasolini, B., Gianfranceschi, M., McCroskey, L.M., and Hatheway, C.L. (1986). Two cases of type E infant botulism caused by neurotoxigenic *Clostridium butyricum* in Italy. J. Infect. Dis. *154*, 201–206.

Aureli, P., Franciosa, G., and Fenicia, L. (2002). Infant botulism and honey in Europe: a commentary. Pediatr. Infect. Dis. J. *21*, 866–868.

Austin, J., Blanchfield, B., Ashton, E., Lorange, M., Proulx, J.F., Trinidad, A., and Winther, W. (1999). Botulism in Canada. Summary for 1997. Can. Commun. Dis. Rep. *25*, 121–122.

Autio, T., Hielm, S., Miettinen, M., Sjöberg, A.-M., Aarnisalo, K., Björkroth, J., Mattila-Sandholm, T., and Korkeala, H. (1999). Sources of *Listeria monocytogenes* contamination in a cold-smoked rainbow trout processing plant detected by pulsed-field gel electrophoresis typing. Appl. Environ. Microbiol. *65*, 150–155.

Bakheit, A.M.O., Ward, C.D., and McLellan, C.L. (1997). Generalized botulism-like syndrome after intramuscular injections of botulinum toxin type A: a report of two cases. J. Neurol. Neurosur. Ps. *62*, 198.

Baker, D.A., Genigeorgis, C., and Garcia, G. (1990). Prevalence of *Clostridium botulinum* in seafood and significance of multiple incubation temperatures for determination of its presence and type in fresh retail fish. J. Food Protect. *53*, 668–673.

Barash, J.R., and Arnon, S.S. (2004). Dual toxin-producing strain of *Clostridium botulinum* type Bf isolated from a California patient with infant botulism. J. Clin. Microbiol. *42*, 1713–1715.

Barash, J.R., Tand, T.W., and Arnon, S.S. (2005). First case of infant botulism caused by *Clostridium baratii* type F in California. J. Clin. Microbiol. *43*, 4280–4282.

Bhandari, M., Campbell, K.D., Collins, M.D., and East, A.K. (1997). Molecular characterization of the clusters of genes encoding the botulinum neurotoxin complex in *Clostridium botulinum* (*Clostridium argentinense*) type G and nonproteolytic *Clostridium botulinum* type B. Curr. Microbiol. *35*, 207–214.

Björkroth, K.J., and Korkeala, H.J. (1997). Use of rRNA gene restriction patterns to evaluate lactic acid bacterium contamination of vacuum-packaged sliced cooked whole-meat product in a meat processing plant. Appl. Environ. Microbiol. *63*, 448–453.

Bossi, P., Tegnell, A., Baka, A., Werner, A., van Loock, F., Hendriks, J., Maidhof, H., and Gouvras, G. (2004). Bichat guidelines for the clinical management of botulism and bioterrorism-related botulism. Eurosurveill. Monthly *9*, 31–32.

Boyer, A., Girault, C., Bauer, F., Korach, J.-M., Salomon, J., Moirot, E., Leroy, J., and Bonmarchand, G. (2001). Two cases of food-borne botulism type E and review of epidemiology in France. Eur. J. Clin. Microbiol. Infect. Dis. *20*, 192–195.

Braconnier, A., Broussolle, V., Perelle, S., Fach, P., Nguyen-The, C., and Carlin, F. (2001). Screening for *Clostridium botulinum* type A, B, and E in cooked chilled foods containing vegetables and raw material using polymerase chain reaction and molecular probes. J. Food Protect. *64*, 201–207.

Brett, M.M., McLauchlin, J., Harris, A., O'Brien, S., Black, N., Forsyth, R.J., Roberts, D., and Bolton, F.J. (2005). A case of infant botulism with a possible link to infant formula milk powder: evidence for the presence of more than one strain of *Clostridium botulinum* in clinical specimens and food. J. Med. Microbiol. *54*, 769–776.

Brin, M., and Irvine, R. (2007). 2007 update: new products and novel emerging therapeutic targets. In In Proceedings of 44th Interagency Botulism Research Coordinating Committee Meeting, Monterey, California, USA, p. 35.

Campbell, K.D., Collins, M.D., and East, A.K. (1993a). Nucleotide sequence of the gene coding for *Clostridium botulinum* (*Clostridium argentinense*) type G neurotoxin: genealogical comparison with other clostridial neurotoxins. Biochim. Biophys. Acta *1216*, 487–491.

Campbell, K.D., East, A.K., Thompson, D.E., and Collins, M.D. (1993b). Studies on the large subunit ribosomal RNA genes and intergenic spacer regions of non-proteolytic *Clostridium botulinum* types B, E and F. Res. Microbiol. *144*, 171–180.

Cann, D.C., Wilson, B.B., Hobbs, G., Shewan, J.M., and Johannsen, A. (1965). The incidence of *Clostridium botulinum* type E in fish and bottom deposits in the North Sea and off the coast of Scandinavia. J. Appl. Bacteriol. *28*, 426–430.

Cann, D.C., Wilson, B.B., and Hobbs, G. (1968). Incidence of *Clostridium botulinum* in bottom deposits in British coastal waters. J. Appl. Bact. *31*, 511–514.

Carlin, F., Broussolle, V., Perelle, S., Litman, S., and Fach, P. (2004). Prevalence of *Clostridium botulinum* in food raw materials used in REPFEDs manufactured in France. Int. J. Food Microbiol. *91*, 141–145.

Carter, A.T., Mason, D.R., Paul, C.J., Austin, J.W., and Peck, M.W. (2007). Microarray analysis of proteolytic *Clostridium botulinum*. In Proceedings of 4th Conference on Functional Genomics of Gram-Positive Microorganisms, Pisa, Italy, P24.

Centers for Disease Control and Prevention. (1998). Botulism in the United States (1899–1996). Handbook for Epidemiologists, Clinicians, and Laboratory Workers (U.S. Department of Health, Education, and Welfare, CDC, Atlanta, GA).

Chen, Y., Korkeala, H., Aarnikunnas, J., and Lindström, M. (2007). Sequencing the botulinum neurotoxin and related genes in *Clostridium botulinum* type E strains reveals *orfx3* and a novel type E neurotoxin subtype. J. Bacteriol. *189*, 8643–8650.

Chertow, D.S., Tan, E.T., Maslanka, S.E., Schulte, J., Bresnitz, E.A., Weisman, R.S., Bernstein, J., Marcus, S.M., Kumar, S., Malecki, J., Sobel, J., and Braden, C.R. (2006). Botulism in 4 adults following cosmetic injections with an unlicensed, highly concentrated botulinum preparation. JAMA *296*, 2476–2479.

Chia, J.K., Clark, J.B., Ryan, C.A., and Pollack, M. (1986). Botulism in an adult associated with food-borne intestinal infection with *Clostridium botulinum*. N. Engl. J. Med. *315*, 239–240.

Clemmens, M.R., and Bell, L. (2007). Infant botulism presenting with poor feeding and lethargy: a review of 4 cases. Pediatr. Emerg. Care. *23*, 492–494.

Craven, K.E., Ferreira, J.L., Harrison, M.A., and Edmonds, P. (2002). Specific detection of *Clostridium botulinum* types A, B, E, and F using the polymerase chain reaction. J. AOAC Int. *85*, 1025–1028.

Creti, R., Fenicia, L., and Aureli, P. (1990). Occurrence of *Clostridium botulinum* in the soil of the vicinity of Rome. Curr. Microbiol. *20*, 317–321.

Dahlenborg, M., Borch, E., and Rådström, P. (2001). Development of a combined selection and enrichment PCR procedure for *Clostridium botulinum* types B, E, and F and its use to determine prevalence in faecal samples from slaughtered pigs. Appl. Environ. Microbiol. *67*, 4781–4788.

Dahlenborg, M., Borch, E., and Rådström, P. (2003). Prevalence of *Clostridium botulinum* types B, E and F in faecal samples from Swedish cattle. Int. J. Food Microbiol. *82*, 105–110.

Dahlsten, E., Korkeala, H., Somervuo, P., and Lindström, M. (2008). PCR assay for differentiating between Group I (proteolytic) and Group II (nonproteolytic) strains of *Clostridium botulnum*. Int. J. Food Microbiol. *124*, 108–111.

Das Gupta, B.R., and Sugiyama, H. (1972). Role of a protease in natural activation of *Clostridium botulinum* neurotoxin. Infect. Immun. *6*, 587–590.

Dhaked, R.K., Sharma, S.K., Parida, M.M., and Singh, L. (2002). Isolation and characterization of *Clostridium botulinum* type E from soil of Gwalior, India. J. Nat. Toxins *11*, 49–56.

Dineen, S.S., Bradshaw, M., and Johnson, E.A. (2003). Neurotoxin gene clusters in *Clostridium botulinum* type A strains: sequence comparison and evolutionary implications. Curr. Microbiol. *46*, 345–352.

Dineen, S.S., Bradshaw, M., Karasek, C.E., and Johnson, E.A. (2004). Nucleotide sequence and transcriptional analysis of the type A2 neurotoxin gene cluster in *Clostridium botulinum*. FEMS Microbiol. Lett. *235*, 9–16.

Duff, J.T., Wright, G.G., and Yarinsky, A. (1956). Activation of *Clostridium botulinum* type E toxin by trypsin. J. Bacteriol. *72*, 455–460.

East, A.K., and Collins, M.D. (1994). Conserved structure of genes encoding components of botulinum neurotoxin complex M and the sequence of the gene coding for nontoxic component in nonproteolytic *Clostridium botulinum* type F. Curr. Microbiol. *29*, 69–77.

East, A.K., Bhandari, M., Stacy, J.M., Campbell, K.D., and Collins, M.D. (1996a). Contrasting organization of the cluster of genes encoding components of the botulinum neurotoxin complex in proteolytic *Clostridium botulinum* types A, B and F. Int. J. Syst. Bacteriol. *46*, 1105–1112.

East, A.K., Bhandari, M., Stacy, J.M., Campbell, K.D., and Collins, M.D. (1996b). Organization and phylogenetic interrelationships of genes encoding components of the botulinum toxin complex in proteolytic *Clostridium botulinum* types A, B, and F: evidence of chimeric sequences in the gene encoding the non-toxic nonhemagglutinin component. Int. J. Syst. Bacteriol. *46*, 1105–1112.

East, A.K., Richardson, P.T., Allaway, D., Collins, M.D., Roberts, T.A., and Thompson, D.E. (1992). Sequence of the gene encoding type F neurotoxin of *Clostridium botulinum*. FEMS Microbiol. Lett. *96*, 225–230.

East, A.K., Stacy, J.M., and Collins, M.D. (1994). Cloning and sequencing of a hemagglutinin component of the botulinum neurotoxin complex encoded by *Clostridium botulinum* type A and B. Syst. Appl. Microbiol. *17*, 306–312.

Eklund, M.W., and Poysky, F. (1967). Incidence of *Cl. botulinum* type E from the Pacific Coast of the United States. In Botulism 1966, M. Ingram, and T. A. Roberts, eds. (London, UK: Chapman and Hall), pp. 49–55.

Eklund, M.W., and Poysky, F. (1974). Interconversion of type C and D strains of *Clostridium botulinum* by specific bacteriophages. Appl. Microbiol. *27*, 251–258.

Eklund, M.W., Poysky, F.T., Mseitif, L.M., and Strom, M.S. (1988). Evidence for plasmid-mediated toxin and bacteriocin production in *Clostridium botulinum* type G. Appl. Environ. Microbiol. *54*, 1405–1408.

Eklund, M.W., Poysky, F.T., and Reed, S.M. (1972). Bacteriophage and the toxigenicity of *Clostridium botulinum* type D. Nature (London) New Biol. *235*, 16–17.

Eklund, M.W., Poysky, F.T., and Wieler, D.I. (1967a). Characteristics of *Clostridium botulinum* type F isolated from the Pacific Coast of the United States. Appl. Microbiol. *15*, 1316–1323.

Eklund, M.W., Wieler, D.I., and Poysky, F.T. (1967b). Outgrowth and toxin production of nonproteolytic type B *Clostridium botulinum* at 3.3 to 5.6°C. J. Bacteriol. *93*, 1461–1462.

Fach, P., Gibert, M., Griffais, R., Guillou, J.P., and Popoff, M.R. (1995). PCR and gene probe identification of botulinum neurotoxin A-, B-, E-, F-, and G-producing

Clostridium spp. and evaluation in food samples. Appl. Environ. Microbiol. *61*, 389–392.

Fach, P., Hauser, D., Guillou, J.P., and Popoff, M.R. (1993). Polymerase chain reaction for the rapid identification of *Clostridium botulinum* type a strains and detection in food samples. J. Appl. Bacteriol. *75*, 234–239.

Fach, P., Perelle, S., Dilasser, F., Grout, J., Dargaignaratz, L.B., Gourreau, J.-M., Carlin, F., Popoff, M.R., and Broussolle, V. (2002). Detection by PCR-enzyme-linked inmmunoassay of *Clostridium botulinum* in fish and environmental samples from a coastal area in Nothern France. Appl. Environ. Microbiol. *68*, 5870–5876.

Fenicia, L., Franciosa, G., Pourshaban, G.M., and Aureli, P. (1999). Intestinal toxemia botulism in two young people, caused by *Clostridium butyricum* type E. Clin. Infect. Dis. *29*, 1381–1387.

Fenicia, L., Anniballi, F., and Aureli, P. (2007a). Intestinal toxemia botulism in Italy, 1984–2005. Eur. J. Clin. Microbiol. Infect. Dis. *26*, 385–394.

Fenicia, L., Anniballi, F., De Medici, D., Delibato, E., and Aureli, P. (2007b). SYBR green real-time PCR method to detect *Clostridium botulinum* type A. Appl. Environ. Microbiol. *73*, 2891–2896.

Franciosa, G., Fenicia, L., Pourshaban, M., and Aureli, P. (1997). Recovery of a strain of *Clostridium botulinum* producing both neurotoxin A and neurotoxin B from canned macrobiotic food. Appl. Environ. Microbiol. *63*, 1148–1150.

Franciosa, G., Ferreira, J.L., and Hatheway, C.L. (1994). Detection of type A, B, and E botulism neurotoxin genes in *Clostridium botulinum* and other *Clostridium* species by PCR: Evidence of unexpressed type B toxin genes in type A toxigenic organisms. J. Clin. Microbiol. *32*, 1911–1917.

Franciosa, G., Floridi, F., Maugliani, A., and Aureli, P. (2004). Differentiation of the gene clusters encoding botulinum neurotoxin type A complexes in *Clostridium botulinum* type A, Ab, and A(B) strains. Appl. Environ. Microbiol. *70*, 7192–7199.

Franciosa, G., Maugliani, A., Floridi, F., and Aureli, P. (2006). A novel type A2 neurotoxin gene cluster in *Clostridium botulinum* strain Mascarpone. FEMS Microbiol. Lett. *261*, 88–94.

Freedman, M., Armstrong, R.M., Killian, J.M., and Bolland, D. (1986). Botulism in a patient with jejunoileal bypass. Ann. Neurol. *20*, 641–643.

Fujii, N., Kimura, K., Yokosawa, N., Oguma, K., Yashiki, T., Takeshi, K., Ohyama, T., Isogai, E., and Isogai, H. (1993a). Similarity in nucleotide sequence of the gene encoding nontoxic component of botulinum toxin produced by toxigenic *Clostridium butyricum* strain BL6340 and *Clostridium botulinum* type E Mashike. Microbiol. Immunol. *37*, 395–398.

Fujii, N., Kimura, K., Yokosawa, N., Yashiki, T., Tsuzuki, K., and Oguma, K. (1993b). The complete nucleotide sequence of the gene encoding the nontoxic component of *Clostridium botulinum* type E progenitor toxin. J. Gen. Microbiol. *139*, 79–86.

Fujinaga, Y., Inoue, K., Shimazaki, S., Tomochika, K., Tsuzuki, K., Fujii, N., Watanabe, T., Ohyama, T., Takeshi, K., Inoue, K., and Oguma, K. (1994). Molecular construction of *Clostridium botulinum* type C progenitor toxin and its gene organization. Biochem. Biophys. Res. Commun. *205*, 1291–1298.

Fujinaga, Y., Inoue, K., Watanabe, S., Yokota, K., Hirai, Y., Nagamachi, E., and Oguma, K. (1997). The haemagglutinin of *Clostridium botulinum* type C progenitor toxin plays an essential role in binding of toxin to the epithelial cells of guinea pig small intestine, leading to the efficient absorption of the toxin. Microbiol. *143*, 3841–3847.

Gao, Q.Y., Huang, Y.F., Wu, J.G., Liu, H.D., and Xia, H.Q. (1990). A review of botulism in China. Biomed. Environ. Sci. *3*, 326–336.

Gauthier, M., Cadieux, B., Austin, J.W., and Blais, B.W. (2005). Cloth-based hybridization array system for the detection of *Clostridium botulinum* types A, B, E, and F neurotoxin genes. J. Food Prot. *68*, 1477–1483.

Giménez, D.F. (1984). *Clostridium botulinum* subtype Ba. Zentralbl. Bakteriol. Parasitenkd. Infektionskr. Hyg. Abt. 1 Orig. Reihe A *257*, 68–72.

Giménez, D.F., and Ciccarelli, A.S. (1970). Studies on strain 84 of *Clostridium botulinum*. Zentralbl. Bakteriol. *215*, 212–220.

Graham, A.F., Mason, D.R., Maxwell, F.J., and Peck, M.W. (1997). Effect of pH and NaCl on growth from spores of nonproteolytic *Clostridium botulinum* at chill temperature. Lett. Appl. Microbiol. *24*, 95–100.

Grimont, F., and Grimont, P.A. (1986). Ribosomal ribonucleic acid gene restriction patterns as potential taxonomic tools. Ann. Inst. Pasteur. Microbiol. *137B*, 165–175.

Gupta, A., Sumner, C.J., Castor, M., Maslanka, S., and Sobel, J. (2005). Adult botulism type F in the United States, 1981–2002. Neurology *65*, 1694–1700.

Hall, J.D., McCroskey, L.M., Pincomb, B.J., and Hatheway, C.L. (1985). Isolation of an organism resembling *Clostridium barati* which produces type F botulinal toxin from an infant with botulism. J. Clin. Microbiol. *21*, 654–655.

Harvey, S.M., Sturgeon, J., and Dassey, D.E. (2002). Botulism due to *Clostridium baratii* type F toxin. J. Clin. Microbiol. *40*, 2260–2262.

Hatheway, C.L. (1995). Botulism: The present status of the disease. Curr. Top. Microbiol. Immunol. *195*, 55–75.

Hatheway, C.L., and McCroskey, L.M. (1987). Examination of feces and serum for diagnosis of infant botulism in 336 patients. J. Clin. Microbiol. *25*, 2334–2338.

Hatheway, C.L., McCroskey, L.M., Lombard, G.L., and Dowell, V.R.Jr. (1981). Atypical toxin variant of *Clostridium botulinum* type B associated with infant botulism. J. Clin. Microbiol. *14*, 607–611.

Hauschild, A.H.W., and Dodds, K.L. (1993) *Clostridium botulinum*. Ecology and control in foods (New York, USA: Marcel Dekker).

Hauschild, A.H., and Gauvreau, L. (1985). Food-borne botulism in Canada, 1971–84. Can. Med. Assoc. J. *133*, 1141–1146.

Hauser, D., Eklund, M.W., Boquet, P., and Popoff, M.R. (1994). Organization of the botulinum neurotoxin C1 gene and its associated non-toxic protein genes in

Clostridium botulinum C 468. Mol. Gen. Genet. *243*, 631–640.

Hayes, S., Craig, J.M., and Pilcher, K.S. (1970). The detection of *Clostridium botulinum* type E in smoked fish products in the Pacific Northwest. Can. J. Microbiol. *16*, 207–209.

Henderson, I., Whelan, S.M., Davis, T.O., and Minton, N.P. (1996). Genetic characterisation of the botulinum toxin complex of *Clostridium botulinum* strain NCTC 2916. FEMS Microbiol. Lett. *140*, 151–158.

Hielm, S., Björkroth, J., Hyytiä, E., and Korkeala, H. (1998a). Genomic analysis of *Clostridium botulinum* group II by pulsed-field gel electrophoresis. Appl. Environ. Microbiol. *64*, 703–708.

Hielm, S., Björkroth, J., Hyytiä, E., and Korkeala H. (1998b). Prevalence of *Clostridium botulinum* in Finnish trout farms: pulsed-field gel electrophoresis typing reveals extensive genetic diversity among type E isolates. Appl. Environ. Microbiol. *64*, 4161–4167.

Hielm, S., Björkroth, J., Hyytiä, E., and Korkeala, H. (1999). Ribotyping as an identification tool for *Clostridium botulinum* strains causing human botulism. Int. J. Food Microbiol. *47*, 121–131.

Hielm, S., Hyytiä, E., Andersin, A. B., and Korkeala H. (1998c). A high prevalence of *Clostridium botulinum* type E in Finnish freshwater and Baltic Sea sediment samples. J. Appl. Microbiol. *84*, 133–137.

Hielm, S., Hyytiä, E., Ridell, J., and Korkeala, H. (1996). Detection of *Clostridium botulinum* in fish and environmental samples using polymerase chain reaction. Int. J. Food. Microbiol. *31*, 357–365.

Hill, K.K. (2008). Personal communication, email 30 May 2008.

Hill, K.K., Barash, J., Smith, T., Helma, C., Ticknor, L., Smith, L., and Arnon, S. (2007a). AFLP analysis of combined *Clostridium botulinum* strain collections from USAMRIID and the California infant botulism program: A first report. In Proceedings of 44th Interagency Botulism Research Coordinating Committee Meeting, Monterey, California, USA, p. 154.

Hill, K.K., Smith, T.J., Helma, C.H., Ticknor, L.O., Foley, B.T., Svensson, R.T., Brown, J.L., Johnson, E.A., Smith, L.A., Okinaka, R.T., Jackson, P.J., and Marks, J.D. (2007b). Genetic diversity among botulinum neurotoxin-producing clostridial strains. J. Bacteriol. *189*, 818–832.

Hoffman, R.E., Pincomb, B.J., Skeels, M.R., and Burkhart, M.J. (1982). Type F infant botulism. Am. J. Dis. Child. *136*, 270–271.

Holdeman, L.V., and Brooks, J.B. (1970). Variation among strains of *Clostridium botulinum*. In Proceedings of the 1st U.S. – Japan Conference of Toxic Microorganisms, M. Herzberg, ed. (Washington, DC, USA: U. S. Government Printing Office), pp. 278–286.

Holzer, V.E. (1962). Botulism from inhalation. Med. Klin. *57*, 1735–1738.

Huss, H.H. (1980). Distribution of *Clostridium botulinum*. Appl. Environ. Microbiol. *39*, 764–769.

Huss, H.H., Pedersen, A., and Cann, D.C. (1974). The incidence of *Clostridium botulinum* in Danish trout farms. I. Distribution in fish and their environment. J. Food Technol. *9*, 445–450.

Hutson, R.A., Collins, M.D., East, A.K., and Thompson, D.E. (1994). Nucleotide sequence of the gene coding for non-proteolytic *Clostridium botulinum* type B neurotoxin: comparison with other clostridial neurotoxins. Curr. Microbiol. *28*, 101–110.

Hutson, R.A., Thompson, D.E., and Collins, M.D. (1993). Genetic interrelationships of saccharolytic *Clostridium botulinum* types B, E and F and related clostridia as revealed by small-subunit rRNA gene sequences. FEMS Microbiol. Lett. *108*, 103–110.

Hyytiä, E., Björkroth, J., Hielm, S., and Korkeala, H. (1999). Characterisation of *Clostridium botulinum* groups I and II by randomly amplified polymorphic DNA analysis and repetitive element sequence-based PCR. Int. J. Food Microbiol. *48*, 179–189.

Hyytiä, E., Hielm, S., Björkroth, J., and Korkeala H. (1998a). Biodiversity of *Clostridium botulinum* type E strains isolated from fish and fishery products. Appl Environ. Microbiol. *65*, 2057–2064.

Hyytiä, E., Hielm, S., and Korkeala, H. (1998b). Prevalence of *Clostridium botulinum* type E in Finnish fish and fishery products. Epidemiol. Infect. *120*, 245–250.

Hyytiä-Trees, E., Lindström, M., Schalch, B., Stolle, A., and Korkeala, H. (1999). *Clostridium botulinum* type E in Bavarian fish. Arch. Lebensmittelhyg. *50*, 79–82.

Hyytiä-Trees, E., Skyttä, E., Mokkila, M., Kinnunen, A., Lindström, M., Lähteenmäki, L., Ahvenainen, R., and Korkeala, H. 2000. Safety evaluation of sous vide-processed products with respect to nonproteolytic *Clostridium botulinum* by use of challenge studies and predictive microbiological models. Appl. Environ. Microbiol. *66*, 223–229.

Insalata, N.F., Witzeman, S.J., Fredericks, G.J., and Sunga, F.C.A. (1969). Incidence study of spores of *Clostridium botulinum* in convenience foods. Appl. Microbiol. *17*, 542–544.

Isacsohn, M., Cohen, A., Steiner, P., Rosenberg, P., and Rudensky, B. (1985). Botulism intoxication after surgery in the gut. Israel J. Med. Sci. *21*, 150–153.

Ito, K.A., Seslar, D.J., Mercer, W.A., and Meyer, K.F. (1967). The thermal and chlorine resistance of *Clostridium botulinum* types A, B, and E spores. In Botulism 1966, M. Ingram, and T. A. Roberts, ed. (London, UK: Chapman and Hall), pp. 108–122.

Johannsen, A. (1962). Förekomst och utbredning av *Cl. botulinum* typ E med särskilt hänsyn till Öresundsområdet. Nord. Vet. Med. *14*, 441–474.

Johannsen, A. (1963). *Clostridium botulinum* in Sweden and the adjacent waters. J. Appl. Bacteriol. *26*, 43–47.

Johnson, E.A., Tepp, W.H., Bradshaw, M., Gilbert, R.J., Cook, P.E., and McIntosh, E.D. (2005). Characterization of *Clostridium botulinum* strains associated with an infant botulism case in the United Kingdom. J. Clin. Microbiol. *43*, 2602–2607.

Johnson, R.O., Clay, S.A., and Arnon, S.S. (1979). Diagnosis and management of infant botulism. Am. J. Dis. Child. *133*, 586.

Kakinuma, H., Maruyama, H., Yamakawa, K., Nakamura, S., and Takahashi, H. (1997). Application of nested polymerase chain reaction for the rapid diagnosis of infant botulism type B. Acta Paediatr. Jpn. *39*, 346–348.

Kalka-Moll. W.M. (2007). Personal communication, email 16 October 2007.

Kalka-Moll, W.M., Aurbach, U., Schaumann, R., Schwarz, R., and Seifert, H. (2007). Wound botulism in injection drug users. Emerg. Infect. Dis. *13*, 942–943.

Keet, C.A., Fox, C.K., Margeta, M., Marco, E., Shane, A.L., Dearmond, S.J., Strober, J.B., and Miller, S.P. (2005). Infant botulism, type F, presenting at 54 hours of life. Pediatr. Neurol. *32*, 193–196.

Keto-Timonen, R., Nevas, M., and Korkeala, H. (2005). Efficient DNA fingerprinting of *Clostridium botulinum* types A, B, E, and F by amplified fragment length polymorphism analysis. Appl. Environ. Microbiol. *71*, 1148–1154.

Keto-Timonen, R., Heikinheimo, A., Eerola, E., and Korkeala, H. (2006). Identification of *Clostridium* species and DNA fingerprinting of *Clostridium perfringens* by amplified fragment length polymorphism analysis. J. Clin. Microbiol. *44*, 4057–4065.

Kimura, K., Fujii, N., Tsuzuki, K., Murakami, T., Indoh, T., Yokosawa, N., Takeshi, K., Syuto, B., and Oguma, K. (1990). The complete nucleotide sequence of the gene coding for botulinum type C1 toxin in the c-st phage genome. Biochem. Biophys. Res. Commun. *171*, 1304–1311.

Kirma, N., Ferreira, J.L., and Baumstark, B.R. (2004). Characterization of six type A strains of *Clostridium botulinum* that contain type B toxin gene sequences. FEMS Microbiol. Lett. *231*, 159–164.

Kitamura, M., Sakaguchi, S., and Sakaguchi, G. (1969). Significance of 12S toxin of *Clostridium botulinum* type E. J. Bacteriol. *98*, 1173–1178.

Klarmann, D. (1989). The detection of *Clostridium botulinum* in fecal samples of cattle and swine and in the raw material and animal meal of different animal body rendering plants. Berl. Munch. Tierarztl. Wochenschr. *102*, 84–86.

Kongsaengdao, S., Samintarapanya, K., Rusmeechan, S., Wongsa, A., Pothirat, C., Permpikul, C., Pongpakdee, S., Puavilai, W., Kateruttanakul, P., Phengtham, U., Panjapornpon, K., Janma, J., Piyavechviratana, K., Sithinamsuwan, P., Deesomchok, A., Tongyoo, S., Vilaichone, W., Boonyapisit, K., Mayotarn, S., Piya-Isragul, B., Rattanaphon, A., Intalapaporn, P., Dusitanond, P., Harnsomburana, P., Laowittawas, W., Chairangsaris, P., Suwantamee, J., Wongmek, W., Ratanarat, R., Poompichate, A., Panyadilok, H., Sutcharitchan, N., Chuesuwan, A., Oranrigsupau, P., Sutthapas, C., Tanprawate, S., Lorsuwansiri, J., Phattana, N; Thai Botulism Study Group. (2006). An outbreak of botulism in Thailand: clinical manifestations and management of severe respiratory failure. Clin. Infect. Dis. *43*, 1247–2356.

Korkeala, H., Stengel, G., Hyytiä, E., Vogelsang, B., Bohl, A., Wihlman, H., Pakkala, P., and Hielm, S. (1998). Type E botulism associated with vacuum-packaged hot-smoked whitefish. Int. J. Food Microbiol. *43*, 1–5.

Kubota, T., Shirakawa, S., Kozaki, S., Isogai, E., Isogai, H., Kimura, K., and Fujii, N. (1996) Mosaic type of the nontoxic-nonhemagglutinin component gene in *Clostridium botulinum* type A strain isolated from infant botulism in Japan. Biochem. Biophys. Res. Commun. *244*, 843–848.

Kubota, T., Yonekura, N., Hariya, Y., Isogai, E., Isogai, H., Amano, K., and Fujii, N. (1998). Gene arrangement in the upstream region of *Clostridium botulinum* type E and *Clostridium butyricum* BL6340 progenitor toxin genes is different from that of other types. FEMS Microbiol. Lett. *158*, 215–221.

Leclair, D., Pagotto, F., Farber, J.M., Cadieux, B., and Austin, J.W. (2006). Comparison of DNA fingerprinting methods for use in investigation of type E botulism outbreaks in the Canadian Arctic. J. Clin. Microbiol. *44*, 1635–1644.

Lin, W.J., and Johnson, E.A. (1995). Genome analysis of *Clostridium botulinum* type A by pulsed-field gel electrophoresis. Appl. Environ. Microbiol. *61*, 4441–4447.

Lindström, M., and Korkeala, H. (2006). Laboratory diagnostics of botulism. Clin. Microbiol. Rev. *19*, 298–314.

Lindström, M., Hielm, S., Nevas, M., Tuisku, S., and Korkeala, H. (2004). Proteolytic *Clostridium botulinum* type B in the gastric content of patient with type E botulism due to whitefish eggs. Food-borne Path. Dis. *1*, 53–57.

Lindström, M., Keto, R., Markkula, A., Nevas, M., Hielm, S., and Korkeala, H. (2001). Multiplex PCR assay for detection and identification of *Clostridium botulinum* types A, B, E, and F in food and fecal material. 2001. Appl. Environ. Microbiol. *67*, 5694–5699.

Lindström, M., Kiviniemi, K., and Korkeala, H. (2006a). Hazard and control of group II (non-proteolytic) *Clostridium botulinum* in modern food processing – a review. Int. J. Food Microbiol. *108*, 92–104.

Lindström, M., Nevas, M., Hielm, S., Lähteenmäki, L., Peck, M.W., and Korkeala, H. (2003). Thermal inactivation of nonproteolytic *Clostridium botulinum* type E spores in model fish media and in vacuum-packaged hot-smoked vacuum-packaged fish products. Appl. Environ. Microbiol. *69*, 4029–4035.

Lindström, M., Vuorela, M., Hinderink, K., Korkeala, H., Dahlsten, E., Raahenmaa, M., and Kuusi, M. (2006b). Botulism associated with vacuum-packed smoked whitefish in Finland, June-July 2006. Euro Surveill. *11*, E060720.3. Available at: http://www.eurosurveillance.org/ew/2006/060720.asp#3.

Lindström, M., Somervuo, P., Kiviniemi, K., Hinderink, K., Nevas, M., Auvinen, P., and Korkeala, H. (2007). Genomic analysis of two major Nordic Group I *Clostridium botulinum* populations reveals different arsenic resistance. In Proceedings of 4th Conference on Functional Genomics of Gram-Positive Microorganisms, Pisa, Italy, P31.

Lundén, J.M., Autio, T.J., and Korkeala, H.J. (2002). Transfer of persistent *Listeria monocytogenes* contamination between food-processing plants associated with a dicing machine. J. Food Prot. *65*, 1129–1133.

Lynt, R.K., Kautter, D.A., and Solomon, H.M. (1981). Heat resistance of proteolytic *Clostridium botulinum* type F in phosphate buffer and crabmeat. J. Food Sci. *47*, 204.

Lynt, R.K., Kautter, D.A., and Solomon, H.M. (1983). Effect of delayed germination by heat-damaged spores on estimates of heat resistance of *Clostridium botulinum* types E and F. J. Food Sci. *48*, 226–229.

Lövenklev, M., Holst, E., Borch, E., and Rådström, P.(2004). Relative neurotoxin gene expression in *Clostridium botulinum* type B, determined using quantitative reverse transcription-PCR. Appl. Environ. Microbiol. *70*, 2919–2927.

Marshall, K.M., Bradshaw, M., Pellett, S., and Johnson, E.A. (2007). Plasmid encoded neurotoxin genes in *Clostridium botulinum* serotype A subtypes. Biochem. Biophys. Res. Commun. *361*, 49–54.

Marvaud, J.C., Gibert, M., Inoue, K., Fujinaga, Y., Oguma, K., and Popoff, M.R. (1998). botR/A is a positive regulator of botulinum neurotoxin and associated non-toxin protein genes in *Clostridium botulinum* A. Mol. Microbiol. *29*, 1009–1018.

Matsumura, T., Jin, Y., Kabumoto, Y., Takegahara, Y., Oguma, K., Lencer, W. I., and Fujinaga, Y. (2007). The HA proteins of botulinum toxin disrupt intestinal epithelial intercellular junctions to increase toxin absorption. Cell Microbiol. E-pub ahead of print at doi:10.1111/j.1462–5822.2007.01048.x.

McCroskey, L.M., and Hatheway, C.L. (1988). Laboratory findings in four cases of adult botulism suggest colonization of the intestinal tract. J. Clin. Microbiol. *26*, 1052–1054.

McCroskey, L.M., Hatheway, C.L., Fenicia, L., Pasolini, B., and Aureli, P. (1986). Characterization of an organism that produces type E botulinal toxin but which resembles *Clostridium butyricum* from the faeces of an infant with type E botulism. J. Clin. Microbiol. *23*, 201–202.

McCroskey, L.M., Hatheway, C.L., Woodruff, B.A., Greenberg, J.A., and Jurgenson, P. (1991). Type F botulism due to neurotoxigenic *Clostridium baratii* from and unknown source in an adult. J. Clin. Microbiol. *29*, 2618–2620.

McKinney, M.W., Levett, P.N., and Haylock, R.W. (1993). Cloning of a DNA sequence unique to *Clostridium botulinum* group I by selective hybridization. J. Clin. Microbiol. *31*, 1845–1849.

Meng, X., Yamakawa, K., Zou, K., Wang, X., Kuang, X., Lu, C., Wang, C., Karasawa, T., and Nakamura, S. (1999). Isolation and characterisation of neurotoxigenic *Clostridium butyricum* from soil in China. J. Med. Microbiol. *48*, 133–137.

Merivirta, L.O., Lindström, M., Björkroth, K.J., and Korkeala, H.J. (2006). The prevalence of *Clostridium botulinum* in European river lamprey (*Lampetra fluviatilis*) in Finland. Int. J. Food Microbiol. *109*, 234–237.

Mezaki, T., Kaji, R., Kohara, N., and Kimura, J. (1996). Development of general weakness in a patient with amyotrophic lateral sclerosis after focal botulinum toxin injection. Neurology *46*, 845–846.

Miettinen, M.K., Siitonen, A., Heiskanen, P., Haajanen, H., Björkroth, K.J., and Korkeala, H.J. (1999). Molecular epidemiology of an outbreak of febrile gastroenteritis caused by *Listeria monocytogenes* in cold-smoked rainbow trout. J. Clin. Microbiol. *37*, 2358–2360.

Montecucco, C., and Schiavo, G. (1993). Tetanus and botulism neurotoxins: a new group of zinc proteases. Trends Biochem. Sci. *18*, 324–327.

Montecucco, C., and Schiavo, G. (1994). Mechanism of action of tetanus and botulinum neurotoxins. Mol. Microbiol. *13*, 1–8.

Moriishi, K., Koura, M., Fujii, N., Fujinaga, Y., Inoue, K., Syuto, B., and Oguma, K. (1996). Molecular cloning of the gene encoding the mosaic neurotoxin, composed of parts of botulinum neurotoxin types C1 and D, and PCR detection of this gene from *Clostridium botulinum* type C organisms. Appl. Environ. Microbiol. *62*, 662–667.

Mulleague, L., Bonner, S.M., Samuel, A., Nichols, P., Khan, M., Shaw, S., and Gruning, T. (2002). Wound botulism in drug addicts in the United Kingdom. Anaesthesia *57*, 301–302.

Myllykoski, J., Nevas, M., Lindström, M., and Korkeala, H. (2006). The detection and prevalence of *Clostridium botulinum* in pig intestinal samples. Int. J. Food Microbiol. *110*, 172–177.

Mäde, D., Trümper, K., and Stark, R. (2000). Nachweis von *Clostridium botulinum* in Honig duch Polymerase-Kettenreaktion. Arch. Lebensmittelhyg. *51*, 57–80.

Nakajima, H., Inoue, K., Ikeda, T., Fujinaga, Y., Sunagawa, H., Takeshi, K., Ohyama, T., Watanabe, T., Inoue, K., and Oguma, K. (1998). Molecular composition of the 16S toxin produced by a *Clostridium botulinum* type D strain, 1873. Microbiol. Immunol. *42*, 599–605.

Nakano, H., Kizaki, H., and Sakaguchi, G. (1994). Multiplication of *Clostridium botulinum* in dead honey-bees and bee pupae, a likely source of heavy contamination of honey. Int. J. Food Microbiol. *21*, 247–252.

Nevas, M., Hielm, S., Lindström, M., Horn, H., Koivulehto, K., and Korkeala, H. (2002). High prevalence of *Clostridium botulinum* types A and B in honey samples detected by polymerase chain reaction. Int. J. Food Microbiol. *72*, 45–52.

Nevas, M., Lindström, M., Hautamäki, K., Puoskari, S., and Korkeala, H. (2005a). Prevalence and diversity of *Clostridium botulinum* types A, B, E and F in honey produced in the Nordic countries. Int. J. Food Microbiol. *105*, 145–151.

Nevas, M., Lindström, M., Hielm, S., Björkroth, K.J., Peck, M.W., and Korkeala, H. (2005b). Diversity of proteolytic *Clostridium botulinum* strains, determined by pulsed-field gel electrophoresis approach. Appl. Environ. Microbiol. *71*, 1311–1317.

Nevas, M., Lindström, M., Hörman, A., Keto-Timonen, R., and Korkeala, H. (2006). *Clostridium botulinum* in the production environment of honey. Environ. Microbiol. *8*, 1085–1094.

Nevas, M., Lindström, M., Virtanen, A., Hielm, S., Kuusi, M., Arnon, S.S., Vuori, E., and Korkeala, H. (2005c). Infant botulism acquired from household dust presenting as sudden infant death syndrome. J. Clin. Microbiol. *43*, 511–513.

Ohishi, I., and Sakaguchi, G. (1977). Activation of botulinum toxins in the absence of nicking. Infect. Immun. *17*, 402–407.

Ohyama, T., Watanabe, T., Fujinaga, Y., Inoue, K., Sunagawa, H., Fujii, N., Inoue, K., and Oguma, K. (1995). Characterization of nontoxic-nonhemagglutinin component of the two types of progenitor toxin (M and L) produced by *Clostridium botulinum* type D CB-16. Microbiol. Immunol. *39*, 457–465.

Pace, P.J., Krumbiegel, E.R., Angelotti, R., and Wisniewski, H.J. (1967a). Demonstration and isolation of *Clostridium botulinum* types from whitefish chubs collected at fish smoking plants of the Milwaukee area. Appl. Microbiol. *15*, 877–884.

Passaro, D.J., Werner, S.B., McGee, J., MacKenzie, W.R., and Vugia, D. J. (1998). Wound botulism associated with black tar heroin among injecting drug users. JAMA *279*, 859–863.

Paul, C.J., Tran, S., Tam, K.J., and Austin, J.W. (2007). A unique restriction site in the *flaA* gene allows rapid differentiation of group I and group II *Clostridium botulinum* strains by PCR-restriction fragment length polymorphism analysis. J. Food Prot. *70*, 2133–2139.

Paul, C.J., Twine, S.M., Tam, K.J., Mullen, J.A., Kelly, J.F., Austin, J.W., and Logan, S.M. (2007). Flagellin diversity in *Clostridium botulinum* groups I and II: a new strategy for strain identification. Appl. Environ. Microbiol. *73*, 2963–2975.

Peck, M.W. (1997). *Clostridium botulinum* and the safety of refrigerated processed foods of extended durability. Trends Food Sci. Technol. *8*, 186–192.

Peck, M.W. (2006). *Clostridium botulinum* and the safety of minimally heated, chilled foods: an emerging issue? 2006. J. Appl. Microbiol. *101*, 556–570.

Peck, M.W., Fairbairn, D.A., and Lund, B.M. (1992a). Factors affecting growth from heat-treated spores of non-proteolytic *Clostridium botulinum*. Lett. Appl. Microbiol. *15*, 152–155.

Peck, M.W., Fairbairn, D.A., and Lund, B.M. (1992b). The effect of recovery medium on the estimated heat-inactivation of spores of non-proteolytic *Clostridium botulinum*. Lett. Appl. Microbiol. *15*, 146–151.

Peck, M.W., Fairbairn, D.A., and Lund, B. M. (1993). Heat-resistance of spores of non-proteolytic *Clostridium botulinum* estimated on medium containing lysozyme. Lett. Appl. Microbiol. *16*, 126–131.

Polulet, S., Hauser, D., Quanz, M., Niemann, H., and Popoff, M.R. (1992). Sequences of the botulinal neurotoxin E derived from *Clostridium botulinum* type E (strain Beluga) and *Clostridium butyricum* (strain ATCC 43181 and ATCC 43755). Biochem. Biophys. Res. Commun. *183*, 107–113.

Pourshafie, M., Vahdani, P., and Popoff, M. (2005). Genotyping *Clostridium botulinum* toxinotype A isolates from patients using amplified rDNA restriction analysis. J. Med. Microbiol. *54*, 933–936.

Raffestin, S., Dupuy, B., Marvaud, J.C., and Popoff, M.R. (2005). BotR/A and TetR are alternative RNA polymerase sigma factors controlling the expression of the neurotoxin and associated protein genes in *Clostridium botulinum* type A and *Clostridium tetani*. Mol. Microbiol. 55, 235–249.

Raphael, B.H., and Andreadis, J.D. (2007). Real-time PCR detection of the nontoxic nonhemagglutinin gene as a rapid screening method for bacterial isolates harboring the botulinum neurotoxin (A-G) gene complex. J. Microbiol. Methods. E-pub ahead of print at doi: 10.1016/j.mimet.2007.09.016.

Roberts, T.A., and Smart, J.L. (1976). The occurrence and possible growth of *Clostridium* spp. in vacuum-packed bacon, with particular Reference to *Clostridium perfringens* (*welchii*) and *Clostridium botulinum*. J. Food Technol. *11*, 229–244.

Rouhbakhsh-Khaleghdoust, A. (1975). The incidence of *Clostridium botulinum* type E in fish and bottom deposits in the Caspian Sea coastal waters. Pahlavi Med. J. *6*, 550–556.

Sakaguchi, Y., Hayashi, T., Kurokawa, K., Nakayama, K., Oshima, K., Fujinaga, Y., Ohnishi, M., Ohtsubo, E., Hattori, M., and Oguma, K. (2005). The genome sequence of *Clostridium botulinum* type C neurotoxin-converting phage and the molecular mechanisms of unstable lysogeny. Proc. Natl. Acad. Sci. USA. *102*, 17472–17477.

Schmidt, C.F. (1964). Spores of *C. botulinum*: Formation, resistance, germination. In Botulism, K. H. Lewis and K. Cassel, Jr., ed. (Cincinnati, Ohio, USA: U. S. Department of Health, Education, and Welfare, Public Health Service), pp. 69–82.

Schmidt, C.F., Lechowich, R.V., and Folinazzo, J.F. (1961). Growth and toxin production by type E *Clostridium botulinum* below 40°F. J. Food Sci. *26*, 626–630.

Sciacchitano, C.J., and Hirshfield, I.N. (1996). Molecular detection of *Clostridium botulinum* type E neurotoxin gene in smoked fish by polymerase chain reaction and capillary electrophoresis. J. AOAC Int. *79*, 861–865.

Scott, V.N., and Bernard, D.T. (1982). Heat resistance of spores of non-proteolytic type B *Clostridium botulinum*. J. Food Protect. *45*, 909.

Sebaihia, M., Peck, M.W., Minton, N.P., Thomson, N.R., Holden, M.T., Mitchell, W.J., Carter, A.T., Bentley, S.D., Mason, D.R., Crossman, L., Paul, C.J., Ivens, A., Wells-Bennik, M.H., Davis, I. J., Cerdeno-Tarraga, A.M., Churcher, C., Quail, M.A., Chillingworth, T., Feltwell, T., Fraser, A., Goodhead, I., Hance, Z., Jagels, K., Larke, N., Maddison, M., Moule, S., Mungall, K., Norbertczak, H., Rabbinowitsch, E., Sanders, M., Simmonds, M., White, B., Whithead, S., Parkhill, J. (2007). Genome sequence of a proteolytic (Group I) *Clostridium botulinum* strain Hall A and comparative analysis of the clostridial genomes. Genome Res. *17*, 1082–1092.

Smelt, J.P.P.M. (1980). Heat resistance of *Clostridium botulinum* in acid ingredients and its signification for the safety of chilled foods. Academic dissertation, University of Utrecht, The Netherlands.

Smith, L.D.S. (1978). The occurrence of *Clostridium botulinum* and *Clostridium tetani* in the soil of the United States. Health Lab. Sci. *15*, 74–80.

Smith, L.D.S., and Hobbs, G. (1974). Genus III: *Clostridium*. In Bergey's Manual of Determinative Bacteriology, 8th ed., R. E. Buchanan and N. E. Gibbons ed. (Baltimore, Maryland, USA: Williams & Wilkins), pp 551–572.

Smith, L.D.S., and Sugiyama, H. (1988). Botulism. The Organism, Its Toxins, the Disease, 2nd ed. (Springfield, Illinois, USA: Charles C. Thomas Publisher).

Smith, T., Hill, K.K., MacDonald, T., Helma, C., Ticknor, L., and Smith, L. (2007). Differentiation of *Clostridium botulinum* BoNT/A strains using multiple-locus variable-number tandem-repeat (VNTR) analysis. In Proceedings of 44th Interagency Botulism Research Coordinating Committee Meeting, Monterey, California, USA, p. 107.

Smith, T.J., Lou, J., Geren, I.N., Forsyth, C.M., Tsai, R., Laporte, S.L., Tepp, W.H., Bradshaw, M., Johnson, E.A., Smith, L.A., and Marks, J.D. (2005). Sequence variation within botulinum neurotoxin serotypes impacts antibody binding and neutralization. Infect. Immun. *73*, 5450–5457.

Sobel, J. (2005). Botulism. Clin. Infect. Dis. *41*, 1167–73.

Sobel, J., Tucker, N., Sulka, A., McLauchlin, J., and Maslanka, S. (2004). Food-borne botulism in the United States, 1990–2000. Emerg. Infect. Dis. *10*, 1606–1611.

Sobel, J., Malavet, M., and John, S. (2007). Outbreak of clinically mild botulism type E illness from home-salted fish in patients presenting with predominantly gastrointestinal symptoms. Clin. Infect. Dis. *45*, 14–16.

Solomon, H.M., and Lilly, T. Jr. (2001). *Clostridium botulinum*. Bacteriological Analytical Manual Online. Available at http://www.cfsan.fda.gov/~ebam/bam-17.html.

Sonnabend, O., Sonnabend, W., Heinzle, R., Sigrist, T., Dirnhofer, R., and Krech, U. (1981). Isolation of *Clostridium botulinum* type G and identification of type G botulinal toxin in humans: report of five sudden unexpected deaths. J. Infect. Dis. *143*, 22–27.

Souayah, N., Karim, H., Kamin, S.S., McArdle, J., and Marcus, S. (2006). Severe botulism after focal injection of botulinum toxin. Neurology *67*, 1855–1856.

Stubmo, C.R. (1973). Thermobacteriology in Food Processing, 2nd edn. (New York, USA: Academic Press).

Suen, J.C., Hatheway, C.L., Steigerwalt, A.G., and Brenner, D.J. (1988). *Clostridium argentinense*, sp. nov: A genetically homogenous group composed of all strains of *Clostridium botulinum* type G and some nontoxigenic strains previously identified as *Clostridium subterminale* or *Clostridium hastiforme*. Int. J. Syst. Bacteriol. *38*, 375.

Szabo, E.A., Pemberton, J.M., Gibson, A.M., Eyles, M.J., and Desmarchelier, P.M. (1994). Polymerase chain reaction for detection of *Clostridium botulinum* types A, B and E in food, soil and infant faeces. J. Appl. Bacteriol. *76*, 539–545.

Taclindo, C., Midura, T. Jr., Nygaard, G.S., and Bodily, H.L. (1967). Examination of prepared foods in plastic packages for *Clostridium botulinum*. Appl. Microbiol. *15*, 426–430.

Takeshi, K., Fujinaga, Y., Inoue, K., Nakajima, H., Oguma, K., Ueno, T., Sunagawa, H., and Ohyama, T. (1996). Simple method for detection of *Clostridium botulinum* type A to F neurotoxin genes by polymerase chain reaction. Microbiol. Immunol. *40*, 5–11.

Thomas, H.A. (1942). Bacterial densities from fermentation tube tests. J. Am. Water Works Assoc. *34*, 572–576.

Thompson, D.E., Brehm, J.K., Oultram, J.D., Swinfield, T.J., Shone, C.C., Atkinson, T., Melling, J. and Minton, N.P. (1990). The complete amino acid sequence of the *Clostridium botulinum* type A neurotoxin, deduced by nucleotide sequence analysis of the encoding gene. Eur. J. Biochem. *189*, 73–81.

Thompson, J.A., Filloux, F.M., Van Orman, C.B., Swoboda, K., Peterson, P., Firth, S.D., Bale, J. F. Jr. (2005). Infant botulism in the age of botulism immune globulin. Neurology. *64*, 2029–2032.

Versalovic, J., Koeuth, T., and Lupski, J.R. (1991). Distribution of repetitive DNA sequences in eubacteria and application to fingerprinting of bacterial genomes. Nucleic Acids Res. *19*, 6823–6831.

Vos, P., Hogers, M., Bleeker, M., Reijans, M., Lee, T. van der, Hornes, M., Frijters, A., Pot, J., Peleman, J., Kuiper, M., Zabeau, M. (1995). AFLP: a new technique for DNA fingerprinting. Nucleic Acids Res. *23*, 4407–4414.

Weber, J.T., Hibbs, R.G. Jr., Darwish, A., Mishu, B., Corwin, A.L., Rakha, M., Hatheway, C.L., el Sharkawy, S., el-Rahim, S.A., al-Hamd, M.F., Sarn, J.E., Blake, P.A., and Tauxe, R.V. (1993). A massive outbreak of type E botulism associated with traditional salted fish in Cairo. J. Infect. Dis. *167*, 451–454.

Werner, S.B., Passaro, D., McGee, J., Schechter, R., and Vugia, D.J. (2000). Wound botulism in California: Recent epidemic in heroin injectors. Clin. Infect. Dis. *31*, 1018–1024.

Whelan, S.M., Elmore, M.J., Bodsworth, N.J., Brehm, J.K., Atkinson, T., Minton, N.P. (1992). Molecular cloning of the *Clostridium botulinum* structural gene encoding the type B neurotoxin and determination of its entire nucleotide sequence. Appl. Environ. Microbiol. *58*, 2345–2354.

Willems, A., East, A.K., Lawson, P.A., and Collins, M.D. (1993). Sequence of the gene coding for the neurotoxin of *Clostridium botulinum* type A associated with infant botulism: comparison with other clostridial neurotoxins. Res. Microbiol. *144*, 547–556.

Williams, J.G., Kubelikt, A.R., Livak, K.J., Rafalski, J.A., and Tingey, S.V. (1990). DNA polymorphism amplified by arbitrary primers are useful as genetic markers. Nucleic. Acids Res. *18*, 6531–6535.

Yamakawa, K., and Nakamura, S. (1992). Prevalence of *Clostridium botulinum* type E and coexistence of *C. botulinum* nonproteolytic type B in the river soil of Japan. Microbiol. Immunol. *36*, 583–591.

Yoon, S.-Y., Chung, G.T., Kang, D.-H., Ryu, C., Yoo, C.-K., and Seong, W.-K. (2005). Application of real-time PCR for quantitative detection of *Clostridium botulinum* type A toxin gene in food. Microbiol. Immunol. *49*, 505–511.

Zhang, L., Lin, W.J., Li, S., and Aoki, K.R. (2003). Complete DNA sequences of the botulinum neurotoxin complex of *Clostridium botulinum* type A-Hall (Allergan) strain. Gene *315*, 21–32.

Zhou, Y., Sugiyama, H., Nakano, H., and Johnson, E.A. (1995). The genes for the *Clostridium botulinum* type G toxin complex are on a plasmid. Infect. Immun. *63*, 2087–2091.

Molecular Variability in *Clostridium difficile* Large Clostridial Toxins

6

Maja Rupnik

Abstract

Clostridium difficile, as all clostridia, is a toxin-producing microorganism and the toxins are the main virulence factors. In early 1980s it was clear that two large toxins are produced by bacterium and epidemiological studies have indicated that strains either produce both toxins (toxin A, TcdA, and toxin B, TcdB) or none of them. Toxigenic strains were usually associated with the disease, while non-toxigenic were not. A few years later, however, this simple situation changed as strains producing only TcdB or strains producing an additional toxin (binary toxin CDT) were described. Such strains with unusual toxin production pattern were subsequently found to have changes in the genomic PaLoc region encoding the toxins TcdA and TcdB. These changes are the basis for a method that distinguish *C. difficile* strains into toxinotypes. The variability of genes coding for large clostridial toxins (LCTs) had consequences in changes of laboratory diagnosis, changes in understanding of the role of both toxins in the pathogenesis, in structure–function relationships and partially in understanding of the evolutions of LCTs.

First variant *C. difficile* strains

First variant strains were recognized because of their toxin B-positive, toxin A-negative phenotype. As mentioned earlier this was unusual because the great majority of strains were A^+B^+.

Already in 1986 during the studies on the influence of defined medium composition on toxin production a single strain was discovered that did not produce TcdA under any conditions used (Haslam *et al.*, 1986). Several years later have two groups independently published analysis of this unique natural mutant, strain 8864 (Lyerly *et al.*, 1992; Borriello *et al.*, 1992).

Soon afterwards description of entire *C. difficile* serogroup (serogroup F) with A^-B^+ phenotype was published (Depitre *et al.*, 1993). It was clear that serogroup F and strain 8864 represent two different types of A^-B^+ strains. The differences in toxin genes *tcdA* and *tcdB* encoding the toxins TcdA and TcdB were analysed in more detail in both A^-B^+ groups by six overlapping PCRs (Rupnik *et al.*, 1997) and changes were described in both groups (Fig. 6.1).

Discovery of toxinotypes

With two groups of A^-B^+ strains and one of them correlated perfectly with serogrouping the next obvious question was: do other forms of variant *C. difficile* toxin genes exist and, if so, do they also correlate with any of >20 known serogroups?

During this time entire 19-kb toxin-coding region called PaLoc or toxigenic element was defined by two different groups (Braun *et al.*, 1996; Hammond and Johnson, 1995). PaLoc contains five genes, two of which are structural genes for two related toxins, *tcdA* and *tcdB*. Another two genes are involved in positive (*tcdR*) and negative (*tcdC*) regulation. The transport function for the last gene (*tcdE*) is so far only putative and is based on its homology with phage holin proteins.

For screening of changes in toxin genes the previously described PCR method amplified putative functional domains: fragment A1/B1

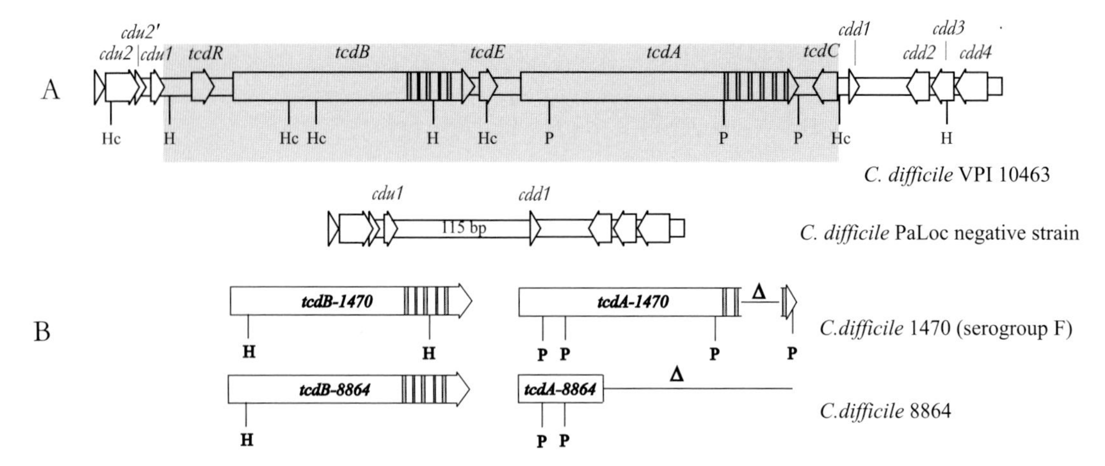

Figure 6.1 Changes in toxin genes *tcdA* and *tcdB* described in first variant *C. difficile* strains. (A) Toxin encoding region PaLoc includes five genes, two large toxin genes (*tcdA* and *tcdB*) with characteristic 3′-end repetitive sequences, and three accessory genes involved in regulation and probably transport (*tcdR*, *E*, *C*). (B) First variant *C. difficile* strains had TcdB-positive and TcdA-negative phenotype. Both had whole length *tcdB* gene that differed from VPI 10463 (reference strain) and *tcdA* gene with larger or smaller deletions.

amplifying catalytic domain, A2/B2 translocation domain and A3/B3 repetitive receptor binding domain (Fig. 6.2). In both known A^-B^+ groups fragments B1, B3 and A3 displayed most variations (Rupnik *et al.*, 1997). Large selection of strains from different serogroups was therefore compared in these three fragments to the reference strain VPI 10463 and eight new variations in PaLoc were found (Rupnik *et al.*, 1998). In variant strains entire PaLoc was screened with 10 overlapping PCRs (Fig. 6.2).

C. difficile toxinotype was defined as a group of strains that has identical changes in PaLoc region encoding the TcdA and TcdB. Variant toxinotypes are named with roman numerals and at the time types I to XXVII are differentiated (Table 6.1). Strains identical to VPI 10463 were defined as toxinotype 0.

Toxinotypes and production of toxins TcdA, TcdB and binary toxin CDT

C. difficile strains can produce three toxins: toxin A (TcdA, enterotoxin) and B (TcdB, cytotoxin) from the group of large clostridial toxins (LCT; Eichel-Streiber *et al.*, 1996) and an unrelated toxin from group of clostridial binary toxins (CDT; Perelle *et al.*, 1997). According to the combination of all three toxins we can differentiate seven toxin production types in *C. difficile*: $A^+B^+CDT^-$, $A^-B^+CDT^-$, $A^-B^+CTD^+$, $A^+B^-CDT^+$, $A^+B^+CDT^+$, $A^-B^-CDT^+$, $A^-B^-CDT^-$. TcdA and TcdB phenotypes are determined with toxin gene amplification and/or toxin production. CDT is usually confirmed only by the presence of functional *cdt* gene(s) but not necessarily with detection of toxin itself.

Usual *C. difficile* strains (toxinotype 0) have phenotype $A^+B^+CDT^-$. Identical phenotype is found in several variant toxinotypes with minor changes in PaLoc (Table 6.1). Most of the variant toxinotypes with significantly changed toxin genes will in addition to TcdA and TcdB produce also binary toxin CDT ($A^+B^+CDT^+$). Other toxin production types found in variant toxinotypes include type $A^-B^+CDT^-$ (toxinotypes VIII, X, XVI, XVII) and some rare types of $A^-B^+CTD^+$ (Stubbs *et al.*, 2000; Rupnik *et al.*, 2003), $A^+B^-CDT^+$ (MacCannell *et al.*, 2006a) and $A^-B^-CDT^+$ strains (Geric *et al.*, 2003).

Toxinotypes and changes in PaLoc

Length differences

Deletions are probably the most prominent changes in PaLoc as some of them can be quite large and easily observed as length polymorphisms of amplified fragments (Fig. 6.3). Most of them are located in repetitive regions of gene

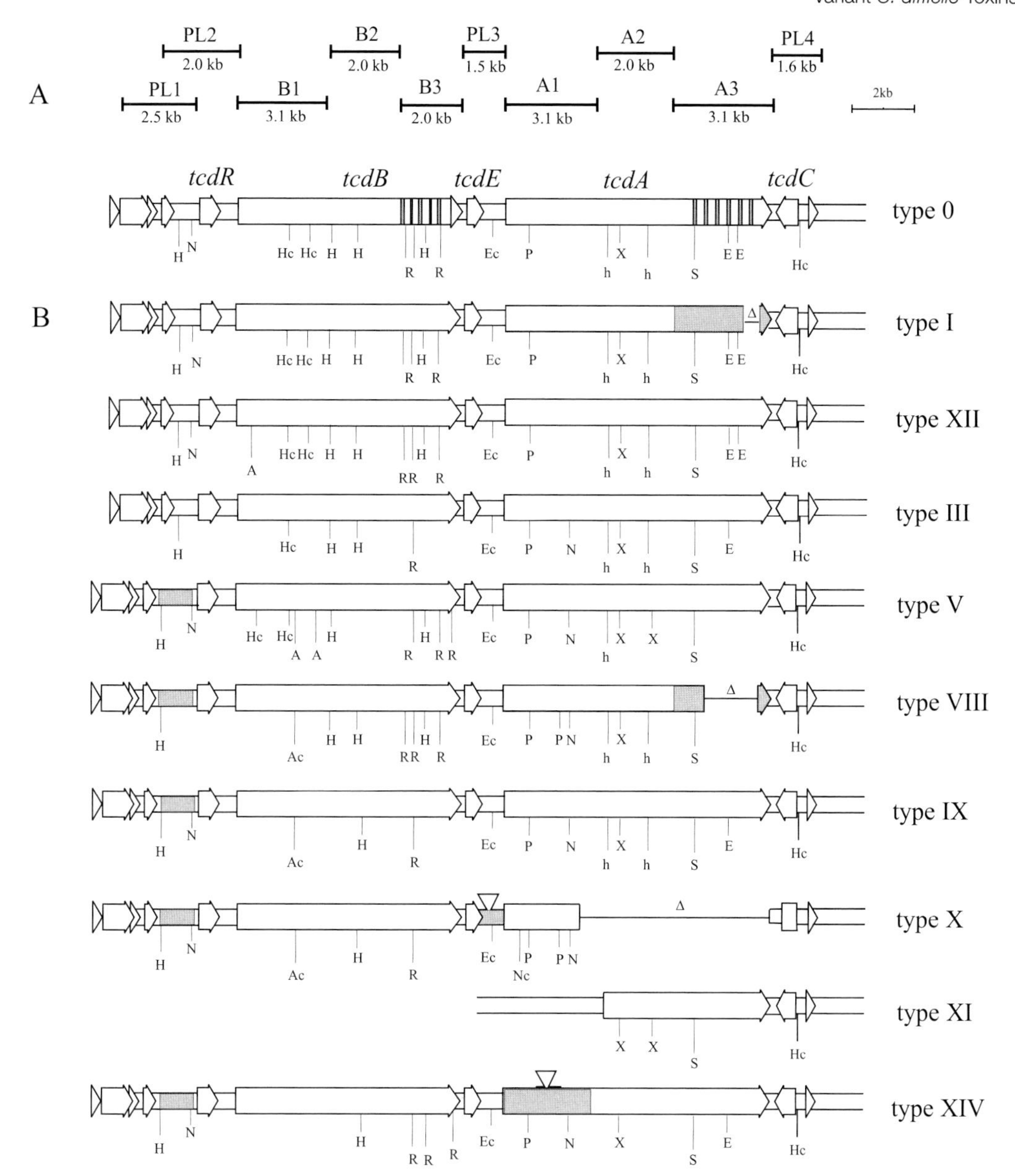

Figure 6.2 Schematic presentation of selected toxinotypes. (A) PaLoc of the reference strain VPI 10463 (toxinotype 0) and PCR fragments used for screening the toxin coding region. (B) PaLoc regions in variant C. difficile strains could have insertions/deletions (shown as gray areas) or heterogeneous sequence (shown as some of the tested restriction sites). Note that in some strains changes are minimal and limited only to one of the PCR fragments (type I, A3; type XI, B1). Other toxinotypes have significantly changed toxin genes and other PaLoc regions. *Acc*I, E – *Eco*RI, Ec – *Eco*RV, H – *Hin*dIII, Hc – *Hin*cII, h – *Hae*III, N – *Nsi*I, Nc – *Nco*I, P – *Pst*I, R – *Rsa*I, S – *Spe*I, X – XbaI.

tcdA or in some cases at 5′-end of *tcdA* (Fig. 6.2). Shorter *tcdA* forms can be found in PaLoc regions identical to the reference strain VPI 10463 (toxinotype 0) like in toxinotypes I, II, XIII, XIX, XX. In other toxinotypes (VIII, X, XVI ...) shorter *tcdA* gene is combined with variant *tcdB* gene and other changes in PaLoc. However, variations in *tcdB* gene are always at the level of polymorphic sequences and no truncated forms of *tcdB* are known so far.

Deletions in *tcdA* can cover from about 100 bp to as much as 5.9 kb in toxinotype X (Soehn *et al.*, 1998). The best studied is approximate 700 bp deletion in *tcdA* of toxinotype VIII (including

Table 6.1 Toxinotypes – the characteristic changes in PaLoc region and toxin production

Toxinotype	Type strain	Toxin production[a]	B1[b]	A3[b]	Type of *tcdC* gene[c]	Type of CPE
0	VPI 10463	A+B+CDT–	1	1	1	D
I	EX 623	A+B+CDT–	1	4	ND	D
II	AC 008	A+B+CDT–	1	3	ND	D
III[d]	SE 844	A+B+CDT+	4	2	2	D
IV	55767	A+B+CDT+	2	2	4	D
V	SE 881	A+B+CDT+	3	8	3	D
VI	51377	A+B+CDT+	3	5d	3	D
VII	57267	A+B+CDT+	3	6d	3	D
VIII	1470	A–B+CDT–	5	7d	1	S
IX	51680	A+ B+CDT+	5	2	1	S
X	8864	A–B+CDT+	5	Neg	Neg	S
XIa	IS 58	A–B–CDT+	Neg	5d	3	None
XIb	R 11402	A–B–CDT+	Neg	8	3	None
XII	IS 25	A+B+CDT–	6	1	1	D
XIII	R 9367	A+B+CDT–	1	9	ND	D
XIV	R 10870	A+B+CDT+	7	2	3	S
XV	R 9385	A+ B+CDT+	7	2	2	S
XVI	SUC36	A–B+CDT+	3	10d	3	D
XVII	J9965	A–B+CDT+	5	Neg	Neg	S
XVIII	GAI00166	A+B+CDT–	1	11	ND	D
XIX	TR13	A+B+CDT–	1	5	ND	D
XX	TR14	A+B+CDT–	1	6	ND	D
XXI	CH6223	A+B+CDT–	5	1	1	S
XXII	CH6143	A+B+CDT+	4	1	1	D
XXIII	8785	A+B+CDT+	5	2	1	S
XXIV	597B	A+B+CDT+	1	1	2	D
XXV	ZZV07-383	A+ B+CDT+	4	12	ND	D
XXVI	ZZV07-382	A+B+CDT–	1	13	ND	D
XXVII	ZZV07-340	A+B+CDT–	1	14	ND	D – low activity

Neg, not amplified; d, deletion (detected already in unrestricted A3 PCR fragment).
[a]A+ and B+ refers to production of toxin TcdA and TcdB; CDT+ refers to the presence of complete CDT locus (production of binary toxin not tested for all strains).
[b]Types of *Hinc/Acc* restrictions for B1 PCR fragment and types of *Eco*RI restrictions for A3 PCR fragment.
[c]Type of *Nco*I restriction type of *tcdC* gene indicating deletions of different lengths.
[d]Within toxinotype III three subtypes can be differentiated (IIIa, IIIb, IIIc) due to the differences in B2 and B3 PCR fragments (Geric *et al*., 2004).

A$^-$B$^+$ strains of serogroup F and X) which was characterized by several groups (Eichel-Streiber *et al.*, 1999; Kato *et al.*, 1999; Sambol *et al.*, 2000).

Very recently deletions in *tcdC* gene encoding a negative regulator have become increasingly important and we can currently differentiate four different types of deletions from 18 bp to 56 bp (Spigaglia and Mastrantonio, 2002; Stare *et al.*, 2007; Curry *et al.*, 2007).

The largest known deletion in PaLoc is found in toxinotype XI, which would have only

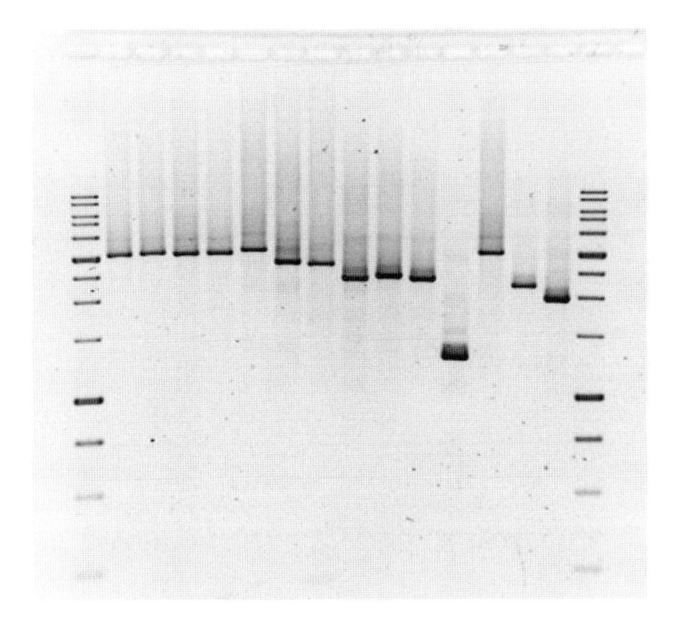

Figure 6.3 Length polymorphisms in *tcdA* repetitive regions. After amplification and *Eco*RI restriction (not shown) 14 different types of A3 could be differentiated. A3 fragment is also the only one that displays visible length differences in uncut PCR fragments (types 5–7, 9–14).

tcdC and two-thirds of *tcdA* left, while *tcdR, B, E* are completely missing (Rupnik *et al.*, 2001).

Insertions are also present and one of them is found at the upstream PaLoc border of several toxinotypes (Fig. 6.2). Toxinotype X has another characteristic >1 kb insertion between *tcdB* and *tcdE* and some authors reported that it contains new ORF (Song *et al.*, 1999).

In 5′-end of *tcdA* a large insertion of about 2000 bp was first described in toxinotype XIV (Mehlig *et al.*, 2001) and thereafter also in other toxinotypes (XVII, XXII and XXIII) (Rupnik *et al.*, 2003; Geric *et al.*, 2004). This large insertion is a mobile element ISTron and is specific only for *C. difficile* (Braun *et al.*, 2000b). It is spliced from mRNA and TcdA produced by these toxinotypes is of usual size.

Changes in nucleotide sequences

Changes in nucleotide sequence are detected as RFLPs of amplified fragments and were confirmed in some toxinotypes by sequencing. Published sequences of whole or partial variant toxin genes include toxinotypes VIII, X and XIV (GenBank numbers Z23277, AF217292, AJ011301, AJ294944, Y10689).

Mutations are not distributed equally over entire length of toxin genes nor are they present in comparable proportions in both toxin genes.

Whole length *tcdB* gene in toxinotypes 0 (VPI 10463) and VIII (strain 1470, serogroup F) shows 94,7% identity at nucleotide level. The majority of mutations is located at 5′-end and the identity in first 1700 bp is only 83% (Eichel-Streiber *et al.*, 1995).

That genes for *tcdB* are much more diverse than *tcdA* genes was first described already on the basis of RFLPs (Rupnik *et al.*, 1998). If published and unpublished sequences (first 1700 bp) of toxinotypes 0, III, IV, VI, VIII, IX and X are compared on the nucleotide level identity from 83% to 98% is observed in *tcdB* and identity from 98.1–98.7% is observed in *tcdA* sequences (Fig. 6.4).

Of non-toxin PaLoc region most is known about *tcdC* variability and 17 variants are described according to deletions and mutations (Curry *et al.*, 2007). This gene has beside deletions of 18–56 bp often also point deletions which sometimes introduce stop codon and premature ending of protein synthesis (MacCannell *et al.*, 2006b; Curry *et al.*, 2007).

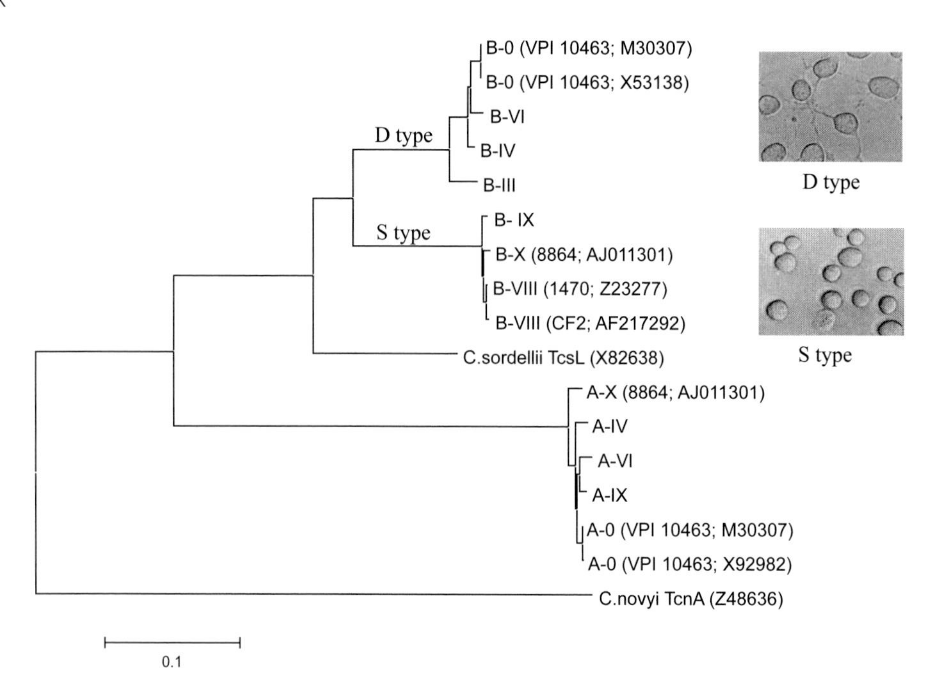

Figure 6.4 Dendrogram of 1700 bp 3′-end sequences of *tcdA* and *tcdB* genes from selected toxinotypes. Large clostridial toxins produced by *C. sordellii* and *C. novyi* are also included. Genes *tcdA* form only one group while *tcdB* genes are much more diverse and group into two clusters which correlate well with cytopathic effect (CPE) caused by TcdB. Two forms of CPE are described: type D is typical for most of the *C. difficile* wild type and variant toxins and type S is caused by *C. sordellii* TcsL and by some of the variant *C. difficile* TcdBs.

Consequences of gene variability on protein structure and function

Deletions in *tcdA* influence the protein size but not the reactivity with antibodies

Deletions in *tcdA* are mostly located in repetitive domain encoding immunodominant part of the TcdA and such toxinotypes with shorter *tcdA* express toxins of lower molecular weight. However, low MW toxins A still react with toxin A-specific antibodies (Blake *et al.*, 2004). Even the large deletion present in A^-B^+ toxinotype VIII does not influence the immunoreactivity (Eichel-Streiber *et al.*, 1999). But it has to be emphasized that reactivity was detected with cloned and recombinantly expressed fragment of truncated repetitive region of *tcdA-VIII*, because no whole-length TcdA is produced by toxinotype VIII strains (see below).

Repeats in *tcdA* can be grouped into short and long forms and in wild type TcdA two long repeats are always separated by several short repeats. Interestingly, even in truncated *tcdA* of toxinotype VIII this succession of long/*n* x short/long/*n* x short ... is conserved suggesting it is important for proper function of the toxin (Ho *et al.*, 2005).

Mutations can result in non-production of TcdA or TcdC

Toxinotype VIII strains do not react with commercial diagnostic TcdA-specific immunodetection kits nor can the TcdA be purified from this type. The reason for absence of the toxin is a nonsense mutation introducing a stop codon and terminating TcdA translation at amino acid position 47 (Eichel-Streiber *et al.*, 1999). Non-production of TcdA and not the deletion in *tcdA* is therefore the reason for non-reactivity of toxinotype VIII in commercial tests specific for TcdA only.

Similarly to toxinotype VIII where deletion was first used as explanation for non-immunoreactivity, much attention was given to the 18-bp deletion in *tcdC* gene. Particularly in the epidemic type NAP1/027 it was suggested to affect functionality of TcdC and could be responsible for increased toxin production. However, changes in regulation is probably due to the point deletion at position 117 which introduce a nonsense codon downstream and result in truncated 65 amino acid TcdC (MacCanell *et al.*, 2006b). Other point mutations can also cause the production of shorter TcdC variants (Curry *et al.*, 2007).

Variant toxins and different substrate recognition patterns

Mutations influence the GTPase pattern used as a substrate and variant pattern was described for TcdB toxins of toxinotype VIII (Chaves-Olarte 1999), toxinotype X (Soehn *et al.*, 1998) and toxinotype XIV (Mehlig *et al.*, 2001). Same as wild type $TcdB_{VPI10463}$ which modifies Rho, Rac and Cdc42 all three variant toxinotypes also use Rac and Cdc42 as a substrate while Rho is modified only by toxinotype XIV. Additionally, all three variant toxinotypes modify Rap, Ral and R-Ras.

Variant toxins and different cytopathic effect

Large clostridial cytotoxins (LCT) can cause two types of cytopathic effects (Fig. 6.4). *C. difficile* toxins A and B typically cause cell rounding with remaining protrusions (difficile type) and *C. sordellii* toxins typically cause complete rounding of the cells. Some variant *C. difficile* TcdBs do have 'sordellii type' instead of 'difficile type' of cytopathic effect (Torres, 1991; Rupnik, 2001; Blake *et al.*, 2004; Chaves-Olarte *et al.*, 1999). The CPE type is influenced by the modified GTPases (Chaves-Olarte *et al.*, 2003) which again depends on the primary differences in toxin coding sequence. Therefore it is not surprising that catalytic domains of variant TcdBs with 'sordellii' type of CPE group together (Fig. 6.4).

Other differences in function of variant toxins

For TcdB, in contrast to TcdA, some immunological differences were observed. TcdBs produced by toxinotypes VIII (serogroup F) and X did not react with some of mAbs specific for reference $TcdB_{VPI10463}$ (Lyerly *et al.*, 1992; Depitre *et al.*, 1993).

For $TcdB_{8864}$ (toxinotype X) also lethal dose in mouse model is reported (6 ng/g), which is much lower than wild type $TcdB_{VPI10463}$ (Lyerly *et al.*, 1992).

Origin of *C. difficile* toxinotypes

Some toxinotypes have only a single representative strain while others include larger number of strains (from approximately five to several hundreds). All strains in given toxinotype correlate well with other molecular typing methods, such as ribotyping (Rupnik *et al.*, 2001; Johnson *et al.*, 2003).

Toxinotypes in which the only difference compared to the reference strain VPI10463 are deletions in *tcdA* are most probably spontaneous mutants. Repeats in *tcdA* form five homology classes (Eichel-Streiber *et al.*, 1992) and could therefore participate in homologous recombination events resulting in deletions or insertions.

In the case of toxinotypes that differ also in nucleotide sequence the number of point mutations is high. The possibility that several strains with identical large number of mutations would emerge independently is unlikely. That a single toxinotype contains clonal or related strains was shown with several molecular methods on the example of toxinotypes VIII and IIIb (Stabler *et al.*, 2006; McDonald *et al.*, 2005; Lemee *et al.*, 2004; Kato *et al.*, 1998; MacCannell *et al.*, 2006b; Curry *et al.*, 2007). Nevertheless, the question is, how have such clones developed and how are different toxinotypes related one to another.

Fig. 6.5 shows two possible models. According to model A all variant toxinotypes have common ancestor from which they have further diversified. Model B represents the development of variant toxinotypes in several independent events. Analysis on genomic relatedness of variant and non-variant strains from serogroups including several toxinotypes (tox−, 0, III, IV, V, VI, VIII, IX) supports the model B (Wozniak *et al.*, 2000).

It is very likely that toxinotype 0 strains were the ancestral line, not only because this type is most numerous at the time, but also, because the older PCR based studies did not detect lot of variant strains (McMillin *et al.*, 1991; Boondeekhun *et al.*, 1993). However, the molecular mechanisms causing increased variability of catalytic

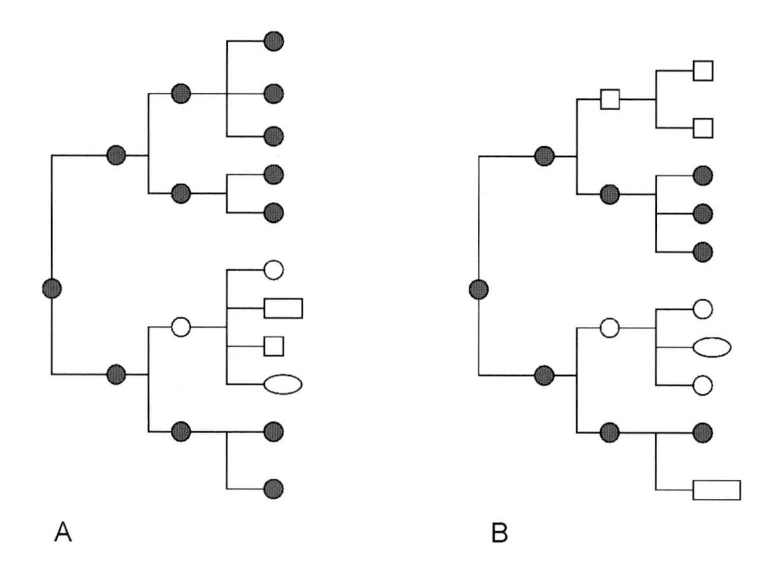

Figure 6.5 Two models of variant strains origin and evolution. (A) All variant *C. difficile* strains (white symbols) could have common ancestor. (B) Different toxinotypes have evolved independently in several lineages from toxinotype 0 strains (black symbols).

parts of the toxin genes and especially the *tcdB* gene are not known at the time. Similarely, it is not explained why only (unrelated) strains with variant PaLoc posses third toxin CDT.

Distribution of *C. difficile* toxinotypes in human and animal hosts

In humans the proportion and types of variant toxinotypes differ among hospitals and among countries.

In strain collections of different sizes proportion of variant strains was 1.8 % (Hungary; Terhes *et al.*, 2004), 13.7% (Italy, Spigaglia and Mastrantonio, 2002), 7.7% (large ribotyped collection; Rupnik *et al.*, 2001) or 21.5 % (large serotyped collection; Rupnik *et al.*, 1998). In a single hospital in a non-outbreak situation the proportion of variant strains can range from 0 to 13% (Rupnik *et al.*, 2003, Goncalves *et al.*, 2004; Geric *et al.*, 2004; Alonso *et al.*, 2005). In EU hospitals the average proportion of variant strains found in 2005 was 20.9% (Barbut *et al.*, 2007). This high percentage is most likely due to local outbreaks of III/027 and VIII strains.

Toxinotypes most often isolated from human patients are III, IV, V, VIII, IX and XII. The same toxinotypes (with exception of type IX and addition of toxinotype XI) were so far also reported from different animal hosts such as horses, pigs, calves and dogs (Braun *et al.*, 2000b; Lefebvre *et al.*, 2006; Keel *et al.*, 2007; Rodriguez-Palacios *et al.*, 2006). Moreover, variant C. *difficile* strains are present also in food (Rodriguez-Palacios *et al.*, 2007). Toxinotype V seems to be particularly adapted to animals. It was found in most of the animals in highest proportion and is present all over the world.

Toxinotypes and laboratory diagnostics

In routine laboratories diagnostics was largely based on direct detection of toxin(s) with commercial immunological tests. Before description of A^-B^+ strains commercial tests were specific only for toxin A. Only minor proportion of laboratories did perform culture or used more sensitive toxin detection test with cell culture cytotoxicity.

After A^-B^+ strains were detected in outbreaks and in severe PMC cases (see below) the need for upgraded toxin test became evident. To date most of the suppliers offer different commercial tests that detect both toxins, TcdA and TcdB.

Some toxinotypes could be associated with disease of increased severity and their recognition in routine lab could be of some advantage. With classical diagnostic methods (culture and/or toxin test) variant C. *difficile* strains are detected but not recognized as variants. Possibilities to distinguish the variant C. *difficile* strains from ordinary ones in routine laboratory include (Rupnik, 2001): (1) detection of changed toxin production as all strains that produce only TcdB ($A^{-}B^{+}$) belong to variant toxinotypes, (2) detection of atypical cytopathic effect as strains that cause 'sordellii-like' effect on cultured cells belong to variant toxinotypes, (3) detection of binary toxin genes as strains that produce binary toxin or have binary toxin gene most likely belong to variant toxinotypes, (4) detection of polymorphic *tcdC* gene, (5) ribotyping as toxinotypes correlate with toxinotypes. The most efficient, but a time-consuming method, for recognition of variant strains is amplification and analysis of B1-, A3- and *tcdC* PCR fragments. With this method all known as well as new variant strains could be detected (www.mf.uni-mb.si/mikro/tox).

Toxinotypes and human disease

Toxinotypes do not correlate with a specific form of the disease. Whether the presence of variant toxin gene *tcdA* or *tcdB* influences the strain virulence can be answered only with constructing appropriate mutants (exchanging wild type toxin gene with a variant one). Only recently have the appropriate methods become available, therefore no data have been published so far.

Toxinotypes are isolated from all known forms of C. *difficile*-associated diseases. Currently, two toxinotypes are associated with outbreaks and severe forms of disease, toxinotype IIIb and toxinotype VIII (McDonald *et al.*, 2005; Kuijper *et al.*, 2006). Data on the variant toxinotypes causing community-acquired disease are very limited. In the USA, toxinotypes IX and XIV/XV have been reported (Chernak *et al.*, 2005), and toxinotype III/027 was community associated in some Canadian regions (MacCannell *et al.*, 2006b).

Already before emergence of type 027 were other non-027 strains from toxinotype III often published in association with PMC (Rupnik *et al.*, 1998; Geric *et al.*, 2004).

Toxinotype VIII strains can be isolated from asymptomatic neonates and children (Depitre *et al.*, 1993) but can also cause outbreaks and severe PMC cases (Alfa *et al.*, 2000; Kuijper *et al.*, 2001; Barbut *et al.*, 2002; Drudy *et al.*, 2007; Johnson *et al.*, 2001; Limaye *et al.*, 2000; Komatsu *et al.*, 2003).

Emergence of $A^{-}B^{+}$ strains in association with severe infection was a surprise and has changed our understanding of the role of the toxins in pathogenesis. Early animal model studies have suggested TcdA to be primary virulence factor (Lyerly *et al.*, 1985). In these studies TcdA alone caused the symptoms while animals given pure TcdB did not develop the disease. The symptoms developed only if TcdB was applied together with subtoxic doses of TcdA. However, $A^{-}B^{+}$ strains also cause (severe) disease in the absence of TcdA. Whether polymorphic TcdB produced by such strains has some of the *in vivo* TcdA functions or toxinotype VIII strains just have genetic background with additional virulence factors needs to be clarified.

Summary

C. *difficile* shows large variability in region encoding two main virulence factors, toxins TcdA and TcdB. Such variations can affect the toxin production or could result in production of toxins with altered properties. Variant strains used to be less prevalent in human population but their isolation rate is increasing. Interestingly, other clostridia producing LCTs also had some variations (Green *et al.*, 1996) and a new form of LCT toxin without repetitive regions was described recently in C. *perfringens* (Amimoto *et al.*, 2007).

References

Alfa, M.J., Kabani A, Lyerly, D., Moncrief, S., Neville, L.M., Al-Barrak, A., Harding, G.K., Dyck, B., Olekson, K., and Embil, J.M. (2000). Characterization of a toxin A-negative, toxin B-positive strain of *Clostridium difficile* responsible for a nosocomial outbreak of *Clostridium difficile*-associated diarrhea. J. Clin. Microbiol. 38, 2706–2714.

Alonso, R., Martin, A., Pelaez, T., Marin, M., Rodriguez-Creixems, M., and Bouza, E. (2005). Toxigenic status of *Clostridium difficile* in a large Spanish teaching hospital. J. Med. Microbiol. 54, 159–162.

Amimoto, K., Taichi, N., Eiji, O., and Mitsugu, S. (2007). A novel toxin homologous to large clostridial cytotoxins found in culture supernatant of *Clostridium perfringens* type. Microbiology 153, 1198–1206.

Barbut, F., Lalande, V., Burghoffer, B., Thien, H.V., Grimprel, E., and Petit, J.C. (2002). Prevalence and genetic characterization of toxin A variant strains of

Clostridium difficile among adults and children with diarrhea in France. J. Clin. Microbiol. *40*, 2079–2083.

Barbut, F., Mastrantonio, P., Delmee, M., Brazier, J., Kuijper, E., and Poxton, I.; on behalf of the European Study Group on Clostridium difficile (ESGCD). (2007). Prospective study of *Clostridium difficile* infections in Europe with phenotypic and genotypic characterisation of the isolates. Clin. Microbiol. Infect. *13*, 1048–1057.

Blake, J.E., Mitsikosta, F., and Metcalfe, M.A. (2004). Immunological detection and cytotoxic properties of toxins from toxin A-positive, toxin B-positive *Clostridium difficile* variants. J. Med. Microbiol. *53*, 197–205.

Boondeekhun, H.S., Gurtler, V., Odd, M.L., Wilson, V.A., and Mayall B.C. (1993). Detection of *Clostridium difficile* enterotoxin gene in clinical specimens by the polymerase chain reaction. J. Med. Mikrobiol. *38*, 384–387.

Borriello, S.P., Wren, B.W., Hyde, S., Seddon, S.V., Sibbons, P., Krishna, M.M., Tabaqchali, S., Manek, S., and Price, A.B. (1992). Molecular, immunological, and biological characterization of a toxin A-negative, toxin B-positive strain of *Clostridium difficile*. Infect. Immun. *60*, 4192–4199.

Braun, M., Herholz, C., Straub, R., Choisat, B., Frey, J., Nicolet, J., and Kuhnert, P. (2000a). Detection of the ADP-ribosyltransferase toxin gene (*cdt*A) and its activity in *Clostridium difficile* isolates from Equidae. FEMS Microbiol. Lett. *184*, 29–33.

Braun, V., Hundsberger, T., Leukel, P., Sauerborn, M., and Eichel-Streiber, C. (1996). Definition of the single integration site of the pathogenicity locus in *Clostridium difficile*. Gene *181*, 29–38.

Braun, V., Mehlig, M., Moos, M., Rupnik, M., Kalt, B., Mahony, D.E., and Eichel-Streiber, C. (2000b). A chimeric ribozyme in *Clostridium difficile* combines features of group I introns and insertion elements. Mol. Microbiol. *36*, 1447–1459.

Chaves-Olarte, E., Low, P., Freer, E., Norlin, T., Weidmann, M., Eichel-Streiber, C., and Thelestam, M. (1999). A novel cytotoxin from *Clostridium difficile* serogroup F is a functional hybrid between two other large clostridial cytotoxins. J. Biol. Chem. *274*, 11046–11052.

Chaves-Olarte, E., Freer, E., Parra, A., Guzman-Verri, C., Moreno, E., and Thelestam, M. (2003). R-Ras glucosylation and transient RhoA activation determine the cytopathic effect produced by toxin B variants from toxin A-negative strains of *Clostridium difficile*. J. Biol. Chem. *278*, 7956–7963.

Chernak, E., Johnson, C.C., Weltman, A., McDonald, L.C., Wiggs, L., Killgore, G., Thompson, A., LeMaile-Williams, M., Tan, E., Lewis, F.M. (2005). Severe *Clostridium difficile*–associated disease in populations previously at low risk-four states. MMWR 54. 1201–1205.

Curry, S.R., Marsh, J.W., Muto, C.A., O'Leary, M.M., Pasculle, A.W., and Harrison, L.H. (2007). *tcdC* genotypes associated with severe TcdC truncation in an epidemic clone and other strains of *Clostridium difficile*. J. Clin. Microbiol. 45, 215–221.

Depitre, C., Delmée, M., Avesani, V., L'Haridon, R., Roels, A., Popoff, M., and Corthier, G. (1993). Serogroup F strains of *Clostridium difficile* produce toxin B but not toxin A. J. Med. Microbiol. *38*, 434–441.

Drudy, D., Harnedy, N., Fanning, S., Hannan, M., and Kyne, L. (2007). Emergence and control of fluoroquinolone-resistant, toxin A-negative, toxin B-positive *Clostridium difficile*. Infect. Control. Hosp. Epidemiol. *28*, 932–940.

Eichel-Streiber, C., Laufenberg-Feldmann, R., Sartingen, S., Shulze, J., and Sauerborn, M. (1992). Comparative sequence analysis of the *Clostridium difficile* toxins A and B. Mol. Gen. Genet. *233*, 260–268.

Eichel-Streiber, C., Meyer zu Heringdorf, D., Habermann, E., and Sartingen, S. (1995). Closing in on the toxic domain through analysis of a variant *Clostridium difficile* cytotoxin B. Mol. Microbiol. *17*, 313–321.

Eichel-Streiber, C., Boquet, P., Sauerborn, M., and Thelestam, M. (1996). Large clostridial cytotoxins – a family of glycosyltransferases modifying small GTP-binding proteins. Trends Microbiol. *4*, 375–382.

Eichel-Streiber, C., Zec-Pirnat, I., Grabnar, M., and Rupnik, M. (1999). A nonsense mutation abrogates production of a functional enterotoxin A in *Clostridium difficile* toxinotype VIII strains of serogroups F and X. FEMS Microbiol. Lett. *178*, 163–168.

Geric, B., Johnson, S., Gerding, D.N., Grabnar, M., and Rupnik, M. (2003). Frequency of binary toxin genes among *Clostridium difficile* strains that do not produce large clostridial toxins. J. Clin. Microbiol. *41*, 5227–5232.

Geric, B., Rupnik, M., Gerding, D.N., Grabnar, M., and Johnson, S. (2004). Distribution of *Clostridium difficile* variant toxinotypes and strains with binary toxin genes among clinical isolates in an American hospital. J. Med. Microbiol. 53, 887–894.

Goncalves, C., Decre, D., Barbut, F., Burghoffer, B., and Petit, J.C. (2004). Prevalence and characterization of a binary toxin (actin-specific ADP-ribosyltransferase) from *Clostridium difficile*. J. Clin. Microbiol. *42*, 1933–1939.

Green, G.A., Schué, V., Girardot, R., and Monteil, H. (1996). Characterization of an enteritoxin-negative, cytotoxin-positive strain of *Clostridium sordellii*. J. Med. Microbiol. *44*, 60–64.

Hammond, G.A., Johnson, J.L. (1995). The toxinogenic element of *Clostridium difficile* strain VPI 10463. Microbial Pathog. *19*, 203–213.

Haslam, S.C., Ketley, J.M., Mitchell, T.J., Stephen, J., Burdon, D.W., and Candy, D.C.A. (1986). Growth of *Clostridium difficile* and production of toxins A and B in complex and defined media. J. Med. Microbiol. *21*, 293–297.

Ho, J.G., Greco, A., Rupnik, M., Ng, K.K. (2005). Crystal structure of receptor-binding C-terminal repeats from Clostridium difficile toxin A. Proc. Natl. Acad. Sci. USA *102*, 18373–18378.

Johnson, S., Kent, S.A., O'Leary, K.J., Merrigan, M.M., Sambol, S.P., Peterson, L.R., and Gerding, D.N. (2001). Fatal pseudomembranous colitis associated with a variant *Clostridium difficile* strain not detected by toxin A immunoassay. Ann. Intern. Med. *135*, 434–438.

Johnson, S., Sambol, S.P., Brazier, J.S., Delmee, M., Avesani, V., Merrigan, M.M., and Gerding, D.N. (2003). International typing study of toxin A-negative, toxin B-positive *Clostridium difficile* variants. J. Clin. Microbiol. *41*, 1543–1547.

Kato, H., Kato, N., Watanabe, K., Iwai, N., Nakamura, H., Yamamoto, T., Suzuki, K., Kim, S.M., Chong, Y., and Wasito, E.B. (1998). Identification of toxin A-negative, toxin B-positive *Clostridium difficile* by PCR. J. Clin. Microbiol. *36*, 2178–2182.

Kato, H., Kato, N., Katow, S., Maegawa, T., Nakamura, S., and Lyerly, D.M. (1999). Deletions in the repeating sequences of the toxin A gene of toxin A-negative, toxin B-positive *Clostridium difficile* strains. FEMS Microbiol. Lett. *175*, 197–203.

Komatsu, M., Kato, H., Aihara, M., Shimakawa, K., Iwasaki, M., Nagasaka, Y., Fukuda, S., Matsuo, S., Arakawa, Y., Watanabe, M., and Iwatani, Y. (2003). High frequency of antibiotic-associated diarrhea due to toxin A-negative, toxin B-positive *Clostridium difficile* in a hospital in Japan and risk factors for infection. Eur. J. Clin. Microbiol. Infect. Dis. *22*, 525–529.

Kuijper, E.J., de Weerdt, J., Kato, H., Kato, N., van Dam, A.P., van der Vorm, E.R., Weel, J., van Rheenen, C., and Dankert, J. (2001). Nosocomial outbreak of *Clostridium difficile*-associated diarrhoea due to a clindamycin-resistant enterotoxin A-negative strain. Eur. J. Clin. Microbiol. Infect. Dis. *20*, 528–534.

Kuijper, E.J., Coignard, B., and Tull, P. (2006). Emergence of *Clostridium difficile*-associated disease in North America and Europe. Clin. Microbiol. Infect. *12* (Suppl 6), 2–18.

Keel, K., Brazier, J.S., Post, K.W., Weese, S., and Songer, J.G. (2007). Prevalence of PCR ribotypes among *Clostridium difficile* isolates from pigs, calves, and other species. J. Clin. Microbiol. *45*, 1963–1964.

Lefebvre, S.L., Arroyo, L.G., and Weese, J.S. (2006). Epidemic *Clostridium difficile* strain in hospital visitation dog. Emerg. Infect. Dis. *12*, 1036–1037.

Lemee, L., Dhalluin, A., Pestel-Caron, M., Lemeland, J.F., and Pons, J.L. (2004). Multilocus sequence typing analysis of human and animal *Clostridium difficile* isolates of various toxigenic types. J. Clin. Microbiol. *42*, 2609–2617.

Limaye, A.P., Turgeon, D.K., Cookson, B.T., and Fritsche, T.R. (2000). Pseudomembranous colitis caused by a toxin A(-) B(+) strain of *Clostridium difficile*. J. Clin. Microbiol. *38*, 1696–1697.

Lyerly, D.M., Saum, K.E., MacDonald, D.K., and Wilkins, T.D. (1985). Effects of *Clostridium difficile* given intragastrically to animals. Infect. Immun. *47*, 349–352.

Lyerly, D.M., Barroso, L.A., Wilkins, T.D., Depitre, C., and Corthier, G. (1992). Characterization of a toxin A-negative, toxin B-positive strain of *Clostridium difficile*. Infect. Immun. *60*, 4633–4639.

MacCannell, D.R., Louie, T.J., Rupnik, M., Krulicki, W., Armstrong, G., Emery, J., Ward, L., and Lye, T. (2006a). Characterization of a novel, TcdB-deficient NAP1 variant strain of *Clostridium difficile*. Presented at the 46th ICAAC Conference, San Francisco, September 2006.

MacCannell, D.R., Louie, T.J., Gregson, D.B., Laverdiere, M., Labbe, A.C., Laing A., and Henwick, S. (2006b). Molecular analysis of *Clostridium difficile* PCR ribotype 027 isolates from eastern and western Canada. J. Clin. Microbiol. *44*, 2147–2152.

McDonald, L.C., Killgore, G.E., Thompson, A., Owens, R.C., Kazakova, S.V., Sambol, S.P., Johnson, S., and Gerding, D.N. (2005). An epidemic, toxin gene–variant strain of *Clostridium difficile*. N. Engl. J. Med. *353*, 2433–2441.

McMillin, D.E., Muldrow, L.L., Leggette, S.J., Abdulahi, Y., and Ekanemesang, U.M. (1991). Molecular screening of *Clostridium difficile* toxins A and B genetic determinants and identification of mutant strains. FEMS Microbiol. Lett. *78*, 75–80.

Mehlig, M., Moos, M., Braun, V., Kalt, B., Mahony, D.E., and Eichel-Streiber, C. (2001). Variant toxin B and a functional toxin A produced by *Clostridium difficile* C34. FEMS Microbiol. Lett. *198*, 171–176.

Perelle, S., Gibert, M., Bourlioux, P., Corthier, G., and Popoff, M.R. (1997). Production of a complete binary toxin (actin-specific ADP-ribosyltranferase) by *Clostridium difficile* CD196. Infect. Immun. *65*, 1402–1407.

Rodruiges-Palacios, A., Stempfli, H., Duffield, T., Peregrine, A.S., Trotz-Williams, L.A., Arrojo, L.G., Brazier, J.S., and Weese, J.S. (2006). *Clostridium difficile* PCR ribotypes in calves, Canada. Emerg. Infect. Dis. *12*, 1730–1736.

Rodriguez-Palacios, A., Staempfli, H.R., Duffield, T., and Weese, J.S. (2007). Clostridium difficile in retail ground meat, Canada. Emerg. Infect. Dis. *13*, 485–487.

Rupnik, M., Braun, V., Soehn, F., Janc, M., Hofstetter, M., Laufenberg-Feldmann, R., and Eichel-Streiber, C. (1997). Characterization of polymorphisms in the toxin A and B genes of *Clostridium difficile*. FEMS Microbiol. Lett. *148*, 197–202.

Rupnik, M., Avesani, V., Janc, M., Eichel-Streiber, C., and Delmee, M. (1998). A novel toxinotyping scheme and correlation of toxinotypes with serogroups of *Clostridium difficile* isolates. J. Clin. Microbiol. *36*, 2240–2247.

Rupnik, M., Brazier, J.S., Duerden, B.I., Grabnar, M., and Stubbs, S.L. (2001). Comparison of toxinotyping and PCR ribotyping of *Clostridium difficile* strains and description of novel toxinotypes. Microbiology *147*, 439–447.

Rupnik, M. (2001). How to detect *Clostridium difficile* variant strains in a routine laboratory. Clin. Microbiol. Infect. *7*, 417–420.

Rupnik, M., Kato, N., Grabnar, M., and Kato, H. (2003). New types of toxin A-negative, toxin B-positive strains among *Clostridium difficile* isolates from Asia. J. Clin. Microbiol. *41*, 1118–1125.

Sambol, S.P., Merrigan, M.M., Lyerly, D., Gerding, D.N., and Johnson, S. (2000). Toxin gene analysis of a variant strain of *Clostridium difficile* that causes human clinical disease. Infect. Immun. *68*, 5480–5487.

Soehn, F., Wagenknecht-Wiesner, A., Leukel, P., Kohl, M., Weidmann, M., Eichel-Streiber, C., and Braun, V. (1998). Genetic rearrangements in the pathogenicity locus of Clostridium difficile strain 8864-implications for transcription, expression and enzymatic activity of toxins A and B. Mol. Gen. Genet. *258*, 222–232.

Song, K.P., Ow, S.E., Chang, S.Y., and Bai. X.L. (1999). Sequence analysis of a new open reading frame located in the pathogenicity locus of Clostridium difficile strain 8864. FEMS Microbiol. Lett. *180*, 241–248.

Spigaglia, P., and Mastrantonio, P. (2002). Molecular analysis of the pathogenicity locus and polymorphism in the putative negative regulator of toxin production (TcdC) among *Clostridium difficile* clinical isolates. J. Clin. Microbiol. *40*, 3470–3475.

Stabler, R.A., Gerding, D.N., Songer, J.G., Drudy, D., Brazier, J.S., Trinh, H.T., Witney, A.A., Hinds, J., and Wren, B.W. (2006). Comparative phylogenomics of *Clostridium difficile* reveals clade specificity and microevolution of hypervirulent strains. J. Bacteriol. *188*, 7297–7305.

Stare, B.G., Delmee, M., and Rupnik, M. (2007). Variant forms of the binary toxin CDT locus and *tcdC* gene in *Clostridium difficile* strains. J. Med. Microbiol. *56*, 329–335.

Stubbs, S., Rupnik, M., Gibert, M., Brazier, J., Duerden, B., and Popoff, M. (2000). Production of actin-specific ADP-ribosyltransferase (binary toxin) by strains of *Clostridium difficile*. FEMS Microbiol. Lett. *186*, 307–312.

Terhes, G., Urban, E., Soki, J., Hamid, K.A., and Nagy, E. (2004). Community-acquired *Clostridium difficile* diarrhea caused by binary toxin, toxin A, and toxin B gene-positive isolates in Hungary. J. Clin. Microbiol. *42*, 4316–4318.

Torres, J.F. (1991). Purification and characterisation of toxin B from a strain of *Clostridium difficile* that does not produce toxin A. J. Med. Mikrobiol. 35, 40–44.

Wozniak, G., Trontelj, P., and Rupnik, M. (2000). Genomic relatedness of *Clostridium difficile* strains from different toxinotypes and serogroups. Anaerobe *6*, 261–267.

Comparative Genomics of *Clostridium difficile* 7

Lisa F. Dawson, Richard A. Stabler and Brendan W. Wren

Abstract

The recent emergence of hypervirulent strains of *Clostridium difficile* and their ability to spread across continents has caused alarm in both hospitals and the community. This has drawn attention away from other important pathogenic *C. difficile* strains, which are responsible for significant morbidity and mortality. Little is known about the genetic diversity of these strains and their less pathogenic counterparts. The recent publication of the genome sequence of strain 630 and advances in both microarray and mutagenesis technologies promises to revolutionize our understanding of the pathogenesis and population dynamics of *C. difficile*. This chapter summarizes the salient findings of the 630 genome sequence and includes phylogenetic analysis of *C. difficile* strains from diverse origins.

Introduction

The importance of the human pathogen *Clostridium difficile* has been brought to the fore recently because of the large outbreaks of *C. difficile* associate disease (CDAD). The transcontinental spread of the hypervirulent strains responsible for outbreaks in Europe, USA and Canada (Loo *et al.*, 2005; McDonald *et al.*, 2005; Warny *et al.*, 2005) coupled with the severity of the disease, means a better understanding of the basic pathogenesis and epidemiology of *C. difficile* is a public and political imperative. The emergent strain, denoted 027 by ribotyping is hypervirulent and resistant to most antibiotics including fluoroquinolones. The rapid spread of the 027 strain has detracted attention from other virulent strains of *C. difficile*, such as 053 and A^-B^+ strains, which are epidemic in Europe and Korea respectively. In recent years, there have been increasing numbers of strains, from several countries with truncated versions of toxin A or toxin B (e.g. A^-B^+/A^-B^-) (Geric *et al.*, 2006; Toyokawa *et al.*, 2003). Additionally epidemiological and clinical studies have shown the emergence of community acquired CDAD (CA-CDAD), in addition to the more widely noted nosocomial transmission. This may be of evolutionary importance, specifically as the toxinotype V strains (A^+B^+), which are predominantly restricted to animals, have been identified as the cause of disease in a number of community acquired infections (Limbago, 2007), suggesting a link between contaminated food/meat products and disease.

It is essential to understand the evolution of both epidemic and non-epidemic strains of *C. difficile*, with respect to their source and genetic constitution. Greater understanding of virulence determinants of strains, in evolutionary terms would provide valuable insights into the pathogenicity of the emergent strains. With this in mind we initiated the *C. difficile* 630 genome sequence in 1998. However, for the genome comparison of larger sets of bacterial strains, microarray-based technology is required that can rapidly provide a bird's eye view of the genome content. During the sequencing of strain 630 we simultaneously constructed a *C. difficile* microarray for expression analysis and whole genome comparisons. The use of comparative genomics combined with robust methods for data analysis (comparative phylogenomics) will form the basis for understanding the population structure of *C. difficile* that may lead to the development of novel intervention strategies to reduce the burden of CDAD. Such studies using a wide variety of clinical, animal and environmental isolates can identify key genes that could be used to differentiate strains from different sources. In this chapter we will review the salient comparative features of the 630 genome sequence and set this into context with the newly developed comparative phylogenomics approach relating to the identification of determinants relevant to strain source and the evolution of virulence.

Strain identification in *C. difficile*

The current methods of typing *C. difficile* provide a valuable resource to clinical laboratories in terms of strain identification; however the information gained is limited in terms of analysing genetic and phenotypic differences between strains and subtypes. Current typing methods, outlined elsewhere in this book, vary in terms of effectiveness and discriminatory powers, which are increased when methods are used concordantly. The subtyping of various strains proves more difficult and in the case of the hypervirulent 027 strains, additional methods are currently being developed. The recent publication of the genome sequence of strain 630 may provide the basis for the further development of typing methods, particularly for differentiating subtypes.

Impact of the genome sequence of 630

The recent publication of the genome sequence of strain 630 (Sebaihia *et al.*, 2006) has been a milestone in *C. difficile* research. The project was conceived in 1998 when a collaboration of scientists augmented the sequencing of the first clostridial species. A fully virulent, highly transmissible, multidrug-resistant strain, 630 was chosen; it was isolated in Zurich, Switzerland from a patient with severe pseudomembranous colitis and proved an ideal strain for analysis due to the disease severity and the ferocity with which it spread.

C. difficile strain 630 has a circular chromosome of 4,290,252 bp and a plasmid of 7881 bp. The chromosome encodes 3776 predicted coding sequences (CDSs), the majority of which (82.1%) are encoded on the leading strand, which is common among low GC Gram-positive bacteria. Comparative analysis with other sequenced clostridial species was performed using reciprocal FASTA analysis, which revealed a low degree of inter-strain homology: *C. acetobutylicum* (Nolling *et al.*, 2001), *C. botulinum*, *C. perfringens* and *C. tetani* share only 15% of *C. difficile* CDSs. The conserved clostridial CDSs mainly encode essential functions, but also encoded potential virulence factors.

Mobile genetic elements

The genome sequence revealed a number of unusual genetic attributes particular to *C. difficile*. The plasticity of the genome may be linked to the high percentage of mobile genetic elements, including conjugative transposons, prophage, IS elements and unusually for a bacterium, IStrons.

Transposons

There are seven conjugative transposons, which harbour the majority of the antibiotic resistance genes in *C. difficile* strain 630 (Sebaihia *et al.*, 2006), including a proportion of the genes that could be important in the development of vancomycin resistance, which exhibits 58–77% identity across the vancomycin resistance locus from *Enterococcus faecalis* (Sebaihia *et al.*, 2006). This may be a particular concern as vancomycin is the current drug of choice use to treat CDAD, alongside metronidazole, whose transient resistance has been noted in *C. perfringens* and *C. difficile* (Hell, 2007; Pituch *et al.*, 2007). CTn3 (Tn5397) in strain 630 was particularly interesting because this transposon harbours a group II intron, which is able to mediate its own splicing from mRNA and is able to mobilize into specific sites within the genome (Mills *et al.*, 1997). This is of particular relevance as the Minton laboratory are in the process of using a derivative of the group II intron from *Lactococcus lactis* to develop a ClosTron system, designed to produce specific gene inactivations in *C. difficile*.

IStrons

The genome contains eight complete and several partial IStrons, which are chimeras between a group I intron and an insertion sequence (IS); this enables the IStrons to insert into the DNA (via the IS) and the group I intron allows the IStron to excise from the mRNA transcript, thus restoring the gene function at the site of insertion. The complete IStrons were all located in CDSs, whereas all the incomplete IStrons were located in DNA intervening CDSs. Interestingly, an IStron was identified in *tcdA*, which encodes toxin A, however, the IStron is successfully spliced out of the primary transcript, therefore the production of TcdA is uninterrupted (Sebaihia *et al.*, 2006).

CRISPR elements

Another unusual observation was the presence of 10 CRISPR elements, otherwise known as clustered regularly interspaced short palindromic repeats. This is among the highest identified for any bacterium to date. Although the functions of these elements are unknown, the spacer regions in the CRISPR elements contain sequence homology to bacteriophage and plasmids (Bolotin *et al.*, 2005; Mojica *et al.*, 2005; Pourcel *et al.*, 2005). These are thought to be important in resistance to prophage infections (Barrangou *et al.*, 2007). These elements are generally associated with *cas* genes, which are thought to be linked to CRISPR mediated resistance or 'immunity' (Makarova *et al.*, 2006), mediated via RNA interference (RNAi). These have been termed CRISPR-cas system (CASS), which act as a defence system against phages and plasmids in the same manner as the eukaryotic RNA interference (RNAi). Although the *cas* genes have not been identified as orthologues of the eukaryotic system, it is hypothesized that they function as prokaryotic-small interfering RNA (psiRNA), whereby the psiRNA basepairs with the target RNA to prevent translation (Makarova *et al.*, 2006). This mechanism of resistance or 'immunity' may be of importance in the development of gene inactivation in *C. difficile*. Until recently the well-defined method of plasmid insertion and incorporation by homologous recombination described in a number of other bacteria, had proved fruitless in *C. difficile*. It is possible that these 10 CRISPR elements may have played a role in interfering with plasmid transfer in *C. difficile*.

Skin elements

Another intriguing genetic element present in the 630 genome is 'skin', a prophage like element which encodes a sporulation specific sigma factor SigK, whose excision at the onset of sporulation is important for effective sporulation (Haraldsen and Sonenshein, 2003).

Potential virulence determinants identified by sequencing

Antibiotic resistance determinants

The 630 strain harbours a wide range of antibiotic resistance genes, including those encoding tetracycline and erythromycin resistance, carried on Tn5397 and Tn5398 respectively. Other antibiotic resistance genes include those encoding lantibiotic, daunorubicin, bacitracin, nogalmycin, streptogramin A acetyltransferase, tellurite and beta-lactamase resistance (Sebaihia *et al.*, 2006). Interestingly, the skin element also encodes teicoplanin resistance (Sebaihia *et al.*, 2006).

Pathogenicity Locus (PaLoc)

The pathogenesis of *C. difficile* is largely due to the production of two exotoxins, designated A (TcdA) and B (TcdB), which form part of the 19.6 kb pathogenicity locus termed PaLoc. The toxins are glucosyltransferases, known to modify the actin cytoskeleton of the intestinal epithelial cells, via the covalent glucosylation of Rho, Rac and Cdc42 (Just *et al.*, 2001; Just *et al.*, 1995). The level of toxin production can vary greatly between strains and seems to be influenced by environmental conditions (Matamouros *et al.*, 2007). Several publications have shown that transcriptional regulation of the PaLoc is driven by a number of different environmental stimuli, including growth phase (Dupuy and Sonenshein, 1998; Hundsberger *et al.*, 1997), temperature (Karlsson *et al.*, 2003), antibiotics (Nakamura *et al.*, 1982), nutrient sources (Yamakawa *et al.*, 1994) and amino acids (Karlsson *et al.*, 2000). There have been several interesting observations made regarding sequence differences in the PaLoc among different C. *difficile* strains, which will be discussed in 'Differences in toxin sequence', below. In addition to the cytotoxins A and B, some strains also produce a binary toxin, encoded by *cdtA*/*cdtB*, which forms a two-subunit actin specific ADP-ribosyltransferase (Perelle *et al.*, 1997).

Cell wall and S-layer

A number of cell wall associated proteins have been discovered in C. *difficile*, which are thought to be involved in virulence and immune modulation. The most widely studied are the S-layer proteins, which are thought to be an important interface between pathogenic bacteria and the host (Cerquetti *et al.*, 2000). The S-layer of C. *difficile* is unique, in that, it is formed from two surface layer proteins, the low-molecular-weight (LMW) and the high-molecular-weight (HMW) proteins, which are encoded by a single gene *slpA* (Calabi *et al.*, 2001). The transcript of this gene contains a signal peptide sequence, alongside the LMW and HMW regions, these proteins are post-translationally cleaved in two rounds, firstly to remove the signal sequence after localization of the HMW and LMW proteins to the cell wall, then internally to release the LMW and HMW proteins, where they form a paracrystalline structure, external to the cell wall. These are thought to be involved in virulence, as the LMW and HMW proteins are thought to modulate cytokine production; however, the mechanism of this modulation is unclear as both proinflammatory cytokines (IL-1β, IL-6 and IL-12p70) and regulatory cytokines (IL-10) have been induced *in vitro* by C. *difficile* (Ausiello *et al.*, 2006). The S-layer, and in particular the LMW peptide is used in serotyping; however, sequence diversity has been identified in the *slpA* gene which varies between ribotypes and can therefore also be used as a typing method of C. *difficile* clinical isolates (Eidhin *et al.*, 2006).

There are 28 *slpA* paralogues in the 630 genome, which contain a HMW domain, including *cwp66* (an adhesin), which has been shown to play a role in adherence to host cells (Waligora *et al.*, 2001). Another example of a characterized *slpA* homologue is *cwp84*, which is a surface associated protein with homology to a serine protease (Savariau-Lacomme *et al.*, 2003), also thought to be involved in virulence of C. *difficile*.

Flagella and adhesins

Recent literature suggests that a number of virulence factors are involved in the adherence of the bacteria to the gut epithelial cells, not just the cell wall proteins, but also flagella, collagen binding proteins and fibronectin-binding proteins.

The flagella of C. *difficile* are peritrichous and have been shown to be involved in adherence to mouse caecum; flagellate organisms were found to bind with a 10-fold higher efficiency to mouse caecum than their aflagellate counterparts (Tasteyre *et al.*, 2001). This is of particular interest as whole genome comparisons revealed that several loci are absent/highly divergent in several of the seventy five strains tested (Stabler *et al.*, 2006) highlighted in 'Non-conserved genes', below.

The genome sequence also highlighted a number of genes which appear to contain extracellular matrix binding domains, namely collagen binding (CD2831) and fibronectin binding (CD2592 and CD2797) genes (Sebaihia *et al.*, 2006). Fibronectin is a multi-domain glycoprotein, attached to the basement membranes of cells, to which fibronectin binding proteins of the bacteria can bind, therefore the collagen and fibronectin-binding proteins may be important in adherence of C. *difficile* to host epithelial cells.

The role of another surface protein, the sortase should not be overlooked. Proteins displayed on the cell surface are most likely moved into position by sortases (encoded by CD2718). These are surface anchor proteins, which cleave surface polypeptides at a conserved LPXTG motif and anchor them to pentaglycine cross-bridges in peptidoglycan (Pallen *et al.*, 2001). Such surface-located proteins are also thought to be important in survival and immune modulation.

Survival in the gastrointestinal tract

C. *difficile* is almost unique in its ability to produce the phenolic compound *p*-cresol. In a two-stage process, tyrosine is oxidized to the intermediate *para*-hydroxyphenyl acetate, which is subsequently decarboxylated to form *p*-cresol (D'Ari and Barker, 1985). The production of *p*-cresol is thought to be linked to virulence and may provide an advantage for *C. difficile* over the gut flora, thus aiding the colonization by *C. difficile* (Sebaihia *et al.*, 2006).

Nutrient adaptation

There were a wide range of genes identified in *C. difficile* strain 630 responsible for both carbohydrate and nitrogen metabolism, which suggest that they are able to use an extended range of carbohydrates for metabolism, which is consistent with those available in the gastrointestinal tract, including the use of ethanolamine, abundant in the hosts diet (Sebaihia *et al.*, 2006).

Genomic analysis by microarray

In recent years, DNA–DNA microarray analysis has been utilized to rapidly assess the genetic diversity of bacterial species. In recent studies on other bacterial genera, microarray data, analysed with complex mathematical modelling, such as Bayesian phylogeny has proved a powerful tool for assessing the genetic diversities between bacterial populations (Champion *et al.*, 2005; Dorrell *et al.*, 2005), which, can then be further refined and related by comparisons with host adaptation, ecological niche and diseases phenotype.

Clostridium difficile microarray

The sequencing of strain 630 provided the opportunity to develop and produce a *C. difficile* 630 specific microarray. The design and production of the array was undertaken at the Bacterial Microarray Group at St George's, London (BUGS). The microarrays are spotted PCR products, the primers were 20–25 bp in length and were designed in accordance with the BUGS protocols (Rozen and Skaletsky, 2000; Stabler *et al.*, 2006). The amplicon size ranges from 50 to 800 bp with an optimum size of 600 bp for each of the chromosomally encoded predicted coding DNA sequences (CDSs) from strain 630 (Stabler *et al.*, 2006). Additionally, multiple reporters were designed to some genes; including eight for *tcdA*, seven for *tcdB*, three for *cdtA*, four for *cdtB* and two for each gene involved in S-layer formation.

Comparative phylogenomics study

Strain selection

75 strains were chosen for phylogenetic analysis, these included hypervirulent outbreak strain, ribotype 027 from Europe, Canada and the USA, showing the transcontinental spread of the disease These were pre-classified hypervirulent by several typing methods, including toxinotyping, Pulse Field Gel Electrophoresis and PCR-ribotyping. Additional strains in the study include current outbreak sub-types, as well as historic isolates and non-epidemic isolates (Table 7.1), the toxin variable samples included A^+B^+, A^-B^+ and A^-B^- strains. Among the human isolates, a wide range of disease outcomes was covered, from the severe pseudomembranous colitis to mild diarrhoea, as well as strains from a variety of hosts, including bovine, swine, equine and murine isolates (Stabler *et al.*, 2006).

Table 7.1 Strain information of *C. difficile* isolates used in the phylogenetic analysis

	Number	Designation
Human	21	Hypervirulent A^+B^+
	13	A^+B^+
	14	A^-B^+
	7	A^-B^-
Animal	7	Bovine
	6	Swine
	4	Equine
	3	Murine

Microarray data collation
In order to determine the phylogenetic relatedness of *C. difficile* strains, a pipeline of analysis was designed (Fig. 7.1). To analyse the data, a certain number of transformations were performed, culminating in a binary output format, which is essential for the Bayesian phylogenetic algorithm, MrBayes (MrBayes v3.1.1, http://morphbank.ebc.uu.se/mrbayes/info.php) (Champion *et al.*, 2005). The original microarray data was collected and analysed in BlueFuse (BlueGnome, UK) and GeneSpring (Agilent Technologies, USA), after which, the CDSs in each strain were marked as present, divergent and absent by the use of GACK software (Howard *et al.*, 2006). GACK calculated an EPP (Estimated Probability of Presence) value for each gene. This output was then used to run the Bayesian phylogenetic algorithm (Champion *et al.*, 2005), which uses a four-chain Markov chain Monte Carlo (MCMC) modelling system. MrBayes was run for a total of a million iterations, which were statistically assessed for convergence. Phylogeny inference was based on a conservative estimation of gene loss, the output of which can be seen in Fig. 7.2.

Analysis using a Bayesian based algorithm uncovered four distinct and statistically significant clades (see Fig. 7.2), comprising a hypervirulent clade, a toxin A^-B^+ clade, and two clades containing both human and animal isolates (Fig. 7.2). The core gene set over all four clades was surprisingly only 19.7%; however, this core gene set was significantly larger within specific clades. HY clade exhibited a 49% core gene set, whereas A^-B^+ clade showed a 36.2%, HA1 clade 41.9% and HA2 clade 47.9% core gene sets (Stabler *et al.*, 2006).

Genetic diversity among isolates
HY clade: Of the 21 hypervirulent isolates, 20 clustered into the HY clade, with only BI-9 as an outlier, located in HA1 clade. The hypervirulent outbreak strains, including the 027 strains R20291 from a particularly severe CDAD outbreak at Stoke Mandeville hospital in the UK, R20352 from Canada and R20928 from the USA clustered tightly together. However, the Canadian and USA outbreak strains appeared to be more closely related than the Stoke Mandeville strain.

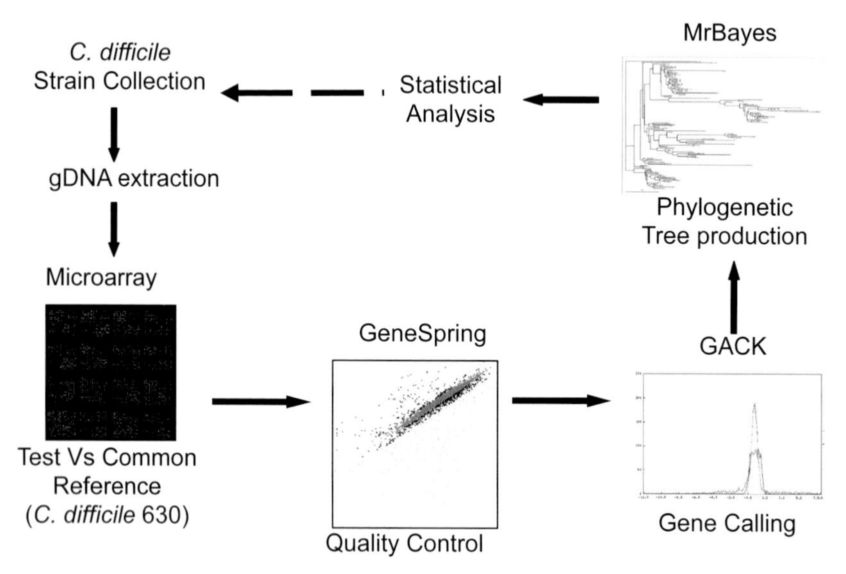

Figure 7.1 Phylogenomics pipeline. Graphical representation of steps involved in production of a phylogenetic tree. Genomic DNA from each sample was hybridized to a BUGS *C. difficile* 630 microarray with *C. difficile* 630 genomic DNA (gDNA) as control. Fluorescent intensities were calculated by BlueFuse and GeneSpring was used to calculate intensity ratio's and to remove low quality data points. GACK converted ratio data to binomial present/absent data. MrBayes used the binary data to construct a putative phylogeny utilizing a Bayesian algorithm. The resultant tree was statistically tested for robustness. Further *C. difficile* strains can be fed into the pipeline and strain selection may be influenced by current phylogeny prediction.

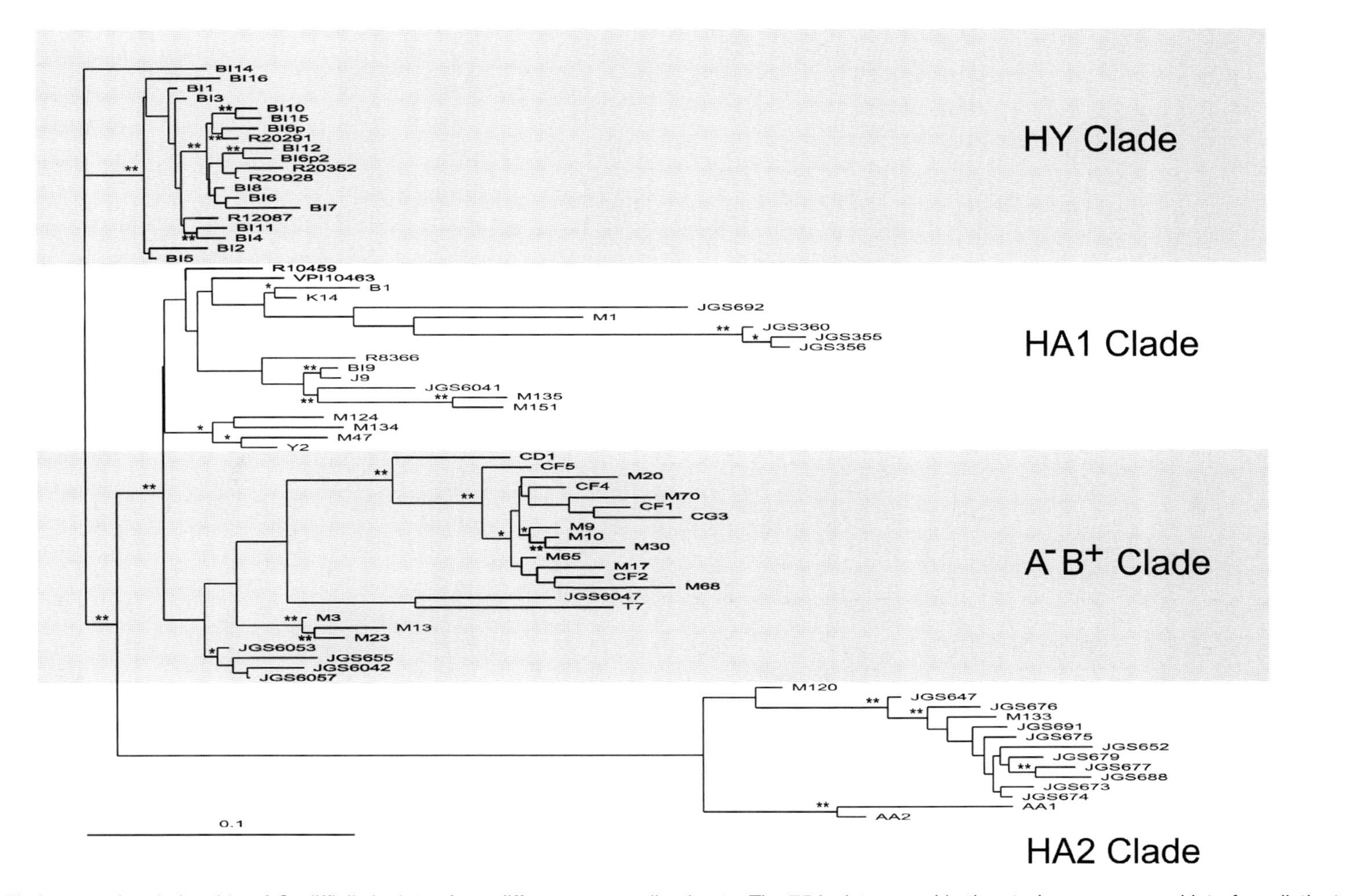

Figure 7.2 Phylogenomic relationship of *C. difficile* isolates from different mammalian hosts. The 75 isolates used in the study were grouped into four distinct clades, using Bayesian analysis of microarray data. The four clades consist of a hypervirulent clade (HY), a toxin variable clade (A^-B^+), and two clades comprising both human and animal isolates (HA1 and HA2) (Stabler *et al.*, 2006). ** represents 100% ($p = 1$) of all phylogenics showing a given topology and * $p \geq 0.98$.

A⁻B⁺ clade: All the 14 A⁻B⁺ isolates clustered into the aptly named A⁻B⁺ clade, along with a subclade of A⁻B⁻ strains. The A⁻B⁺ strains were from a diverse geographical background, with strains from Eire, UK and USA, once again confirming the transcontinental spread of genetically uniform *C. difficile* strains.

HA1 and HA2 clades: The HA1 and HA2 clades contained a mixture of both animal and human isolates (see Fig. 7.2). The most common UK isolate, ribotype 001 (R8366) (Rahmati *et al.*, 2005) clustered into the HA1 clade. The particularly interesting observation was the clustering of various animal isolates: the bovine and murine isolates were tightly grouped in the HA2 and HA1 clades respectively, whereas the porcine and equine strains were distributed throughout the HA1, HA2 and A⁻B⁺ clades, in which they were intermixed with the human isolates (see Fig. 7.2). Interestingly, the human 017 and 027 strains have also been isolated from calves, suggesting that cattle may be a reservoir for CDAD (Rodriguez-Palacios *et al.*, 2006). This may be of particular relevance, as a number of recent publications have noted a possible zoonotic link of *C. difficile* as a potential food borne pathogen

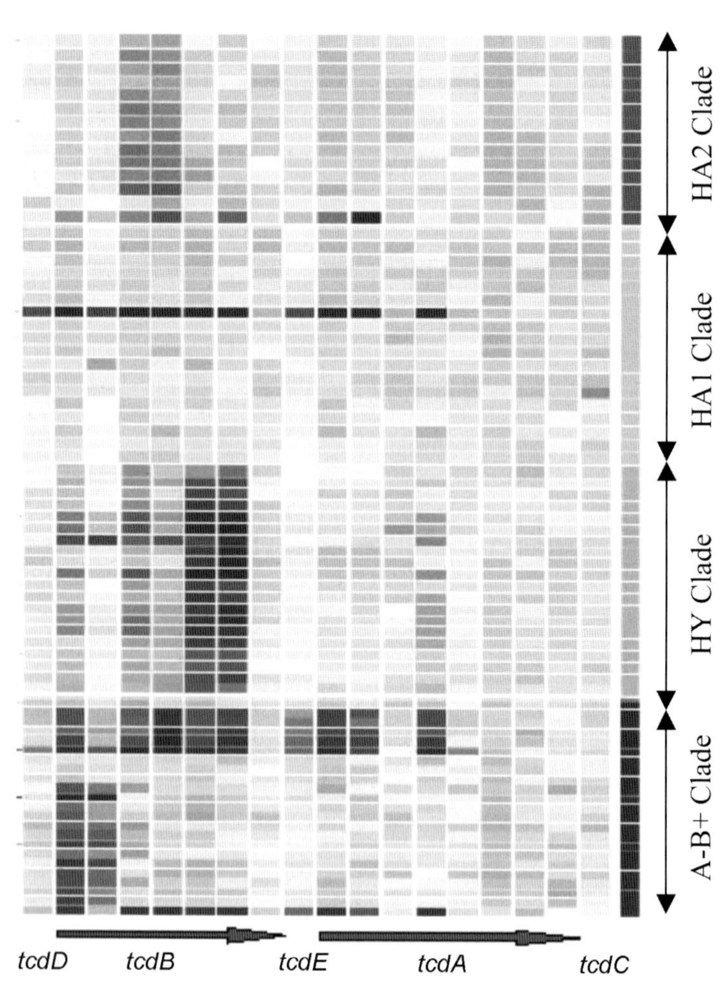

Figure 7.3 A selected gene map of the toxin pathogenicity locus (PaLoc). The horizontal colour bar indicates a single strain in the comparative strain hybridization, and the horizontal lines represent presence (light grey) or absence/high divergence (dark grey) of each gene. The four clades are represented by arrows on the right vertical axis, labelled A⁻B⁺, HY clade, HA1 and HA2. The genes in the PaLoc are listed below the heat map, and the horizontal bars indicate competitive genomic hybridizations to the probes on the microarray by a single strain (Stabler *et al.*, 2006).

(Rupnik, 2007). This may have significant implications in possible origins of *C. difficile* and may have implications for disease control strategies. Analysing these four distinct clades revealed a number of potential deletions or sequence divergences, which are discussed below:

Differences in toxin sequence

The microarray investigation revealed potential divergence in the PaLoc, where probes were unable to bind to the N-terminal region of *tcdB* in all the hypervirulent strains (Fig. 7.3), with the exception of BI-9 (Stabler *et al.*, 2006). Further analysis of the *tcdB* region in the Stoke Mandeville strain (R20291) revealed a high level of diversity in the binding domain (Stabler *et al.*, 2008), suggesting TcdB may have different binding capacities, which was observed by Merrigan *et al.* in human intestinal epithelial cells (Merrigan, 2007).

Expression of the toxin genes is mediated via TcdR, an alternative sigma factor present in the PaLoc. Recent publications indicate that the regulation of the PaLoc and therefore toxin production is a complex multifactorial process, which involves a negative regulator TcdC. A number of groups have noted mutations in *tcdC*, a missense mutation C184T, and a 39-bp deletion 341–379, and an 18-bp deletion 330–347 (Spigaglia and Mastrantonio, 2002) these were annotated genotype *tcdC-A*, *tcdC-B* and *tcdC-C* respectively; however, these mutations were not detected by the *tcdC* microarray probe, which covered almost the entire length of the gene, as the detection of small deletions/single basepair mutations is not possible with a PCR product microarray (Stabler *et al.*, 2006).

The deletions in *tcdC* are interesting yet counterintuitive, as they do not appear to play a role in de-repression of the PaLoc (Matamouros *et al.*, 2007), however it has been noted that the majority of strains containing the 18 bp deletion, also contain the 1 bp deletion, which results in the formation of a truncated protein, which would be non-functional, therefore the 1-bp deletion could be partially responsible for increased toxin production. However the hypervirulence of the 027 strains cannot be solely attributed to the *tcdC* deletions and frameshift mutation, which has opened up the debate on the mode of hypervirulence in the recent outbreak strains. A recent publication analysing $A^{-}B^{+}$ strains of *C. difficile* has shown different specificity of Toxin B for effector molecules in the host epithelial cells, suggesting that the cytotoxic effect of the different strains may be linked to subtle differences in the substrate specificity of the toxins (Huelsenbeck *et al.*, 2007).

One observation regarding the $A^{-}B^{+}$ strains was rather unexpected, the first two *tcdB* probes did not hybridize on the microarray; however, this can easily be explained, as experimental data on CF2 strain revealed a number of point mutations in the 5′ region of *tcdB* (Sambol *et al.*, 2000).

Determinants relating to virulence and niche adaption

There were a number of genetic differences observed among the four clades, these included several genetic islands potentially relating to virulence and niche adaptation, including antibiotic resistance, motility, adhesion and enteric metabolism. Importantly, it has been noted that gene deletions may contribute to increased virulence and pathoadaption, as seen in *Shigella* spp., in which the loss of lysine decarboxylase increased pathogenicity (Day *et al.*, 2001).

Variable genes/gene cassettes

Flagellae have been implicated in virulence, nevertheless a large number of strains from all four clades analysed showed deletions/divergence in their motility loci, more particularly in the HY clade. There are two flagella loci (CD0226–0240 and CD0245–0271) in 630, which are separated by an inter-flagella locus, with unknown function (CD0241–0244). Only seven of the sixty two strains contained the full strain 630 complement of the flagella loci, including three murine (JGS355, JGS356 and JGS360), two equine (JGS692 and JGS6047), and two human strains (CD1 and T7) (Fig. 7.4), all other strains, with the exception of the $A^{-}B^{+}$ strains have lost the first locus (CD0226–0240) and the inter-flagella locus (CD0214–0244). The $A^{-}B^{+}$ strains are only missing CD0252 and CD0255 (Stabler *et al.*, 2006) (Fig. 7.3). Interestingly, the three loci relating to flagella biosynthesis appear to be absent/highly divergent from the HA2 clade

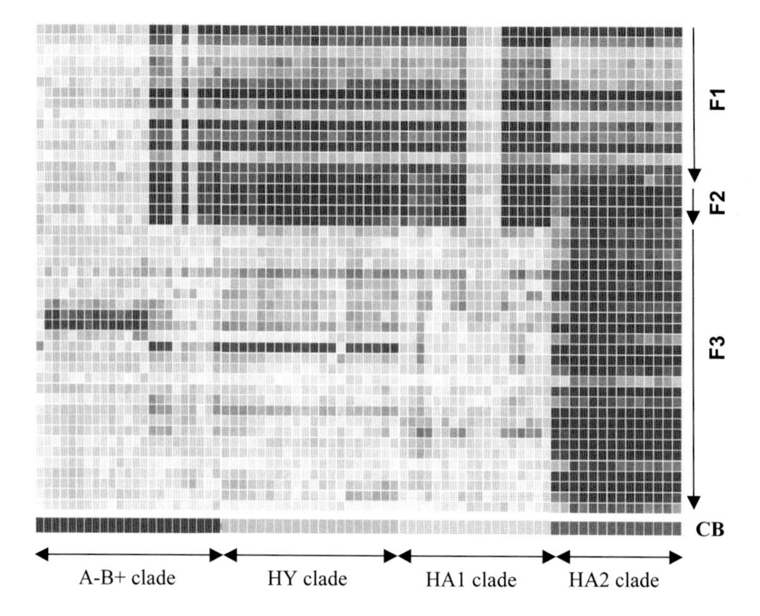

Figure 7.4 Selected gene map for flagellin-associated genes. The vertical colour bar indicates a single strain in the comparative strain hybridization, and the horizontal lines represent presence (light grey) or absence/high divergence (dark grey) of each gene. The flagella loci start from CD0226 located at top, through to CD0271 at bottom, where F1 indicates flagellar locus CD0226–0240, F2 indicates inter-flagella loci CD0241-0244, F3 indicates second flagellar locus CD0245-0271. The clade blocks (CB) represent the four clades in Fig. 7.2; labelled A^-B^+ clade, HY clade, HA1 clade and the HA2 clade.

(Fig. 7.2). This diversity of flagella genes among these isolates maybe important in virulence and immune evasion.

Antibiotic resistance appears for the most part to be transposon mediated; of the 36 potential drug resistance associated genes, the majority were common to all strains, however, absent/divergent genes were generally clustered into a specific clade/subclade (Stabler *et al.*, 2006), for example, the lantibiotic resistance loci (CD0643–46 and CD1349–52) were absent from all the HA2 clade exclusively. The daunorubicin resistance ABC transporter gene (*CD0456*) was absent from all HA2 strains and the majority of the HA1 strains (with the exception of B1, K14 and JGS692) (Stabler *et al.*, 2006).

There are a number of genes whose absent/divergence is specific to the A^-B^+ strains, including genes such as putative sodium solute symporter (CD3575), a membrane protein (CD3574), some hypothetical proteins (CD3573 and CD0590*) and a putative ABC transporter (CD0654*) (*gene present in CF5).

Conserved genes

The majority of the core genes encoded housekeeping functions, such as metabolic, biosynthetic, cellular and regulatory processes. However, there were a number of potentially interesting virulence determinants, which were conserved among all the strains suggesting that they are conserved and required for virulence. These included a capsule operon (CD2769-2780), a type IV pilus (CD3294-3297), fibronectin-binding proteins (CD1304 and 2592), collagen binding protein (CD2831), a sortase (CD2718), the sortase pseudogene (CD3146), along with the decarboxylase genes (hpdA, B and C), which convert pHPA to *p*-cresol.

Ethanolamine degradation (see *Nutrient adaptation*) appears to be encoded on a genomic island which resembles an element that was acquired by horizontal gene transfer, as the final gene in the cluster encodes a site-specific recombinase (Sebaihia *et al.*, 2006). This genomic island was completely intact in all strains except the three murine strains and the human non-toxigenic isolates AA1 and AA2 (Stabler *et al.*, 2006). The conservation of ethanolamine utilization may be important in adaptation to the gastrointestinal lifestyle.

Among the conserved genes there were examples that were conserved in certain sub-clades; CTn2 and CTn5 are conjugative transposons, which were partially present in more recent 027, but absent from historic BI (1–5 inclusive), as well as BI9, 11, 14, 16 and R12087 (the original 027 strain). Their roles are unknown, but they may encode an ABC transporter and exhibits similarity to MatE, a drug/protein antiporter respectively (Stabler *et al.*, 2006), therefore implicating a role of CTn2 and CTn5 in virulence/adaptation.

Future perspectives

In an ideal world phylogenomic analysis of diverse *C. difficile* strains would involve the analysis of the full genome sequences of all strains. These are currently not available, but the genome sequence for the Stoke Mandeville outbreak (R20291) and a Canadian 027 strain are near completion. The release of these sequences will enable the direct comparison between 630 and two 027 strains. However, the recent development of the next generation DNA sequencing technology (e.g. Solexa and 454 sequencing platforms) will provide enormous possibilities for whole genome analysis.

In the meantime we will have to rely on improved microarrays constituting the genes from a broader sweep of *C. difficile* genomes (a pan-*C. difficile* array) and improved methods for data interpretation and analysis. One current project includes the development of a microarray to include the genes present in both the 630 and R20291 strains. The differential analysis of R20291 and 630 will enable novel genes from R20291 to be added to the array to study potential evolution and horizontal gene transfer of R20291 specific genes in other *C. difficile* isolates. The emergence of a novel technology for the production of microarrays, ink jet *in situ* synthesis (IJISS) will revolutionize the design and production of microarrays, and enable them to rapidly adapt to ever increasing sequence information. The IJISS arrays are synthesized *in situ*, base by base. The flexibility with the design and probe size means that additional genes can be added easily to the array at the synthesis stage of an individual slide, and the number of probes per genes can be increased, within the density limits of reporters on an array. The construction of a pan-genome microarray, containing CDSs from a number of different strains and subtypes, will enable closer phenotypic and genotypic comparisons. The limitations with the current *C. difficile* microarrays are that genes unique to strains other than 630 are absent from the array, however, it has opened a number of avenues for functional analysis, which is essential to further understand pathogenesis. The microarrays provide a platform, from which functional analysis can be performed, particularly since an exciting and groundbreaking technique using the modified group II intron system from *Lactococcus lactis* (Mills *et al.*, 1997) has enabled the construction of defined *C. difficile* mutants. At present these are limited to 630 strain, however research is under way to increase the number of strains into which this system can be applied, thus increasing the potential for inter-strain functional analysis. This, coupled with the development of a vastly improved animal model of infection for *C. difficile* (Douce, 2007), is set to revolutionize our basic understanding of this enigmatic pathogen.

References

Ausiello, C.M., Cerquetti, M., Fedele, G., Spensieri, F., Palazzo, R., Nasso, M., Frezza, S., and Mastrantonio, P. (2006). Surface layer proteins from *Clostridium difficile* induce inflammatory and regulatory cytokines in human monocytes and dendritic cells. Microbes Infect. *8*, 2640–2646.

Barrangou, R., Fremaux, C., Deveau, H., Richards, M., Boyaval, P., Moineau, S., Romero, D.A., and Horvath, P. (2007). CRISPR provides acquired resistance against viruses in prokaryotes. Science *315*, 1709–1712.

Bolotin, A., Quinquis, B., Sorokin, A., and Ehrlich, S.D. (2005). Clustered regularly interspaced short palindrome repeats (CRISPRs) have spacers of extrachromosomal origin. Microbiology *151*, 2551–2561.

Calabi, E., Ward, S., Wren, B., Paxton, T., Panico, M., Morris, H., Dell, A., Dougan, G., and Fairweather, N. (2001). Molecular characterization of the surface layer proteins from *Clostridium difficile*. Mol. Microbiol. *40*, 1187–1199.

Cerquetti, M., Molinari, A., Sebastianelli, A., Diociaiuti, M., Petruzzelli, R., Capo, C., and Mastrantonio, P. (2000). Characterization of surface layer proteins from different *Clostridium difficile* clinical isolates. Microbial Pathogenesis *28*, 363–372.

Champion, O.L., Gaunt, M.W., Gundogdu, O., Elmi, A., Witney, A.A., Hinds, J., Dorrell, N., and Wren, B.W. (2005). Comparative phylogenomics of the food-borne pathogen Campylobacter jejuni reveals genetic markers predictive of infection source. Proc. Natl. Acad. Sci. USA *102*, 16043–16048.

D'Ari, L., and Barker, H.A. (1985). p-Cresol formation by cell-free extracts of *Clostridium difficile*. Arch. Microbiol. *143*, 311–312.

Day, W.A., Jr., Fernandez, R.E., and Maurelli, A.T. (2001). Pathoadaptive mutations that enhance virulence: genetic organization of the cadA regions of *Shigella* spp. Infect. Immun. *69*, 7471–7480.

Dorrell, N., Hinchliffe, S.J., and Wren, B.W. (2005). Comparative phylogenomics of pathogenic bacteria by microarray analysis. Curr Opin Microbiol *8*, 620–626.

Douce, G. (2007). Refinement of the syrian golden hamster model of *Clostridium difficile*, Paper presented at: 2nd international *Clostridium difficile* symposium (Maribor, Slovenia).

Dupuy, B., and Sonenshein, A.L. (1998). Regulated transcription of *Clostridium difficile* toxin genes. Mol. Microbiol. *27*, 107–120.

Eidhin, D.N., Ryan, A.W., Doyle, R.M., Walsh, J.B., and Kelleher, D. (2006). Sequence and phylogenetic analysis of the gene for surface layer protein, slpA, from 14 PCR ribotypes of *Clostridium difficile*. J. Med. Microbiol. *55*, 69–83.

Geric, B., Carman, R.J., Rupnik, M., Genheimer, C.W., Sambol, S.P., Lyerly, D.M., Gerding, D.N., and Johnson, S. (2006). Binary toxin-producing, large clostridial toxin-negative *Clostridium difficile* strains are enterotoxic but do not cause disease in hamsters. J. Infect. Dis. *193*, 1143–1150.

Haraldsen, J.D., and Sonenshein, A.L. (2003). Efficient sporulation in *Clostridium difficile* requires disruption of the sigmaK gene. Mol. Microbiol. *48*, 811–821.

Hell, M.S., Indra D., Huhulescu A., Maass M., Allerberger F. (2007). A cluster of *Clostridium difficile* associated disease – caused by a ribotype other than 027 – in a University hospital in Austria, 2006. Paper presented at: 2nd International *Clostridium difficile* symposium (Maribor, Slovenia).

Howard, S.L., Gaunt, M.W., Hinds, J., Witney, A.A., Stabler, R., and Wren, B.W. (2006). Application of comparative phylogenomics to study the evolution of *Yersinia enterocolitica* and to identify genetic differences relating to pathogenicity. J. Bacteriol. *188*, 3645–3653.

Huelsenbeck, J., Dreger, S., Gerhard, R., Barth, H., Just, I., and Genth, H. (2007). Difference in the cytotoxic effects of toxin B from *Clostridium difficile* strain VPI 10463 and toxin B from variant *Clostridium difficile* strain 1470. Infect. Immun. *75*, 801–809.

Hundsberger, T., Braun, V., Weidmann, M., Leukel, P., Sauerborn, M., and von Eichel-Streiber, C. (1997). Transcription analysis of the genes tcdA-E of the pathogenicity locus of *Clostridium difficile*. Eur. J. Biochem. *244*, 735–742.

Just, I., Hofmann, F., Genth, H., and Gerhard, R. (2001). Bacterial protein toxins inhibiting low-molecular-mass GTP-binding proteins. Int. J. Med. Microbiol. *291*, 243–250.

Just, I., Selzer, J., Wilm, M., von Eichel-Streiber, C., Mann, M., and Aktories, K. (1995). Glucosylation of Rho proteins by *Clostridium difficile* toxin B. Nature *375*, 500–503.

Karlsson, S., Dupuy, B., Mukherjee, K., Norin, E., Burman, L.G., and Akerlund, T. (2003). Expression of *Clostridium difficile* toxins A and B and their sigma factor TcdD is controlled by temperature. Infect. Immun. *71*, 1784–1793.

Karlsson, S., Lindberg, A., Norin, E., Burman, L.G., and Akerlund, T. (2000). Toxins, butyric acid, and other short-chain fatty acids are coordinately expressed and down-regulated by cysteine in *Clostridium difficile*. Infect. Immun. *68*, 5881–5888.

Limbago, B., Thopmson A.D., Killgore G.E., Hannett G., Havill N., Mickelson S., Lathrop S., Jones T,F., Park M., Cronquist A., Harriman K.H., McDonald L.C., Angulo F.J., (2007). Isolation and characterization of Clsotridium difficile responsible for communtiy-associated disease, Paper presented at: 2nd international *Clostridium difficile* symposium (Maribor, Slovenia).

Loo, V.G., Poirier, L., Miller, M.A., Oughton, M., Libman, M.D., Michaud, S., Bourgault, A.M., Nguyen, T., Frenette, C., Kelly, M. (2005). A predominantly clonal multi-institutional outbreak of *Clostridium difficile*-associated diarrhea with high morbidity and mortality. N. Engl. J. Med. *353*, 2442–2449.

Makarova, K.S., Grishin, N.V., Shabalina, S.A., Wolf, Y.I., and Koonin, E.V. (2006). A putative RNA-interference-based immune system in prokaryotes: computational analysis of the predicted enzymatic machinery, functional analogies with eukaryotic RNAi, and hypothetical mechanisms of action. Biol Direct *1*, 7.

Matamouros, S., England, P., and Dupuy, B. (2007). *Clostridium difficile* toxin expression is inhibited by the novel regulator TcdC. Mol. Microbiol. *64*, 1274–1288.

McDonald, L.C., Killgore, G.E., Thompson, A., Owens, R.C., Jr., Kazakova, S.V., Sambol, S.P., Johnson, S., and Gerding, D.N. (2005). An epidemic, toxin gene-variant strain of *Clostridium difficile*. N. Engl. J. Med. *353*, 2433–2441.

Merrigan, M.M., Roxas, J., Viswanathan, V.K., Fairweather, N., Gerding, D.N., Vedantam.G., (2007). Hypervirulent C. difficle strains have altered surface proteins and increased adherence to human intestinal epithelial cells., Paper presented at: 2nd international *Clostridium difficile* symposium (Maribor, Slovenia).

Mills, D.A., Manias, D.A., McKay, L.L., and Dunny, G.M. (1997). Homing of a group II intron from Lactococcus lactis subsp. lactis ML3. J. Bacteriol. *179*, 6107–6111.

Mojica, F.J., Diez-Villasenor, C., Garcia-Martinez, J., and Soria, E. (2005). Intervening sequences of regularly spaced prokaryotic repeats derive from foreign genetic elements. J. Mol. Evol. *60*, 174–182.

Nakamura, S., Mikawa, M., Tanabe, N., Yamakawa, K., and Nishida, S. (1982). Effect of clindamycin on cytotoxin production by *Clostridium difficile*. Microbiol. Immunol. *26*, 985–992.

Nolling, J., Breton, G., Omelchenko, M.V., Makarova, K.S., Zeng, Q., Gibson, R., Lee, H. M., Dubois, J., Qiu, D., Hitti, J. (2001). Genome sequence and comparative analysis of the solvent-producing bacterium Clostridium acetobutylicum. J. Bacteriol. *183*, 4823–4838.

Pallen, M.J., Lam, A.C., Antonio, M., and Dunbar, K. (2001). An embarrassment of sortases – a richness of substrates? Trends in Microbiology *9*, 97–101.

Perelle, S., Gibert, M., Bourlioux, P., Corthier, G., and Popoff, M.R. (1997). Production of a complete binary toxin (actin-specific ADP-ribosyltransferase) by *Clostridium difficile* CD196. Infect. Immun. *65*, 1402–1407.

Pituch, H., Obuch-Woszczatynski, P., Wultanska, D., van Belkum, A., Meisel-Mikolajczyk, F., and Luczak, M. (2007). Laboratory diagnosis of antibiotic-associated diarrhea: a Polish pilot study into the clinical relevance of *Clostridium difficile* and Clostridium perfringens toxins. Diagnostic Microbiol. Infect. Dis. *58*, 71–75.

Pourcel, C., Salvignol, G., and Vergnaud, G. (2005). CRISPR elements in *Yersinia pestis* acquire new repeats by preferential uptake of bacteriophage DNA, and provide additional tools for evolutionary studies. Microbiology *151*, 653–663.

Rahmati, A., Gal, M., Northey, G., and Brazier, J.S. (2005). Subtyping of *Clostridium difficile* polymerase chain reaction (PCR) ribotype 001 by repetitive extragenic palindromic PCR genomic fingerprinting. J. Hosp. Infect. *60*, 56–60.

Rodriguez-Palacios, A., Stampfli, H.R., Duffield, T., Peregrine, A.S., Trotz-Williams, L.A., Arroyo, L.G., Brazier, J.S., and Weese, J.S. (2006). *Clostridium difficile* PCR ribotypes in calves, Canada. Emerg. Infect. Dis. *12*, 1730–1736.

Rozen, S., and Skaletsky, H. (2000). Primer3 on the WWW for general users and for biologist programmers. Methods Mol. Biol. *132*, 365–386.

Rupnik, M. (2007). Is *Clostridium difficile*-associated infection a potentially zoonotic and food-borne disease? Clin. Microbiol. Infect. *13*, 457–459.

Sambol, S.P., Merrigan, M.M., Lyerly, D., Gerding, D.N., and Johnson, S. (2000). Toxin gene analysis of a variant strain of *Clostridium difficile* that causes human clinical disease. Infect. Immun. *68*, 5480–5487.

Savariau-Lacomme, M.P., Lebarbier, C., Karjalainen, T., Collignon, A., and Janoir, C. (2003). Transcription and analysis of polymorphism in a cluster of genes encoding surface-associated proteins of *Clostridium difficile*. J. Bacteriol. *185*, 4461–4470.

Sebaihia, M., Wren, B.W., Mullany, P., Fairweather, N.F., Minton, N., Stabler, R., Thomson, N.R., Roberts, A.P., Cerdeno-Tarraga, A.M., Wang, H. (2006). The multidrug-resistant human pathogen *Clostridium difficile* has a highly mobile, mosaic genome. Nat. Genet. *38*, 779–786.

Spigaglia, P., and Mastrantonio, P. (2002). Molecular analysis of the pathogenicity locus and polymorphism in the putative negative regulator of toxin production (TcdC) among *Clostridium difficile* clinical isolates. J. Clin. Microbiol. *40*, 3470–3475.

Stabler, R.A., Gerding, D.N., Songer, J.G., Drudy, D., Brazier, J.S., Trinh, H.T., Witney, A.A., Hinds, J., and Wren, B.W. (2006). Comparative phylogenomics of *Clostridium difficile* reveals clade specificity and microevolution of hypervirulent strains. J. Bacteriol. *188*, 7297–7305.

Tasteyre, A., Barc, M.C., Collignon, A., Boureau, H., and Karjalainen, T. (2001). Role of FliC and FliD Flagellar Proteins of *Clostridium difficile* in Adherence and Gut Colonization. Infect. Immun. *69*, 7937–7940.

Toyokawa, M., Ueda, A., Tsukamoto, H., Nishi, I., Horikawa, M., Sunada, A., and Asari, S. (2003). Pseudomembranous colitis caused by toxin A-negative/toxin B-positive variant strain of *Clostridium difficile*. J. Infect. Chemother. *9*, 351–354.

Waligora, A.J., Hennequin, C., Mullany, P., Bourlioux, P., Collignon, A., and Karjalainen, T. (2001). characterization of a cell surface protein of *Clostridium difficile* with adhesive properties. Infect. Immun. *69*, 2144–2153.

Warny, M., Pepin, J., Fang, A., Killgore, G., Thompson, A., Brazier, J., Frost, E., and McDonald, L.C. Toxin production by an emerging strain of *Clostridium difficile* associated with outbreaks of severe disease in North America and Europe. Lancet *366*, 1079–1084.

Yamakawa, K., Kamiya, S., Meng, X. Q., Karasawa, T., and Nakamura, S. (1994). Toxin production by *Clostridium difficile* in a defined medium with limited amino acids. J. Med. Microbiol. *41*, 319–323.

Surface Structures of *C. difficile* and Other Clostridia: Implications for Pathogenesis and Immunity

8

Jenny Emerson and Neil Fairweather

Abstract

The cell wall of *Clostridium difficile* has an architecture typical of other Gram-positive bacteria. A thick peptidoglycan layer lies external to the cell membrane with many associated cell wall proteins. In *C. difficile* two major cell wall proteins constitute the S-layer, a paracrystalline two-dimensional array surrounding the entire cell. The sequences of these S-layer proteins (SLPs) are variable between strains, perhaps reflecting immunological pressures on the cell. The genome sequence reveals a family of proteins with homology to the high molecular weight SLP; each of these proteins has a second unique domain but their functions remain largely uncharacterized. This family of cell wall proteins is also found in some other species, for example *C. botulinum* and *C. tetani*, but not in others such as *C. perfringens*. Some cell wall proteins of *C. difficile*, including the SLPs, have properties that imply an involvement in pathogenesis, particularly in binding to host cell tissues. The cell wall proteins of *C. difficile* may also act as immunogens to induce a partially protective immune response to infection, and may be considered as components of future vaccines against *C. difficile* associated disease.

Introduction

Clostridium difficile is now recognized as a major nosocomial pathogen in man, causing a range of gastrointestinal diseases ranging from diarrhoea to more severe pseudomembranous colitis (Bartlett, 2006). It is also becoming clear that this bacterium can cause disease in a variety of animal species, including pigs, horses and dogs (Songer and Anderson, 2006). However *C. difficile* is not always a pathogen, and indeed it is carried as a commensal in the intestine by a low percentage of healthy adults, with prevalence rising to up to 20% in hospitalized patients. Asymptomatic carriage is even more common in children (Matsuki *et al.*, 2005), perhaps reflecting the different conditions in the immature intestine which might lead to less advantageous environmental conditions for the induction of bacterial pathogenicity. *C. difficile* is now under intensive study in many laboratories, including those aiming to understand in more detail the properties of the organism and how it mediates pathogenesis. Like many clostridia, *C. difficile* has not received the detailed attention from molecular microbiologists that other Gram-positive bacteria have enjoyed, partly due to general difficulties in handling anaerobes and more specifically the lack of genetic tools. However this situation is changing, and in recent years we have learned much about this organism. In this chapter, we review our understanding of the cell wall of *C. difficile* and make comparisons with other clostridia.

The Gram-positive cell wall

It is well known that the cell wall of Gram-positive bacteria differs significantly from its Gram-negative counterpart, most notably by the absence of an outer membrane. The majority of Gram-positive bacteria also possess a thicker peptidoglycan layer than the Gram-negatives. One consequence of this arrangement is that Gram-positive surface proteins, defined as those that are located largely external to the cell membrane, have different architectures to those associated with the outer membrane of Gram-negatives. Gram-positive surface proteins are not required to pass though or interact with an outer membrane in order to be located on the cell surface. Surface proteins in Gram-positive bacteria have evolved to remain associated with the underlying peptidoglycan layer or with other secondary cell wall polymers, by covalent association or by hydrophobic and/or electrostatic interactions. Several excellent reviews outline the mechanisms that Gram-positive bacteria use to secret proteins and to anchor them to the cell wall (Buist *et al.*, 2006; Desvaux *et al.*, 2006; Marraffini *et al.*, 2006; Navarre and Schneewind, 1999). Surface localized proteins in Gram-positive bacteria can be considered as falling into one of two categories: either as components of the cell wall or as components of surface structures or appendages of the cell, such as flagella and fimbriae. A simplified diagrammatic representation of the cell wall of *C. difficile* is given in Fig. 8.1.

Bacterial S-layers

Surface-layers, or S-layers, are paracrystalline arrays of protein forming a complete monolayer coating the outermost surface of bacteria (reviewed in Sara and Sleytr, 2000; Sleytr and Beveridge, 1999). S-layers have been identified in both Gram-positive and Gram-negative Bacteria and are present on almost all Archaea isolated from many different habitats. They are usually composed of a single protein (Sleytr and Messner, 1988). The S-layer protein subunits, SLPs, create a regular lattice that may have square, oblique or hexagonal symmetry with uniform pores of 2–8 nm comprising up to 70% of the surface area. Many purified SLPs retain the ability to spontaneously re-form regular arrays upon removal of the disrupting agent used for their isolation. SLPs are usually 40–170 kDa in size and are weakly acidic with a high proportion of acidic amino acids, low levels of histidine and methionine, and little or no cysteine. Although little sequence homology is seen even between individual strains of some species, some SLPs have been shown to contain one or more S-layer homology (SLH) domains at their N-terminus (Lupas *et al.*, 1994; Sara, 2001). This is a complex motif thought to anchor the protein to the underlying cell wall. The SLH domain has also been found at the C-terminal of cell-associated exoenzymes. S-layers of Gram-positive bacteria are commonly but not always glycosylated, although the glycosylation bears little resemblance to that seen in eukaryotes (Messner and Sleytr, 1991).

S-layers are a huge metabolic investment: it has been estimated that in bacteria replicating every 20 minutes, over 400 SLP subunits must be produced and translocated every second (Sleytr and Messner, 1983). This, together with the ubiquitous nature of S-layers, suggests that they serve an important purpose, although no consistent function has been characterized for all S-layers. The regular arrangement of precisely sized pores may allow the S-layer to serve as a molecular sieve; indeed the S-layer of *Bacillus stearothermophilus* has been shown to demonstrate a sharp cut-off in permeability to

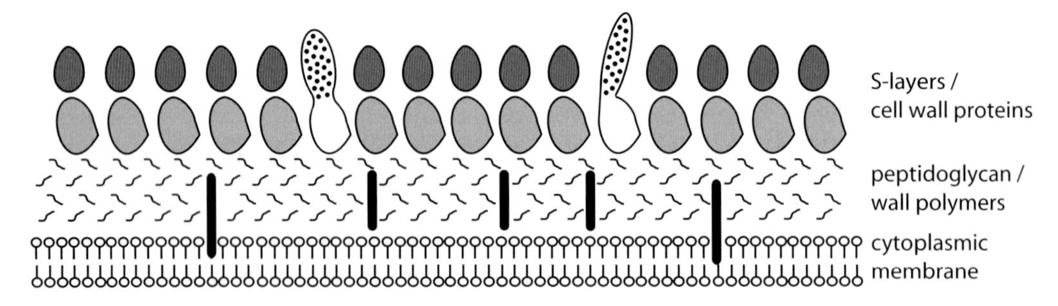

Figure 8.1 A model of the cell wall of *C. difficile*. The cytoplasmic membrane is shown at the bottom of the figure, below the peptidoglycan layer, which is shown containing putative secondary cell wall polymers such as teichoic acids and lipoteichoic acid molecules (Poxton and Cartmill, 1982) (black). The HMW SLP (light grey) and the cell wall binding domains of other proteins, e.g. Cwp2 and Cwp66, (white) are shown above the peptidoglycan layer. The LMW SLP (dark grey) and the surface exposed domains of the other cell wall proteins (spotted) are also shown.

molecules above 30–45 kDa (Sara and Sleytr, 1987). Other functions for the S-layer have been demonstrated, including as cell adhesion factors, virulence factors, anchoring sites for hydrolytic exoenzymes, receptors for phages, or as protection from phagocytosis or from predation by protozoa (Beveridge *et al.*, 1997; Sara and Sleytr, 2000; Sleytr and Beveridge, 1999). However such functions are largely unrelated, and have presumably arisen as a consequence of evolutionary pressures.

The *C. difficile* S-layer

The S-layer of C. *difficile* can be considered, as will be discussed below, to be a member of a family of cell wall proteins (CWPs). It is distinguished from other CWPs by two important properties: (1) it can form a two-dimensional array, and (2) it is expressed at very high levels. The C. *difficile* S-layer is unusual in that it is composed of two SLPs of different sizes and properties (Kawata *et al.*, 1984; McCoubrey and Poxton, 2001). In general, the two proteins each migrate on SDS-polyacrylamide gels at between 35 and 55 kDa, with one protein usually being approximately 36 kDa and the other around 45 kDa. The S-layer may be isolated from vegetative bacteria by a variety of methods including treatment with guanidinium hydrochloride, urea, low pH or EDTA. Several studies have visualized the C. *difficile* S-layer by freeze etch electron microscopy (Cerquetti *et al.*, 2000; Kawata *et al.*, 1984; Masuda *et al.*, 1989). In one study using strain GII071 (Kawata *et al.*, 1984) a single regular array with square symmetry was visualized. When mixed together in equimolar amounts, the two SLPs from strain GAI0714 underwent reassociation and reassembly to form open ended cylinders possessing a regular molecular pattern, as shown by negative staining (Masuda *et al.*, 1989). These studies showed that the regular array was composed of equimolar amounts of the two SLPs, suggesting that both the proteins are required for the formation of the array. A later study of six strains of C. *difficile* revealed the presence of two superimposed, structurally distinct regular arrays (Cerquetti *et al.*, 2000). In each strain, square symmetry was seen in the outer layer, and hexagonal symmetry in the inner S-layer lattice (Cerquetti *et al.*, 2000). Thus C. *difficile* may be able to form two distinct S-layers. Further experiments are required to fully understand how the S-layer is formed, and to reveal its three-dimensional structure.

It is likely that many clostridia produce an S-layer, but few have been characterized in detail. Both C. *botulinum* (Takumi *et al.*, 1992) and C. *tetani* (Takumi *et al.*, 1991) have been shown to produce an S-layer. In the case of C. *tetani*, the S-layer is composed of only one protein of approximately 160 kDa and its gene, termed *slpA*, has been identified (Qazi *et al.*, 2007). Interestingly C. *perfringens*, the other well characterized clostridial pathogen, does not appear to produce such a structure.

Molecular characterization of *C. difficile* SLPs

Variability in the sizes of the major surface proteins of C. *difficile* was reported by several authors (Kawata *et al.*, 1984; McCoubrey and Poxton, 2001; Sharp and Poxton, 1988) before it became clear that the proteins under investigation were in fact the SLPs. These studies showed the variability in the molecular masses of the two SLPs between strains. Molecular analysis has revealed that the two SLPs are derived by processing of a precursor protein, derived from a gene termed *slpA* (Calabi *et al.*, 2001; Karjalainen *et al.*, 2001) (Fig. 8.2). These studies showed that the variability in molecular masses of the SLPs from different strains, seen by SDS-PAGE, is due to DNA sequence differences in the *slpA* genes.

Sequencing shows that the HMW SLP is highly conserved between strains, whereas the LMW SLP is more divergent (Calabi *et al.*, 2001; Eidhin *et al.*, 2006; Karjalainen *et al.*, 2002; Karjalainen *et al.*, 2001; Kato *et al.*, 2005). The divergence of the LMW SLP sequence reflects immunological observations: the HMW SLPs often cross react with antibodies raised against the HMW SLP of another strain, while the LMW SLPs from strains of different serogroups rarely cross react (Calabi *et al.*, 2001).

The *slpA* gene is strongly transcribed during the entire growth phase (Savariau-Lacomme *et al.*, 2003). The HMW SLP shows weak homology to the cell wall binding region of the *N*-acetyl muramoyl-L-alanine amidase from *Bacillus subtilis*; both native SLP and recombinant HMW

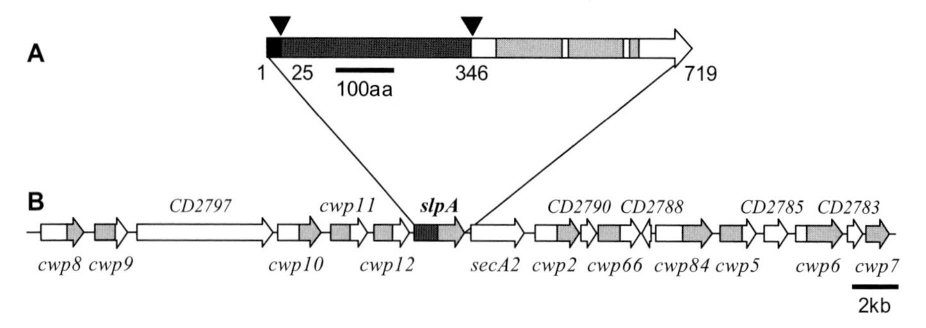

Figure 8.2 (A) Diagrammatic representation of SlpA protein from *C. difficile* 630. Signal sequence is depicted in black, the LMW SLP in dark grey, and the HMW SLP in white with its PF04122 cell wall anchoring domains indicated in light grey. The sites of cleavage to generate the mature proteins are indicated by black triangles. Amino acid residues are numbered. (B) The surface protein gene cluster around *slpA* in *C. diffficile*. The region of each CWP gene encoding the cell wall anchoring region is shown in light grey, and the unique regions and unrelated genes in white.

SLP show amidase activity by zymography (Calabi *et al.*, 2001). The HMW SLP incorporates two complete copies and one partial copy of the Pfam domain PF04122 cell_wall_binding_2. This domain is thought to mediate binding to the underlying bacterial cell wall, possibly through an affinity for teichoic acids, resulting in the anchoring of the protein to the exterior of the bacterium (Hennequin *et al.*, 2001b; Herbold and Glaser, 1975; Kuroda and Sekiguchi, 1991). PF04122 has so far been identified in multiple (typically three) copies in: 19 proteins of *C. tetani* (Bruggemann *et al.*, 2003); *Clostridium kluyveri* (17 proteins); *Desulfitobacterium hafniense* (20–24), *Kineococcus radiotolerans* (9); *Acidothermus cellulolyticus* (4); *B. subtilis* (CwlB and CwbA); and *B. licheniformis* (LytB and LytC). The LMW SLP of *C. difficile* exhibits no significant sequence homology to other proteins in the biological databases, indicating that LMW SLP protein may serve a unique function in *C. difficile*. Despite the low level of sequence homology between the LMW SLPs from different strains of *C. difficile*, one region of homology is seen: a highly conserved GKR motif, located near to the predicted cleavage site that generates the two mature SLPs (Calabi and Fairweather, 2002; Eidhin *et al.*, 2006). This motif may be essential for the recognition by the protease which mediates this cleavage or it may perhaps be involved in formation of the two-dimensional lattice. The identity of this protease is currently unknown.

Little is known concerning the mechanism of secretion and assembly of the S-layer of *C. difficile*. In common with some other Gram-positive pathogens, including *Listeria monocytogenes, Mycobacterium tuberculosis, Streptococcus parasanguis* and *Streptococcus gordonii* (Pallen *et al.*, 2003), *C. difficile* contains two alleles of *secA*, an essential gene encoding an ATPase in the general secretion pathway. In *L. monocytogenes*, *secA2* is required for secretion of several autolysins, and deletion of *secA2* results in attenuation of virulence (Lenz *et al.*, 2003; Lenz and Portnoy, 2002). In *C. difficile* the *secA2* allele is found immediately downstream of *slpA*, and is transcribed in the same direction (Fig. 8.2). While it is tempting to speculate that SecA2 is required for secretion of SlpA, no data is yet available to support this hypothesis.

Despite our knowledge of the molecular properties of the *C. difficile* SLPs, our knowledge of their function remains sparse. It is clear that the SLPs form an S-layer on the outer surface of *C. difficile*, but the function of this structure is unknown. Studies using purified S-layer proteins together with antisera reactive against these proteins have revealed that they may mediate adherence or association with the host cell. Using a cell culture model of adherence, antisera against the 36 kDa LMW SLP reduced adherence of *C. difficile* to Caco2 cells (Karjalainen *et al.*, 2001). In a separate study, the HMW SLP was found to bind to human gastrointestinal biopsy tissue, and anti-HMW SLP antibodies to block bacterial adherence to HEp2 cells, suggesting that this protein may mediate adherence to human cells during infection (Calabi *et al.*, 2002). SLPs have also been shown to mediate adherence to the gut wall in other organisms: the S-layers of

Lactobacillus acidophilus and *L. brevis* are necessary for intestinal adhesion and survival in the gastrointestinal tract (Hynonen *et al.*, 2002; Schneitz *et al.*, 1993). Definitive evidence of the roles of the *C. difficile* SLPs in adherence to host tissues will come from genetic experiments where the creation of gene knock-outs will allow comparison of the properties of isogenic strains containing defined genetic lesions.

Other cell wall proteins of *C. difficile*

One interesting finding to emerge from studies of the S-layer of *C. difficile* was the presence in the *C. difficile* 630 genome of 28 genes that show homology to *slpA* (Calabi *et al.*, 2001; Karjalainen *et al.*, 2001). These proteins, termed cell wall proteins (CWPs), are all found to contain three copies of the PF04122 cell_wall_binding_2 domain, as shown in Fig. 8.3. Structural predictions reveal that these cell wall anchoring regions share a common secondary structural pattern (R. Fagan, personal communication). All CWPs are also found to have an N-terminal signal sequence for Sec-mediated export from the cell. Eleven cwp genes are found clustered around the *slpA* locus (Calabi *et al.*, 2001; Hennequin *et al.*, 2001a) (Fig. 8.2). The six cwp genes located closely downstream of *slpA*, termed *cwp2* to *cwp7*, are conserved between strains and have been shown to be transcribed in cultures grown *in vitro* (Calabi et al., 2001; Sebaihia *et al.*, 2006).

In addition to possessing cell wall anchoring domains, each CWP also contains a unique region of varying size. When a phylogenetic tree is constructed using only the amino acid alignment of the PF04122 domains, those with similar 'unique' regions tended to cluster together (data not shown). The PF00188 SCP domain (putative calcium-chelating serine protease) is encoded by *cwp9, cwp11* and *cwp12*; *cwp12* also encodes a PF07523 Big_3 (bacterial immunoglubulin-like 3) domain. *cwp16* and *cwp17* are found adjacent in the genome, and together with *cwp6* encode proteins incorporating the PF01520 amidase_3 domain. Cwp20 features a PF00144 β-lactamase domain, and Cwp19 features a PF02638 DUF187 (conserved domain of unknown function) domain. *cwp22* encodes a protein with one PF03734 ErkK_YbiS_YhnG domain (conserved bacterial domain – possibly enzymatic) in addition to seven copies of PF01473 cell_wall_binding_1 domain, which

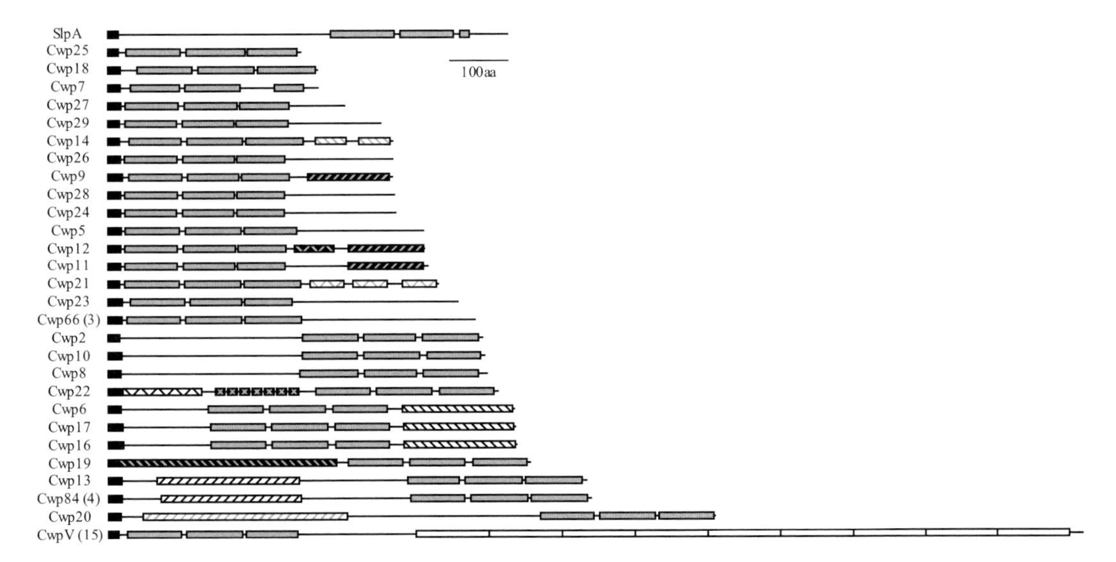

Figure 8.3 The cell wall proteins of *C. difficile*. The 29 putative cell wall proteins from the *C. difficile* 630 genome containing the Pfam domain PF01422 are shown. Pfam domains and repeated sequences are indicated by boxes: ■, signal peptide; ▭, CW_binding_2 PF04122; ▭, SH3_3 PF08239; ▭, SCP PF00188; ▭, Big_3 PF07523; ▭, PepSY PF03413; ▭, ErkK_YbiS_YhnG PF03734; ▭, CW_binding_1 PF01473; ▭, Amidase_3 PF01520; ▭, DUF187 PF02638; ▭, Peptidase_C1 PF00112; ▭, _-lactamase PF00144; ▭, repetitive sequences.

is also found in the repeat region of the large clostridial toxins. Cwp14 incorporates two PF08239 SH3_3 domains; and Cwp21 features three PF03413 PepSY (putative protease inhibitory function) domains. *cwp84* (*cwp4*) and *cwp13* encode proteins incorporating PF00112 peptidase_C1 domains; Cwp84 has been shown to be a cysteine protease, possibly involved in maturation of secreted bacterial proteins or in degradation of extracellular matrix proteins (Janoir *et al.*, 2007; Savariau-Lacomme *et al.*, 2003). Both the cell wall anchoring region and the 'unique' region of Cwp66 (Cwp3) have been shown to be involved in adhesion of *C. difficile* to the gut wall (Hennequin *et al.*, 2001b; Janoir *et al.*, 2004). CwpV (Cwp15) contains three copies of the PF04122 cell wall anchoring domain, but does not feature any other Pfam domains. It is the largest CWP in *C. difficile* 630, and its variable domain consists of multiple repeats.

In an effort to establish which of these CWPs are expressed by *C. difficile*, a proteomic approach was used to identify the proteins present in growing cells (Wright *et al.*, 2005). Two methods of preparing cell wall extracts were employed. Low-pH glycine extracts were prepared yielding the HMW and LMW SLPs as the predominant proteins. This procedure also identified the cell wall proteins Cwp2, Cwp6, Cwp12, CwpV, Cwp24 and Cwp25 as minor components of the glycine extract. All these proteins were visualized as spots with mobilities close to those expected of the mature proteins after signal peptide cleavage, indicating that additional cleavage events had not occurred. One exception was CwpV which was seen as a series of spots of between 47 and 85 kDa. This is consistent with the presence of a number of amino acid repeats in this protein (see Fig. 8.3), and the cleavage of a sequence either within or between the repeats. The second method employed used lysozyme, mutanolysin and lysostaphin to degrade the peptidoglycan layer and release cell wall proteins associated more tightly with the cell wall. Additional proteins released included Cwp84 (Cwp4). Interestingly, the well-characterized adhesin Cwp66 (Cwp3) (Waligora *et al.*, 2001) was not detected, presumably because it is not expressed under the conditions used in these proteomics experiments.

Cell wall proteins in other clostridia

Bioinformatic analysis of the genome sequences of several clostridia reveals that many, but not all, species possess a family of proteins containing the PF04122 domain. Thus in *C. tetani* 19 proteins are found, including the S-layer protein (Qazi *et al.*, 2007), in *C. botulinum* 8 proteins and in *C. kluyveri* 17 proteins (Bruggemann and Gottschalk, 2008). Five other clostridial genomes were analysed, including the pathogen *C. perfringens*, and shown not to encode proteins with the PF04122 domain. This suggests that the PF04122 domain is used by some species as a basis for the formation of cell wall proteins, but that this domain is certainly not universal and can be substituted by other sequences. In *C. beijerinckii*, which possesses only one PF04122 domain, 66 proteins are found containing a second cell wall binding domain, PF01473. This motif may serve to anchor cell wall proteins to underlying cell wall polymers in *C. beijerinckii*, however further work is necessary to clarify this. The PF01473 domain is found most notably in the large clostridial cytotoxins produced by *C. difficile* (TcdA and TcdB), *C. sordellii* (TcsH and TcsL) and *C. novyi* (TcnA). It is also found in glucosyltransferases and glucan-binding proteins of oral streptococci, and in LytA, the major autolysin from *Streptococcus pneumoniae* (Fernandez-Tornero *et al.*, 2001). In the case of the large clostridial cytotoxins, multiple copies of the PF01473 domain are located in the C-terminal segment of the toxin that is required for binding to host tissues. Whether these repeats also have a role in binding to the bacterial cell wall is not known.

Sortase-mediated attachment to the cell wall

The most common mechanism of cell-wall association, present in almost all Gram-positive bacteria, is covalent anchoring to the underlying peptidoglycan (reviewed in Navarre and Schneewind 1999; Marraffini *et al.*, 2006). This is signalled by the LPXTG amino acid motif, which is located close to the C-terminal of the protein and is followed by a hydrophobic region of amino acids and a positively charged tail. It is thought that this region halts the transit of the protein through the Sec machinery and allows recognition of the LPXTG motif by sortase, which cleaves the protein and covalently attaches the threonine residue to an amino group at the end of an un-cross-linked peptidoglycan pentapeptide. In several pathogenic Gram-positive bacteria, many virulence factors are anchored to the underlying cell wall by the sortase mechanism. These include protein A, fibronectin binding proteins and clumping factor of *S. aureus* and the M proteins, protein G and fibronectin binding proteins from *Streptococcus* sp. (Navarre and Schneewind, 1999).

In *C. difficile* 630, only one protein containing the LPXTG motif is found, although a number of other proteins have a sequence motif with a more relaxed specificity, and these may function as sortase substrates. A sortase enzyme, CD2718, is found in the genome suggesting that the sortase system is operational in *C. difficile*. Other clostridia also contain sortase proteins and sortase substrates. This is particularly evident in *C. perfringens* where 13 LPXTG substrates are found in strain 13, while 7 are found in strain SM1010 and 13 in ATCC 13214 (Bruggemann and Gottschalk, 2008). The relatively low number of sortase substrates in some species, e.g. *C. difficile* and *C. tetani*, may reflect the particular niches that these pathogens inhabit. Alternatively perhaps some clostridia have evolved sortase substrates that are not recognized by the conventional algorithms, or distinct pathways mediating secretion are used in these organisms in preference to the sortase.

Flagella and fimbriae

In many bacterial species, flagella can contribute to virulence by mediating motility, chemotaxis and adhesion. It was established over a decade ago that *C. difficile* express flagella (Delmee *et al.*, 1990). In a more recent study, flagella were observed by electron microscopy in 30 out of 47 strains studied (Tasteyre *et al.*, 2000). A recent comparative genomic microarray study highlighted the variation of flagellar gene content in a collection of 75 strains (Stabler *et al.*, 2006). Two loci encoding potential flagellum-associated proteins are found in the genome of strain 630, located on either side of a locus containing genes of unknown function. One of these loci, CD0226–CD0240, contains the *fliC* and *fliD* genes encoding the flagellin subunit and the flagellar cap protein respectively. All toxin $A^{-}B^{+}$ strains retained both flagellar loci, together with the intermediate locus. However the majority of other strains studied did not contain the *fliC*/*fliD* locus, suggesting that chemotaxis and motility are not essential to the virulence of *C. difficile*.

In an earlier study, the variability of the *fliC* and *fliD* genes was studied (Tasteyre *et al.*, 2001b). These genes, which were present in the majority of strains studied including non-flagellated strains, were highly conserved except in the region encoding amino acids exposed to the environment. Further studies, including the behaviour of flagellated and non-flagellated strains in mouse models of infection, suggested that the presence of flagella increased the association of bacteria with tissues (Tasteyre *et al.*, 2001a). Thus flagella may play a role in adherence of bacteria to host tissues, but this is probably not an essential property for virulence.

Many pathogenic bacteria employ fimbriae, also termed pili, to facilitate colonization to host tissues during colonization. While pili are well described and characterized in Gram-negative pathogens, they are less well known in the Gram-positives. However it is now clear that several Gram-positive pathogens express pili, including *Streptococcus pneumoniae* and *Corynebacterium diphtheriae* (Telford *et al.*, 2006). These pili are unlike structures seen in Gram-negative species and are anchored to the underlying peptidoglycan layer *via* a sortase-dependent mechanism. Recently, the presence of another type of pili, the type IV pili, was demonstrated in *C. perfringens* (Varga *et al.*, 2006). These pili are known to mediate a form of motility termed 'twitching motility' in Gram-negative bacteria, most notably in *Pseudomonas aeruginosa* and *Neisseria gonorrhoea* (Mattick, 2002). In *C. perfringens* the production of these pili was associated with an unusual form of motility, which the authors termed gliding motility. Mutations introduced into two key genes, *pilT* and *pilC*, abolished both the surface localization of the pili and the twitching motility of in *C. perfringens*. Interestingly, homologous type IV pili genes are found in nine other clostridia, including the pathogens *C. tetani*, *C. difficile* and *C. botulinum*. It remains to be seen whether these pili are expressed in these strains and if so, whether they have a role in pathogenesis.

The role of surface proteins and toxins in immunity to clostridial diseases

The clostridia are unusual pathogens in that their virulence is largely mediated by exotoxins that are secreted from the bacterium. In the case of *C. tetani* and *C. botulinum*, the toxins are considered to be the sole virulence factors, and immunity to intoxication is achieved by immunization with inactivated toxins (toxoids). A second group of clostridia, the histotoxic clostridia, encompasses *C. perfringens*, *C. histolyticum*, *C. novyi*, *C. septicum*, *C. sordelii* and *C. chauvoei* (reviewed in Stevens and Rood, 2006). These species all produce numerous toxins that are thought to contribute to pathogenesis, although in many cases their exact role of each toxin may be unclear. Although human vaccines against these diseases are not available, for many years veterinary vaccines have been used to prevent tissue infections (Walker, 1992).

The best-characterized toxin of *C. perfringens*, the alpha toxin, is the major virulence factor implicated in clostridial myonecrosis, or gas gangrene (Titball, 2005). It has been shown that recombinant forms of alpha toxin can generate protective immunity in animal models of infection, and therefore this may form the basis of a human vaccine should the need arise (Stevens *et al.*, 2004).

In the case of *C. difficile*, the two large clostridal cytotoxins, TcdA and TcdB, have been the main focus of work aimed at development of a vaccine against disease. Over the last ten years or so, many studies have investigated the protective nature of toxoid vaccines (reviewed in Giannasca and Warny, 2004). Using the hamster model of infection, toxoids of TcdA and TcdB generally provide partial or complete protection against a subsequent challenge with pure toxin, but the degree of protection seen appears to be dependent on the route of immunization. For example, a combined mucosal (rectal) and intramuscular immunization schedule using toxoids of TcdA and TcdB led to high levels of serum-neutralizing antibodies and provided full protection against death and diarrhoea (Giannasca *et al.*, 1999). Novel routes of immunization with *C. difficile* toxins have also been explored, including transdermal (transcutaneous) immunization without the use of needles (Glenn and Kenney, 2006). Mice immunized with detoxified toxin TcdA together with cholera toxin as an adjuvant developed high anti-TcdA IgA and IgG antibodies, which were capable of neutralizing TcdA in a cell based assay (Ghose *et al.*, 2007). Such studies are promising and may lead to novel routes of immunization against clostridial diseases. Several clinical studies have investigated the utility of inactivated toxins TcdA and TcdB as human vaccines. These are based on the correlation between the presence of serum antibodies against TcdA (and TcdB to a lesser extent) and protection against disease and relapse in patients (Aboudola *et al.*, 2003; Kyne *et al.*, 2001). Recent results using a TcdA/TcdB toxoid-based vaccine in a limited number of patients were encouraging (Sougioultzis *et al.*, 2005), suggesting the utility of a *C. difficile* vaccine for prevention or treatment of disease.

Alternative approaches to vaccination against *C. difficile* have been investigated, with a focus on preventing association of the bacterium with gut tissues. To this end, cell wall associated proteins have been used as antigens in experimental animals, and protection assessed by a reduction in bacterial colonization of the gut or by protection against *C. difficile* infection. Using the SLPs from *C. difficile* as an antigen, O'Brien *et al.* (2005) immunized rabbits and used the resulting sera to passively immunize hamsters. The hamsters were then challenged with *C. difficile* and the time taken to develop symptoms compared with control animals which received an unrelated rabbit serum. The survival time in the hamsters given anti-SLP sera was significantly greater than those given the control sera. Further work showed that the anti-S-layer antibodies mediated phagocytosis, which may be related to the ability of the sera to prolong survival time of the animals. In a separate study, Pechine *et al.* (2007) used a mouse model of infection, where the animals were previously administered with a faecal suspension from gnotobiotic mouse harbouring human faecal flora. After immunization mice were challenged with virulent *C. difficile* and the degree of colonization assessed. Immunization via the rectal route with a combination of cell wall antigens, including the SLPs, Cwp84 and flagellar proteins resulted in the greatest reduction in colonization, suggesting a role for multiple cell wall proteins in the colonization process.

Conclusions

In this review we have described the properties of a number of cell wall proteins from clostridia. We have focussed primarily on *C. difficile*, as this species is currently the most important clostridial species in clinical terms. It is clear that the clostridial cell wall proteins are an interesting and diverse group of proteins, with distinct properties compared to their counterparts in other Gram-positive bacteria. Recent advances in genetics will allow further characterization of these proteins, which may have utility as vaccines against clostridial infections.

References

Aboudola, S., Kotloff, K.L., Kyne, L., Warny, M., Kelly, E.C., Sougioultzis, S., Giannasca, P.J., Monath, T.P., and Kelly, C.P. (2003). *Clostridium difficile* vaccine and serum immunoglobulin G antibody response to toxin A. Infect. Immun. *71*, 1608–1610.

Bartlett, J.G. (2006). Narrative review: the new epidemic of *Clostridium difficile*-associated enteric disease. Ann. Intern. Med. *145*, 758–764.

Beveridge, T.J., Pouwels, P.H., Sara, M., Kotiranta, A., Lounatmaa, K., Kari, K., Kerosuo, E., Haapasalo, M., Egelseer, E.M., Schocher, I. (1997). Functions of S-layers. FEMS Microbiol. Rev. *20*, 99–149.

Bruggemann, H., Baumer, S., Fricke, W.F., Wiezer, A., Liesegang, H., Decker, I., Herzberg, C., Martinez-Arias, R., Merkl, R., Henne, A. (2003). The genome sequence of *Clostridium tetani*, the causative agent of tetanus disease. Proc. Natl. Acad. Sci. USA *100*, 1316–1321.

Bruggemann, H., and Gottschalk, G. (2008). Comparative genomics of clostridia: link between the ecological niche and cell surface properties. Ann. NY Acad. Sci. *1125*, 73–81.

Buist, G., Ridder, A.N.J.A., Kok, J., and Kuipers, O.P. (2006). Different subcellular locations of secretome components of Gram-positive bacteria. Microbiology *152*, 2867–2874.

Calabi, E., Calabi, F., Phillips, A.D., and Fairweather, N.F. (2002). Binding of *Clostridium difficile* surface layer proteins to gastrointestinal tissues. Infect. Immun. *70*, 5770–5778.

Calabi, E., and Fairweather, N. (2002). Patterns of sequence conservation in the S-layer proteins and related sequences in *Clostridium difficile*. J. Bacteriol. *184*, 3886–3897.

Calabi, E., Ward, S., Wren, B., Paxton, T., Panico, M., Morris, H., Dell, A., Dougan, G., and Fairweather, N. (2001). Molecular characterization of the surface layer proteins from *Clostridium difficile*. Mol. Microbiol. *40*, 1187–1199.

Cerquetti, M., Molinari, A., Sebastianelli, A., Diociaiuti, M., Petruzzelli, R., Capo, C., and Mastrantonio, P. (2000). Characterization of surface layer proteins from different *Clostridium difficile* clinical isolates. Microb Pathog *28*, 363–372.

Delmee, M., Avesani, V., Delferriere, N., and Burtonboy, G. (1990). Characterization of flagella of *Clostridium difficile* and their role in serogrouping reactions. J. Clin. Microbiol. *28*, 2210–2214.

Desvaux, M., Dumas, E., Chafsey, I., and Hebraud, M. (2006). Protein cell surface display in Gram-positive bacteria: from single protein to macromolecular protein structure, pp. 1–15. FEMS Microbiol. Lett. *256*, 1–15.

Eidhin, D.N., Ryan, A.W., Doyle, R.M., Walsh, J.B., and Kelleher, D. (2006). Sequence and phylogenetic analysis of the gene for surface layer protein, *slpA*, from 14 PCR ribotypes of *Clostridium difficile*. J. Med. Microbiol. *55*, 69–83.

Fernandez-Tornero, C., Lopez, R., Garcia, E., Gimenez-Gallego, G., and Romero, A. (2001). A novel solenoid fold in the cell wall anchoring domain of the pneumococcal virulence factor LytA. Nat. Struct. Biol. *8*, 1020–1024.

Ghose, C., Kalsy, A., Sheikh, A., Rollenhagen, J., John, M., Young, J., Rollins, S.M., Qadri, F., Calderwood, S.B., Kelly, C.P. (2007). Transcutaneous immunization with *Clostridium difficile* toxoid A induces systemic and mucosal immune responses and toxin A-neutralizing antibodies in mice. Infect. Immun. *75*, 2826–2832.

Giannasca, P.J., and Warny, M. (2004). Active and passive immunization against *Clostridium difficile* diarrhea and colitis. Vaccine *22*, 848–856.

Giannasca, P.J., Zhang, Z.X., Lei, W.D., Boden, J.A., Giel, M.A., Monath, T.P., and Thomas, W.D., Jr. (1999). Serum antitoxin antibodies mediate systemic and mucosal protection from *Clostridium difficile* disease in hamsters. Infect. Immun. *67*, 527–538.

Glenn, G.M., and Kenney, R.T. (2006). Mass vaccination: solutions in the skin. Curr. Top. Microbiol. Immunol. *304*, 247–268.

Hennequin, C., Collignon, A., and Karjalainen, T. (2001a). Analysis of expression of GroEL (Hsp60) of *Clostridium difficile* in response to stress. Microb. Pathog. *31*, 255–260.

Hennequin, C., Porcheray, F., Waligora-Dupriet, A., Collignon, A., Barc, M., Bourlioux, P., and Karjalainen, T. (2001b). GroEL (Hsp60) of *Clostridium difficile* is involved in cell adherence. Microbiology *147*, 87–96.

Herbold, D.R., and Glaser, L. (1975). *Bacillus subtilis* N-acetylmuramic acid L-alanine amidase. J. Biol. Chem. *250*, 1676–1682.

Hynonen, U., Westerlund-Wikstrom, B., Palva, A., and Korhonen, T.K. (2002). Identification by flagellum display of an epithelial cell- and fibronectin-binding function in the SlpA surface protein of *Lactobacillus brevis*. J. Bacteriol. *184*, 3360–3367.

Janoir, C., Grenery, J., Savariau-Lacomme, M.P., and Collignon, A. (2004). Characterization of an extracellular protease from *Clostridium difficile*. Pathologie-biologie 52, 444–449.

Janoir, C., Pechine, S., Grosdidier, C., and Collignon, A. (2007). Cwp84, a surface-associated protein of *Clostridium difficile*, is a cysteine protease with degrading activity on extracellular matrix proteins. J. Bacteriol. *189*, 7174–7180.

Karjalainen, T., Saumier, N., Barc, M.-C., Delmee, M., and Collignon, A. (2002). *Clostridium difficile* genotyping based on *slpA* variable region in S-layer gene sequence: an alternative to serotyping. J. Clin. Microbiol. *40*, 2452–2458.

Karjalainen, T., Waligora-Dupriet, A.-J., Cerquetti, M., Spigaglia, P., Maggioni, A., Mauri, P., and Mastrantonio, P. (2001). Molecular and genomic analysis of genes encoding surface-anchored proteins from *Clostridium difficile*. Infect. Immun. *69*, 3442–3446.

Kato, H., Yokoyama, T., and Arakawa, Y. (2005). Typing by sequencing the *slpA* gene of *Clostridium difficile* strains causing multiple outbreaks in Japan. J. Med. Microbiol. *54*, 167–171.

Kawata, T., Takeoka, A., Takumi, M., and Masuda, K. (1984). Demonstration and preliminary characterisation of a regular array in the cell wall of *Clostridum difficile*. FEMS Microbiol. Lett. *24*, 323–328.

Kuroda, A., and Sekiguchi, J. (1991). Molecular cloning and sequencing of a major *Bacillus subtilis* autolysin gene. J. Bacteriol. *173*, 7304–7312.

Kyne, L., Warny, M., Qamar, A., and Kelly, C.P. (2001). Association between antibody response to toxin A and protection against recurrent *Clostridium difficile* diarrhoea. Lancet *357*, 189–193.

Lenz, L.L., Mohammadi, S., Geissler, A., and Portnoy, D.A. (2003). SecA2-dependent secretion of autolytic enzymes promotes *Listeria monocytogenes* pathogenesis. Proc. Natl. Acad. Sci. USA *100*, 12432–12437.

Lenz, L.L., and Portnoy, D.A. (2002). Identification of a second Listeria *secA* gene associated with protein secretion and the rough phenotype. Mol. Microbiol. *45*, 1043–1056.

Lupas, A., Engelhardt, H., Peters, J., Santarius, U., Volker, S., and Baumeister, W. (1994). Domain structure of the *Acetogenium kivui* surface-layer revealed by electron crystallography and sequence-analysis. J. Bacteriol. *176*, 1224–1233.

Marraffini, L.A., DeDent, A.C., and Schneewind, O. (2006). Sortases and the art of anchoring proteins to the envelopes of gram-positive bacteria, pp. 192–221. Microbiol. Mol. Biol. Rev. *70*, 192–221.

Masuda, K., Itoh, M., and Kawata, T. (1989). Characterization and reassembly of a regular array in the cell wall of *Clostridium difficile* GA I4131. Microbiology and Immunology 33, 287–298.

Matsuki, S., Ozaki, E., Shozu, M., Inoue, M., Shimizu, S., Yamaguchi, N., Karasawa, T., Yamagishi, T., and Nakamura, S. (2005). Colonization by *Clostridium difficile* of neonates in a hospital, and infants and children in three day-care facilities of Kanazawa, Japan. Int. Microbiol. *8*, 43–48.

Mattick, J.S. (2002). Type IV pili and twitching motility. Annu. Rev. Microbiol. *56*, 289–314.

McCoubrey, J., and Poxton, I.R. (2001). Variation in the surface layer proteins of *Clostridium difficile*. FEMS Immunol. Med. Microbiol. *31*, 131–135.

Messner, P., and Sleytr, U.B. (1991). Bacterial surface layer glycoproteins. Glycobiology *1*, 545–551.

Navarre, W.W., and Schneewind, O. (1999). Surface proteins of gram-positive bacteria and mechanisms of their targeting to the cell wall envelope. Microbiol. Mol. Biol. Rev. *63*, 174–229.

O'Brien, J.B., McCabe, M.S., Athie-Morales, V., McDonald, G.S.A., Ni Eidhin, D.B., and Kelleher, D.P. (2005). Passive immunisation of hamsters against *Clostridium difficile* infection using antibodies to surface layer proteins. FEMS Microbiol. Lett. *246*, 199–205.

Pallen, M.J., Chaudhuri, R.R., and Henderson, I.R. (2003). Genomic analysis of secretion systems. Current Opinion in Microbiology *6*, 519–527.

Pechine, S., Janoir, C., Boureau, H., Gleizes, A., Tsapis, N., Hoys, S., Fattal, E., and Collignon, A. (2007). Diminished intestinal colonization by *Clostridium difficile* and immune response in mice after mucosal immunization with surface proteins of *Clostridium difficile*. Vaccine *25*, 3946–3954.

Poxton, I.R., and Cartmill, T.D. (1982). Immunochemistry of the cell-surface carbohydrate antigens of *Clostridium difficile*. J. Gen. Microbiol. *128*, 1365–1370.

Qazi, O., Brailsford, A., Wright, A., Faraar, J., Campbell, J., and Fairweather, N. (2007). Identification and characterization of the surface-layer protein of *Clostridium tetani*. FEMS Microbiol. Lett. *274*, 126–131.

Sara, M. (2001). Conserved anchoring mechanisms between crystalline cell surface S-layer proteins and secondary cell wall polymers in Gram-positive bacteria? Trends Microbiol *9*, 47–49; discussion 49–50.

Sara, M., and Sleytr, U.B. (1987). Molecular sieving through S layers of *Bacillus stearothermophilus* strains. J. Bacteriol. *169*, 4092–4098.

Sara, M., and Sleytr, U.B. (2000). S-layer proteins. J. Bacteriol. *182*, 859–868.

Savariau-Lacomme, M.P., Lebarbier, C., Karjalainen, T., Collignon, A., and Janoir, C. (2003). Transcription and analysis of polymorphism in a cluster of genes encoding surface-associated proteins of *Clostridium difficile*. J. Bacteriol. *185*, 4461–4470.

Schneitz, C., Nuotio, L., and Lounatma, K. (1993). Adhesion of *Lactobacillus acidophilus* to avian intestinal epithelial cells mediated by the crystalline bacterial cell surface layer (S-layer). J. Appl. Bacteriol. *74*, 290–294.

Sebaihia, M., Wren, B.W., Mullany, P., Fairweather, N.F., Minton, N., Stabler, R., Thomson, N.R., Roberts, A.P., Cerdeno-Tarraga, A.M., and Wang, H. (2006). The multidrug-resistant human pathogen *Clostridium difficile* has a highly mobile, mosaic genome. Nat. Genet. *38*, 779–786.

Sharp, J., and Poxton, I.R. (1988). The cell wall proteins of *Clostridium difficile*. FEMS Microbiol. Lett. *55*, 99–104.

Sleytr, U.B., and Messner, P. (1983). Crystalline surface layers on bacteria. Annu. Rev. Microbiol. *37*, 311–339.

Sleytr, U.B., and Messner, P. (1988). Crystalline surface layers in procaryotes. J. Bacteriol. *170*, 2891–2897.

Sleytr, U.B., and Beveridge, T.J. (1999). Bacterial S-layers. Trends Microbiol *7*, 253–260.

Songer, J.G., and Anderson, M.A. (2006). *Clostridium difficile*: An important pathogen of food animals. Anaerobe *12*, 1–4.

Sougioultzis, S., Kyne, L., Drudy, D., Keates, S., Maroo, S., Pothoulakis, C., Giannasca, P.J., Lee, C.K., Warny, M., Monath, T.P. (2005). *Clostridium difficile* toxoid vaccine in recurrent *C. difficile*-associated diarrhea. Gastroenterology *128*, 764–770.

Stabler, R.A., Gerding, D.N., Songer, J.G., Drudy, D., Brazier, J.S., Trinh, H.T., Witney, A.A., Hinds, J., and Wren, B.W. (2006). Comparative phylogenomics of *Clostridium difficile* reveals clade specificity and microevolution of hypervirulent strains. J. Bacteriol. *188*, 7297–7305.

Stevens, D.L., and Rood, J.I. (2006). Histotoxic Clostridia. In Gram-positive pathogens, V.A. Fischetti, R.P. Novick, J.J. Ferretti, D.A. Portnoy, and J.I. Rood, eds. (Washington, ASM Press), pp. 715–725.

Stevens, D.L., Titball, R.W., Jepson, M., Bayer, C.R., Hayes-Schroer, S.M., and Bryant, A.E. (2004). Immunization with the C-Domain of alpha-toxin prevents lethal infection, localizes tissue injury, and promotes host response to challenge with *Clostridium perfringens*. J. Infect. Dis. *190*, 767–773.

Takumi, K., Susami, Y., Takeoka, A., Oka, T., and Koga, T. (1991). S layer protein of *Clostridium tetani*: purification and properties. Microbiol. Immunol. *35*, 569–575.

Takumi, K., Ichiyanagi, S., Endo, Y., Koga, T., Oka, T., and Natori, Y. (1992). Characterization, self-assembly and reattachment of S layer from *Clostridium botulinum* type E saroma. Tokushima J. Exp. Med. *39*, 101–107.

Tasteyre, A., Barc, M.C., Collignon, A., Boureau, H., and Karjalainen, T. (2001a). Role of FliC and FliD flagellar proteins of *Clostridium difficile* in adherence and gut colonization. Infect. Immun. *69*, 7937–7940.

Tasteyre, A., Barc, M.C., Karjalainen, T., Dodson, P., Hyde, S., Bourlioux, P., and Borriello, P. (2000). A *Clostridium difficile* gene encoding flagellin. Microbiology *146*, 957–966.

Tasteyre, A., Karjalainen, T., Avesani, V., Delmee, M., Collignon, A., Bourlioux, P., and Barc, M.C. (2001b). Molecular characterization of *fliD* gene encoding flagellar cap and its expression among *Clostridium difficile* isolates from different serogroups. J. Clin. Microbiol. *39*, 1178–1183.

Telford, J.L., Barocchi, M.I.A., Margarit, I., Rappuoli, R., and Grandi, G. (2006). Pili in Gram-positive pathogens. Nat Rev Micro *4*, 509–519.

Titball, R.W. (2005). Gas gangrene: an open and closed case. Microbiology *151*, 2821–2828.

Varga, J.J., Nguyen, V., O'Brien, D.K., Rodgers, K., Walker, R.A., and Melville, S.B. (2006). Type IV pili-dependent gliding motility in the Gram-positive pathogen *Clostridium perfringens* and other Clostridia. Mol. Microbiol. *62*, 680–694.

Waligora, A.J., Hennequin, C., Mullany, P., Bourlioux, P., Collignon, A., and Karjalainen, T. (2001). Characterization of a cell surface protein of *Clostridium difficile* with adhesive properties. Infect. Immun. *69*, 2144–2153.

Walker, P.D. (1992). Bacterial vacines: old and new. Vaccine *10*, 977–990.

Wright, A., Wait, R., Begum, S., Crossett, B., Nagy, J., Brown, K., and Fairweather, N. (2005). Proteomic analysis of cell surface proteins from *Clostridium difficile*. Proteomics *5*, 2443–2452.

Antibiotic Resistance Determinants in *Clostridium difficile*

Paola Mastrantonio and Patrizia Spigaglia

Abstract

Clostridium difficile, the well-known nosocomial pathogen responsible for the majority of antibiotic associated diseases, is increasingly recognized also as the cause of community-associated disease and of enteric disease in animals.

The organism is resistant to several antibiotics and can survive disruption of the normal intestinal flora after antibiotic treatment exploiting this advantage to colonize and cause disease. The study of the mechanisms responsible for resistance have highlighted the presence of mobile genetic elements in the *C. difficile* genome, potentially acquired from other microorganisms. *C. difficile* might be able to disseminate resistance determinants to other species, thus collaborating to the evolution of the antibiotic-resistant patterns that characterize the bacteria circulating worldwide.

Introduction

Clostridium difficile is the most common nosocomial enteric pathogen causing antibiotic-associated diarrhoea and pseudomembranous colitis (Gerding *et al.*, 1995). The clinical manifestations range from asymptomatic carriage to severe diarrhoea and pseudomembranous colitis with toxic megacolon (Bartlett, 2002). Hospital patients receiving antibiotics are the most at risk. Disease occurs after eradication of the patient's intestinal microflora due to the administration of broad-spectrum antibiotics thus creating conditions that favour acquisition and proliferation of *C. difficile* (Kelly *et al.*, 1994).

Third-generation cephalosporins, clindamycin and amoxicillin have been associated with the highest risk for developing antibiotic-associated disease (AAD), but almost all antibiotics can promote the disease (Freeman *et al.*, 1999). Recently, also fluoroquinolones (FQs) have emerged as a major risk for AAD during hospital outbreaks (Pepin *et al.*, 2005).

The increased activity of the new generation fluoroquinolones against anaerobes, and thus against the normal anaerobic gut flora constituting the main barrier to exogenous bacteria, and the increased resistance shown by many recent *C. difficile* strains may explain the role of FQs in inducing *C. difficile*-associated diseases.

Clostridium difficile is generally resistant to cephalosporins and cephamycin antibiotics, frequently resistant to clindamycin and erythromycin, to tetracyclines, chloramphenicol and more recently to fluoroquinolones. Only few strains have been reported worldwide to be resistant or be less susceptible to metronidazole and vancomycin (Barbut *et al.*, 1999; Brazier *et al.*, 2001; Pelaez *et al.*, 2002). The latter are considered the drugs of choice for the treatment of severe cases of *C. difficile*-associated disease. Both the clinical significance and the mechanisms responsible for metronidazole or vancomycin resistance in *C. difficile* treatment are still undefined since the number of failures reported are relatively small as well as the number of strains showing *in vitro* minimum inhibitory concentration (MIC) values over the breakpoint value.

Very recently, The Clinical and Laboratory Standards Institute – Seventh Edition (formerly NCCLS, 2007) published the interpretative categories and minimal inhibitory concentration of antibiotics against anaerobic bacteria including

C. difficile and the acceptable ranges of MICs of control strains.

Resistance to metronidazole

In a paper published in 2002 by Pelaez *et al.*, a rate of resistance of 6.3% at the critical breakpoint (16 mg/l) was reported mainly among isolates from human immunodeficiency virus-infected patients. In some other studies (Barbut *et al.*, 1999; Brazier *et al.*, 2001) only few sporadic resistant strains have been described.

Metronidazole (Mz), a 5-nitroimidazole, is regarded as an important therapeutic agent in the treatment of anaerobic infections (Reysset, 1996). Resistance mechanisms are associated in particular with the presence of *nim* genes. Seven *nim* genes (*nim*A–*nim*G), plasmid or chromosomally encoded, have been described in Bacteroides spp (Lofmark *et al.*, 2005). Carlier *et al.*, 1997 have suggested that these genes encode a nitroimidazole reductase that converts 4- or 5-nitroimidazole to 4- or 5-aminoimidazole, thus avoiding the formation of toxic nitroso radicals essential for antimicrobial activity.

A number of *C. difficile* strains, resistant to Mz, have been analysed so far for the presence of *nim* genes but with negative results (Brazier *et al.*, 2001) so other mechanisms may be involved or other *nim*-like genes not yet detected may exist. Further mechanisms for Mz resistance of anaerobic bacteria have been described: (1) altered membrane proteins affecting permeability, (2) reduced pyruvate/ferredoxin oxidoreductase activity and (3) reduced DNA interactions. Future studies to investigate the resistance mechanism of *C. difficile* to Mz should carefully consider them.

Resistance to vancomycin

Pelaez *et al.*, 2002 reported 3.1% of *C. difficile* strains with decreased susceptibility to vancomycin (MIC values of 8–16 mg/l) since the critical breakpoint is considered 32 mg/l.

No other strains, intermediate or resistant to vancomycin have been reported so far, so the mechanism responsible for this resistance in *C. difficile* has not yet been defined.

Vancomycin (Va) is a glycopeptide antibiotic interfering with cell wall synthesis by binding to the terminal dipeptide D-alanyl-D-alanine region of the pentapeptide precursors of peptidoglycan side chains (Nagarajan, 1991). The final step of this interference is cellular death. Va resistance is widespread among enterococci, especially *E. faecalis* and *E. faecium* (Leclercq *et al.*, 1988; McKessar *et al.*, 2000) and represents one of the most serious problems in hospital infections. Six different gene clusters (*van*A–*van*E and *van*G) have been described in enterococci whereas a *van*B gene was isolated in a strain belonging to Clostridium spp. from the human intestinal flora (Domingo *et al.*, 2005). This report is not surprising because of the high density of clostridia in the human intestine and the possibility of horizontal transfer events among resident bacteria also through conjugative transposons carrying *van*B genes. Future studies looking for the resistance mechanism to Va of *C. difficile* should carefully consider this possibility.

Resistance to macrolide–lincosamide–streptogramin B (MLS) antibiotics

Erythromycin is the first macrolide antibiotic introduced in 1952 against bacterial infections, clindamycin is a semisynthetic derivative of lincomycin and belongs to the lincosamide class. This drug is also potent against anaerobic bacteria. Streptogramin antibiotics are also used, even if more rarely, in clinical practice. These three classes of antibiotics have a similar mode of action inhibiting protein synthesis due to the dissociation of the peptidyl-tRNA molecule from the ribosomes during elongation (Weisblum, 1995). The first mechanism of resistance described was the modification of the 23S r RNA by the adenine-N^6 methyltransferase (Leclercq *et al.*, 1991). Many different methyltransferases have been described in different bacteria and the genes encoding these enzymes are named *erm* (erythromycin ribosome methylation) (Weisblum, 1998). Numerous *erm* genes have been characterized and divided into distinct classes according to their sequence similarity (Roberts *et al.*, 1999). ErmB class of determinants is detected in many different bacterial species suggesting its potential for intergeneric transfer.

The resistance to erythromycin and clindamycin in *C. difficile* is old-dated and was already described at the beginning of the 1980s

(Weisblum, 1995). During the 1990s, clindamycin-resistant strains of *C. difficile* have been reported and their enhanced epidemic potential has been suggested (Johnson *et al.*, 1999).

In *C. difficile* this resistance is mainly due to the alteration of the antibiotic target site catalyzed by an rRNA methyltransferase that is encoded by an *erm*(B) gene located on the transposon Tn*5398* (Mullany *et al.*, 1995). Initial studies suggested a conjugation-like mechanism to transfer the ErmB determinant to other strains of *C. difficile* or to *Staphylococcus aureus* or *Bacillus subtilis* (Mullany *et al.*, 1990). Conjugative transposons are usually located in the bacterial chromosome and can be transferred by a process requiring cell to cell contact. This transposition involves excision of the element, formation of a non-replicating covalently closed circular intermediate which will integrate into the recipient's genome, a processes mediated by transposases and site-specific recombinases encoded by the transposon itself.

Since Tn*5398* encodes no readily identifiable transposition or mobilization protein, it appears to be a mobilizable non-conjugative element. It has been hypothesized that a much larger region, containing Tn*5398*, excises from the chromosome and can be integrated into the recipient chromosome by RecA-dependent homologous recombination.

Farrow *et al.* (2000, 2001) demonstrated in *C. difficile* 630 that Tn*5398* is 9.6 kb in size and carries two copies of *erm*(B) followed by direct repeat sequences. The latter are variants of the two almost identical directly repeated sequences flanking the *erm*(B) gene in *Clostridium perfringens* and it has been hypothesized that *erm*(B) of *C. difficile* originated from duplication of the determinant progenitor of *C. perfringens* followed by recombination events that resulted in two *erm*(B) genes separated by a complete copy of the DR sequence (Fig. 9.1).

Further studies (Farrow *et al.*, 2001; Spigaglia *et al.*, 2002) showed a heterogeneity in both Tn*5398* and *erm*(B) determinants in *C. difficile* strains obtained from different geographic locations. In particular, it has been found that *C. difficile* strains can show the presence of one or two copies of *erm*(B) gene and different arrangements of the surrounding regions.(Fig. 9.1).

Recently, a more frequent circulation of erythromycin-resistant strains, such as the current epidemic PCR ribotype 027, which lack *erm*(B) or other known *erm* genes or efflux mechanisms has been described (Ackermann *et al.*, 2003; Kuijper *et al.*, 2006). Further studies will be necessary to understand the mechanism responsible for resistance in those strains.

Resistance to tetracyclines

Tetracyclines inhibit bacterial protein synthesis by blocking the attachment of the transfer RNA-amino acid to the ribosome. More precisely they are inhibitors of the codon–anticodon interaction. Tetracycline resistance is mainly due to two different mechanisms: (1) the energy-dependent membrane-associated proteins (efflux proteins) which export tetracyline out of the cell; (2) the ribosomal protection proteins which interacting with the ribosome cause an allosteric disruption of the primary tetracycline binding site and allows the release of the tetracycline molecules from the ribosome which returns to its standard conformational state and protein synthesis proceeds. These proteins are encoded by *tet* genes (Roberts, 1996 and 2005).

The majority of tetracycline-resistant *C. difficile* strains harbour *tet*(M) genes encoding for ribosomal protection proteins. This resistance determinant is carried on a conjugative transposon Tn*5397*, as shown in *C. difficile* 630 and was easily transferred to *B. subtilis* and back to *C. difficile* as well as between *C. difficile* strains. The central region of Tn*5397* is closely related to the first conjugative transposon identified in *Enterococcus faecalis* and the most extensively studied Tn*916*, but the integration and excision regions are completely different. In Tn*5397* the gene for the site specific recombinase TndX replaces the Tn*916* genes *int* and *xis* (Wang *et al.*, 2000a).

In 2000, Tn*916* conjugative transposon was found for the first time in a *C. difficile* environmental strain (Wang *et al.*, 2000b).

Only recently, in some tetracycline-resistant clinical isolates the presence of Tn*916*-like elements showing new *tet*(M) alleles have been demonstrated (Spigaglia *et al.*, 2006). In particular, hybridization assays were performed using the amplified products from *tet*(M) and *int* as

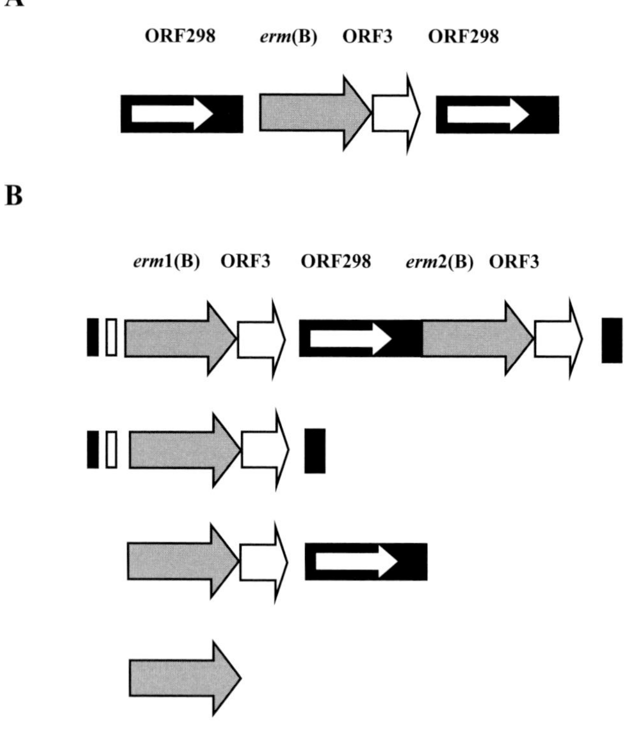

Figure 9.1 Schematic organization of the different arrangements of the Erm(B) determinant in *C. perfringens* CP592 (A) and the different arrangements found in *C. difficile* (B). The approximate extent and organization of the determinants are not necessarily to scale. Regions of nucleotide sequence similarity are indicated by the same shading. The arrows indicate the individual ORFs and their respective direction of transcription. The black rectangles indicate the direct repeat (DR) sequence, whereas the Erm leader peptide is indicated by the white rectangle.

probes for Tn*916* and *tnd*X gene for Tn*5397*. Twenty-eight per cent of the examined strains were positive for *tnd*X and 21% for *int* suggesting the circulation of strains carrying Tn*5397* and Tn*916*, respectively. Sequencing the *tet*(M) gene and the upstream region of eight selected strains as representative for genotypic and phenotypic characteristics, six different *tet*(M) variants were identified: *tet*(M)1–*tet*(M)6. *tet*(M)1 sequence was 100% identical to the *tet*(M) gene found in *Streptococcus agalactiae*, the other alleles did not show complete identity with other *tet*(M) genes already known. All *tet*(M) genes were carried by elements showing a genetic organization very similar to that of the *Enterococcus faecalis* D16 Tn*916*. The phylogenetic tree (Fig. 9.2) seems to suggest that, in C. *difficile*, the acquisition of these Tn*916*-like *tet*(M) determinants from a hypothetical precursor is a relatively more recent event than the acquisition of the Tn*5397* *tet*(M) determinant.

The presence of two Tn*916*-like elements has been found in some C. *difficile* strains. In particular, one of these isolates harboured two *tet*(M) genes: one carried by a Tn*916*-like element, able to transfer to a recipient C. *difficile* strain; the second was genetically linked to an *erm*(B) in a composite element probably unable to conjugate (Spigaglia *et al*., 2007). Molecular analysis of the *tet*(M)–*erm*(B) DNA sequence demonstrated that this region has arisen by recombination of DNA fragments from different plasmids and transposons. This is the first demonstration that C. *difficile* is able to accumulate and maintain antibiotic resistance genes, as observed in other pathogens.

Tn*916* elements are known to be able to transfer antibiotic resistance genes and also virulence determinants among different bacterial species and genera and their presence in C. *difficile* could increase its ability to acquire new virulence traits.

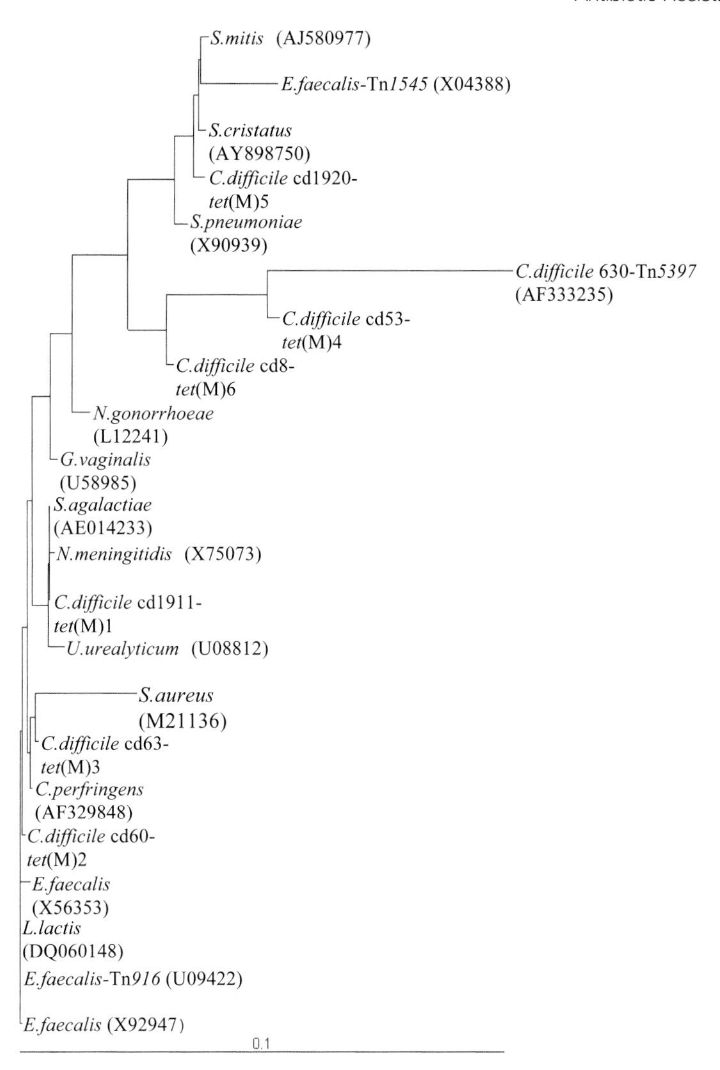

Figure 9.2 Unrooted neighbour-joining phylogenetic tree obtained from the nucleotide multiple alignment of the six *C. difficile tet*(M) variants identified and other 16 *tet*(M) gene sequences available in GenBank. The branch lengths are scaled in proportion to the extent of the change per position as indicated by the scale bar. GenBank accession numbers are in parentheses. (From P. Spigaglia *et al.*, J Antimicrob Chemother 2006, *57*, 1205–1209.)

Resistance to chloramphenicol

As far as the resistance of *C. difficile* to chloramphenicol is concerned, it is mediated, as in other bacteria, by a gene which encodes chloramphenicol acetyltransferases, CAT, (Wren *et al.*, 1988). In this paper, some clinical isolates resistant to chloramphenicol were described and a 0.27 kb *Eco*RV–*Taq*I restriction fragment carrying the chloramphenicol resistance gene (*cat*D) was identified. Southern blot hybridization clearly showed the presence of two copies of the gene in each of the strains examined. In 1989, Rood *et al.* found a significant identity of *cat*D of *C. difficile* with *cat*P of *C. perfringens*. The latter is located on a mobilizable transposon of the Tn*4451* family which resides on a plasmid. In fact, mobilisable transposons are elements that, unlike conjugative elements, do not encode all of the proteins necessary for bacterial conjugation but only some proteins which are able to facilitate the transfer that is dependent upon the conjugation machinery of a co-resident conjugative plasmid or transposon. The sequence of Tn*4451* showed six genes, among them the site-specific *tnp*X gene is a site-specific recombinase gene which is responsible for both excision and insertion and a potential mobilization gene *tnp*Z.

In chloramphenicol-resistant *C. difficile* strains, it has been shown (Lyras *et al.*, 1998) that the *cat*D gene responsible for the resistance is carried on the mobilizable transposons Tn*4453*a and *4453*b which are functionally and structurally related to Tn*4451* from *C. perfringens*, suggesting the transfer of DNA between these two species.

Resistance to fluoroquinolones

Fluoroquinolones are inhibitors of bacterial DNA replication by interacting with either or both of two target enzymes necessary for DNA supercoiling, the DNA gyrases and the topoisomerase IV (Hooper, 2000, 2001).

The first generation of fluoroquinolones such as ciprofloxacin and ofloxacin were inactive against most anaerobic bacteria but the new generation includes products with a broad-spectrum activity such as moxifloxacin and gatifloxacin (Wilcox *et al.*, 2000).

Resistance is generally caused by two mechanisms: (a) mutations in target enzyme encoding genes and (b) increased efflux of the drug or reduced permeability. In the first case, resistance is due to nucleotide mutations occurring in the quinolone resistance determining region (QRDR) of the *gyr*A and *gyr*B genes that encode for the two DNA gyrase subunits, GyrA and GyrB, or in the *par*C and *par*E genes that encode for the topoisomerase IV subunits, ParC and ParE.

Different amino acid substitutions have been detected in FQs resistant *C. difficile* so far (Table 9.1). In particular, Ackermann *et al.* (2001) reported some FQ-resistant *C. difficile* strains which, in particular, contained a mutation at codon 82 (either ACT→ATT or ACT→GTT) in the *gyr*A gene resulting in amino acid exchange: threonine→isoleucine or the threonine→valine, respectively. *C. difficile* Thr82 corresponds to Ser83 in *E. coli*, a position known to play a key role in fluoroquinolone resistance.

In 2002, Dridi *et al.* described the detection of the same amino acid change in French isolates and also of further two in GyrA: Asp71 →Val and Ala118→Thr.

The substitution Asp71 →Val corresponds to a change at position 72 of GyrA in *E. coli* but, although located in QRDR, there is no direct proof that it is involved in the resistance.

The other substitution Ala118→Thr was only occasionally reported in other species and its role in resistance has been demonstrated in *S. aureus* (Fournier *et al.*, 1998).

Interestingly, in the same paper by Dridi *et al.*, two amino acid substitutions in GyrB were also described for the first time: Asp426→Asn and Arg447→Leu. They resulted either identical or located at the same position, respectively, when compared with those described in *E. coli*. In 2006, Drudy *et al.*, described a novel substitution in GyrB: Asp426→Val in FQ highly resistant ToxA$^-$/ToxB$^+$ *C. difficile* strains belonging to the PCR ribotype 017, toxinotype VIII. More recently, the substitution Thr82→Ile has been detected in *C. difficile* PCR ribotype 027 strains isolated in Ireland by the same group (Drudy *et al.*, 2007).

Since the initial studies, no gene with homology to known *par*C genes from other bacterial species have been found, and therefore it was supposed that *C. difficile* lacks genes for topoisomerase IV. This was confirmed by analysing the *C. difficile* 630 genome sequence published by the Sanger Centre, UK (http://www.sanger.ac.uk/Projects/C_difficile/).

Recently, emergence of fluoroquinolones as the predominant risk factor for *C. difficile* –associated diarrhoea has been reported (Pepin *et al.*, 2005). In particular, since March 2003, outbreaks of more severe cases of CDAD associated

Table 9.1 Amino acid substitutions detected in FQ-resistant *C. difficile* isolates

Amino acid substitutions in GyrA	Amino acid substitutions in GyrB
Thr82→Ile/Val	Asp426→Asn/Val
Asp71→Val	Arg447→Leu
Ala11→Thr	

with *C. difficile* PCR ribotype 027, Pulsed Field Gel Electrophoresis type NAP1, toxinotype III, were reported in Canada (Pepin *et al.*, 2004) and in the USA (McDonald *et al.*, 2005), rapidly followed by outbreaks in England (Smith, 2005), in the Netherlands (van Steenberen *et al.*, 2005), in Belgium (Joseph *et al.*, 2005), and in France (Tachon *et al.*, 2006).

A recent paper (Barbut *et al.*, 2007) reports the results obtained from a collaborative study performed to phenotypically and genotypically characterize *C. difficile* strains isolated in Europe with the aim to estimate the incidence of CDAD among hospitalized patients and to determine the point prevalence of the epidemic clone 027.

In that study, FQ-resistant *C. difficile* strains were 37.2% and resistance was more frequent in patients previously treated with FQs (58.7% vs. 31.4%, $P < 0.001$). The results seem to confirm that the recent acquisition of resistance to the newer fluoroquinolones by *C. difficile* and in particular by the hypervirulent *C. difficile* 027 is the major reason for its wide dissemination.

In conclusion, the antibiotic resistance of *C. difficile* strains is mostly due to determinants carried by mobile elements. The recent identification of a large number of mobile genetic elements carrying sequences potentially involved in antimicrobial resistance in the genome of multidrug-resistant *C. difficile* strain 630 (Sebaihia *et al.*, 2006) has highlighted an evolving process that provides this bacterium with potential advantages over the co-resident gut flora.

References

Ackermann, G., Tang, Y.J., Kueper, R., Heisig, P., Rodloff, A.C., Silva, J.J.R., and Cohen, S.H. (2001). Resistance to moxifloxacin in toxigenic *Clostridium difficile* isolates is associated with mutations in *gyr*A. Antimicrob. Agents Chemother. *45*, 2348–2353.

Barbut, F., Decrè, D., Burghoffer, B., Lesage, D., Delisle, F., Lalande, V., Delmèe, M., Avesani, V., Sano, N., Coudert, C., and Petit, J.C. (1999) Antimicrobial susceptibility and serogroups of clinical strains of *Clostridium difficile* isolated in France in 1991 and 1997 Antimicrob. Agents Chemother. *43*, 2607–2611.

Barbut,F., Mastrantonio, P., Delmee, M., Brazier, J., Kuijper, E., Poxton,I., on behalf of the European Study Group on *Clostridium difficile* (*ESGCD*) (2007). Prospective study of *Clostridium difficile* infections in Europe with phenotypic and genotypic characterisation of the isolates. Clin. Microbiol. Infect. *13*, 1048–1057.

Bartlett, J.G. (2002). Clinical practice. Antibiotic-associated diarrhea. N. Engl. J. Med. *346*, 334–339.

Brazier, J.S., Fawley, W., Freeman, J., and Wilcox, M.H. (2001) Reduced susceptibility of *Clostridium difficile* to metronidazole. J. Antimicrob. Chemother. *48*, 741–742.

Carlier, J.P., Seller, N., Rager, M.N., and Reysset, G. (1997) Metabolism of a 5-nitroimidazole in susceptible and resistant isogenic strains of *Bacteroides fragilis*. Antimicrob. Agents Chemother. *41*, 1495–1499.

Clinical and Laboratory Standards Institute. (2007). Methods for Antimicrobial Susceptibility Testing of Anaerobic Bacteria; Approved Standard-Seventh Edition. CLSI document M11-A7, Vol 27, No 2. Wayne, PA 19087 USA.

Domingo, M.C., Huletsky, A., Bernal, A,Giroux, R., Boudreau, D.K., Picard, F.J., and Bergeron, M.G. (2005) Characterization of a Tn*5382*-like transposon containing the *van*B2 gene cluster in a Clostridium strain isolated from human faeces. J. Antimicrob. Chemother. 55, 466–474.

Dridi, L., Tankovic, J., Burghoffer, B., Barbut, F., and Petit, J.C. (2002). *gyr*A and *gyr*B mutations are implicated in cross-resistance to ciprofloxacin and moxifloxacin in *Clostridium difficile*. Antimicrob. Agents Chemother. *46*, 3418–3421.

Drudy, D., Quinn, T., O'Mahony, R., Kyne, L., O'Gaora, P., Fanning, S. (2006). High-level resistance to moxifloxacin and gatifloxacin associated with a novel mutation in *gyr*B in toxin A-negative, toxin B-positive *Clostridium difficile*. J. Antimicrob. Chemother. *58*, 1264–1267.

Drudy, D., Kyne, L., O'Mahony, R., Fanning, S. (2007). *gyr*A mutations in fluoroquinolone-resistant *Clostridium difficile* PCR-027. Emerg. Infect. Dis. *13*, 504–505.

Farrow, K. A., Lyras, D., and Rood, J.I.(2001). Genomic analysis of the erythromycin resistance element Tn5398 from *Clostridium difficile*. Microbiology *147*, 2717–2728.

Farrow, K.A., Lyras, D., and Rood, J.I. (2000). The macrolide-lincosamide-streptogramin B resistance determinant from *Clostridium difficile* 630 contains two *erm*(B) genes. Antimicrob. Agents Chemother. *44*, 411–413.

Fournier, B., Hooper, D.C. (1998). Effects of mutations in GrlA of topoisomerase IV from *S.aureus* on quinolone and coumarin activity. Antimicrob. Agents Chemother. *42*, 2109–2112.

Freeman, J., and Wilcox, M.H. (1999). Antibiotics and *Clostridium difficile*. Microbes Infect. *1*, 777–784.

Gerding, D.N., Johnson, S., Peterson, L.R., Mulligan, M.E., and Silva, J.Jr. (1995). *Clostridium difficile*-associated diarrhea and colitis. Infect. Control Hosp. Epidemiol. *16*, 459–477.

Hooper, D.C. (2000). Mechanisms of action and resistance of older and newer fluoroquinolones. Clin. Infect. Dis *31*, (suppl 2) S24–28.

Hooper, D.C. (2001). Emerging mechanisms of fluoroquinolone resistance. Emerg. Infect. Dis. *7*, 337–341.

Johnson, S., Samore, M.H., Farrow, K.A., Killgore, G.E., Tenover, F.C., Lyras, D., Rood, J.I., DeGirolami, P., Baltch, A., Rafferty, M.E., Pear, S.M., and

Gerding,D.N. (1999). Epidemics of diarrhea caused by a clindamycin-resistant strain of *Clostridium difficile* in four hospitals. N. Engl. J. Med. *341*, 1645–1651.

Joseph, R., Demeyer, D., Vanrenterghem, B., van den Berg, R., Kuijper, E., and Delmee, M. (2005). First isolation of *Clostridium difficile* PCR ribotype 027, toxinotype III in Belgium. Eurosurveill. Weekly 10:EO51020.4. Available from: http://www.eurosurveillance.org/ew/2005/051020.asp#4

Kelly, C.P., Pothoulakis, C., and Lamont, J.T., (1994). *Clostridium difficile* colitis. N. Engl. J. Med. *330*, 257– 262.

Kuijper, EJ., van den Berg, R., Debast, S., Visser, C.E., Veenendaal, D., Troelstra, A., van der Kooi, T., van den Hof, S., and Notermans, D.W. (2006). *Clostridium difficile* ribotype 027, toxinotype III, in the Netherlands. Emerg. Infect. Dis. *12*, 827–830.

Leclercq, R., and Courvalin,P. (1991). Bacterial resistance to macrolide, lincosamide, and streptogramin antibiotics by target modification. Antimicrob. Agents Chemother. *35*, 1267–1272.

Leclercq, R., Derlot, E., and Duval, J. (1988). Plasmid-mediated resistance to vancomycin and teicoplanin in *Enterococcus faecium*. N. Engl. J. Med *319*, 157–161.

Lofmark, S., Fang, H., Hedberg, M., and Edlund, C. (2005). Inducible metronidazole resistance and *nim* genes in clinical *Bacteroides fragilis* group isolates. Antimicrob. Agents Chemother. *49*, 1253–1256.

Lyras, D., Storie, C., Huggins, A.S., Crellin, P.K., Bannam, T.I., and Rood, J.I. (1998). Chloramphenicol resistance in *Clostridium difficile* is encoded on Tn*4453* transposons that are closely related to Tn*4451* from *Clostridium perfringens*. Antimicrob. Agents Chemother. *42*, 1563–1567.

McDonald, L.C., Killgore, G.E., Thompson, A., Owens, R.C. Jr, Khazakova, S.V., Sambol, S.P., Johnson, S., and Gerding, D.N. (2005). An epidemic toxin gene-variant strain of *Clostridium difficile.* N Engl J Med 353,2433–2441.

McKessar, S. J., Berry, A.M., and Bell, J.M. (2000). genetic characterization of *van*G, a novel vancomycin resistance locus of *Enterococcus faecalis*. Antimicrob. Agents Chemother. *44*, 3224–3228.

Mullany, P., Wilks, M., Lamb, I., Clayton, C., Wren, B., and Tabaqchali, S. (1990). Genetic analysis of a tetracycline resistance element from *Clostridium difficile* and its conjugal transfer to and from *Bacillus subtilis*. J. Gen. Microbiol. *136*,1343–1349.

Mullany, P., Wilks, M., and Tabaqchali S. (1995). Transfer of macrolide-lincosamide-streptogramin B (MLS) resistance element from *Clostridium difficile* is linked to a gene homologous with toxin A and is mediated by a conjugative transposon, Tn*5398*. J. Antimicrob. Chemother. *35*, 305–315.

Nagarayan, R. (1991). Antibacterial activities and modes of action of vancomycin and related glycopeptides. Antimicrob. Agents Chemother. *35*, 605–609.

Pelaez, T., Alcalà, L., Alonso, R., Rodriguez-Creixems, M., Garcia-Lechuz, J.M., and Bouza, E. (2002) Reassessment of *Clostridium difficile* susceptibility to metronidazole and vancomycin. Antimicrob. Agents Chemother. *46*, 647–1650.

Pepin, J., Saheb, N., Coulombe, M.A., Alary, M.E., Corriveau, M.P., Authier, S., Leblanc, M., Rivard, G., Bettez, M., Primeau,V., Nguyen, M., Jacob, C.E., and Lanthier, L. (2005). Emergence of fluoroquinolones as the predominant risk factor for *Clostridium difficile*-associated diarrhea: a cohort study during an epidemic in Quebec. Clin. Infect. Dis. *41*, 1254–1260.

Pepin, J., Valiquette, L., Alary, M.E., Villemure, P., Pelletier, A., Forget, K., Pepin, K., and Chouinard, C. (2004). *Clostridium difficile*-associated diarrhoea in a region of Quebec from 1991 to 2003:a changing pattern of disease severity. Canad. Med. Assoc. J. *171*,466–472.

Reysset, G. (1996). Genetics of 5-nitroimidazole resistance in Bacteroides species. Anaerobe *2*, 59–69.

Roberts, M.C. (1996). Tetracycline resistance determinants: mechanisms of action, regulation of expression, genetic mobility, and distribution. FEMS Microbiol. Rev. *19*, 1–24.

Roberts, M.C. (2005). Update on acquired tetracycline resistance genes. FEMS Microbiol. Lett *245*, 195–203.

Roberts, M.C., Sutcliffe, J., Courvalin, P., Jensen, L.B., Rood,J., and Seppala, H. (1999). Nomenclature for macrolide and macrolide-lincosamide-streptogramin B resistance determinants. Antimicrob. Agents Chemother. *43*, 2823–2830.

Rood, J.I., Jefferson, S., Bannam T.L., Wilkie, J.M., Mullany, P., and Wren,B.W. (1989). Hybridization analysis of three chloramphenicol resistance determinants from *Clostridium perfringens* and *Clostridium difficile*. Antimicrob. Agents Chemother. *33*, 1569–1574.

Sebaihia, M., Wren, B.W., Mullany,P., Fairweather, N.F., Minton,N., Stabler, R., *et al.* (2006). The multidrug-resistant human pathogen *Clostridium difficile* has a highly mobile,mosaic genome. Nature Genetics *38*, 779–785.

Smith, A. (2005). Outbreak of *Clostridium difficile* infection in an English hospital linked to hypertoxin-producing strains in Canada and the US. Eurosurveillance 10:050630 http://www.eurosurveillance.org/ew/2005/050630asp#2

Spigaglia, P., and Mastrantonio, P. (2002). Analysis of macrolide-lincosamide-streptograminB (MLS_B) resistance determinant in strains of *Clostridium difficile*. Microb. Drug Res. *8*, 45–53.

Spigaglia, P., Barbanti, F., and Mastrantonio, P. (2006). New variants of the *tet*(M) gene in *Clostridium difficile* clinical isolates harbouring Tn*916*-like elements. J. Antimicrob. Chemother. *57*, 1205–1209.

Spigaglia, P., Barbanti, F., Mastrantonio, P. (2007). Detection of a genetic linkage between genes coding for resistance to tetracycline and erythromycin in Clostridium difficile. Microb Drug Res *13*,90–95.

Tachon, M., Cattoen, C., Blanckaert, K. Poujol, I., Carbonne, A., Barbut, F., Petit, J.C., and Coignard, B. (2006). First cluster of C. *difficile* toxinotype III, PCR ribotype 027 associated disease in France: preliminary report. Eurosurveillance 11:EO60504.1. Available from:http://www.eurosurveillance.org/ew/2006/060504.asp#1

van Steenberen, J., Debast, S., vanCregten, E., vanden Berg, R., Notermans, D., and Kuijper, E. (2005). Isolation of *Clostridium difficile* ribotype 027, toxinotype III in the Netherlands after increase in C. *difficile*-associated diarrhoea. Eurosurveill Weekly 10:EO50714.1. Available from: http://www.eurosurveillance.org/ew/2005/050714.asp#1

Wang, H., Roberts, A.P., Lyras, D., Rood, J.I., Wilks M., and Mullany, P. (2000a). Characterization of the ends and target sites of the novel conjugative transposon Tn*5397* from *Clostridium difficile*: excision and circularization is mediated by the large resolvase, TndX. J. Bacteriol. *182*, 3775–3783.

Wang, H., Roberts, A.P., and Mullany, P. (2000b). DNA sequence of the insertional hot spot of Tn*916* in the *Clostridium difficile* genome and discovery of a Tn*916*-like element in an environmental isolate integrated in the same hot spot. FEMS Microbiol. Lett. *192*, 15–20.

Weisblum, B. (1995) Erythromycin resistance by ribosome modification. Antimicrob. Agents Chemother. *39*, 577–585.

Weisblum, B. (1998) Macrolide resistance. Drug Resist. Update *1*, 29–41.

Wilcox, M.H., Fawley, W., Freeman, J., and Brayson, J. (2000). In vitro activity of new generation fluoroquinolones against genotypically distinct and indistinguishable *Clostridium difficile* isolates. J. Antimicrob Chemother. *46*, 551–555.

Wren, B.W., Mullany, P., Clayton, C., and Tabaqchali, S. (1988). Molecular cloning and genetic analysis of a chloramphenicol acetyltransferase determinant from *Clostridium difficile*. Antimicrob. Agents Chemother. *32*, 1213–1217.

Development of Genetic Knock-out Systems for Clostridia

10

John T. Heap, Stephen T. Cartman, Oliver J. Pennington, Clare M. Cooksley, Jamie C. Scott, Ben Blount, David A. Burns and Nigel P. Minton

Abstract

Despite the medical and industrial importance of the genus *Clostridium*, our understanding of its basic biology lags behind that of its more illustrious counterpart, *Bacillus*. The advent of the genomics era has provided new insights, but full exploitation of the data becoming available is being hindered by a lack of mutational tools for functional genomic studies. Thus, in the preceding decades the number of clostridial mutants generated has been disappointingly low. On the one hand, the absence of effective transposon elements has stymied random mutant generation. On the other hand, the construction of directed mutants using classical methods of recombination-based, allelic exchange has met with only limited success. Indeed, in the majority of clostridial species mutants are largely based on integration of plasmids by a Campbell-like mechanism. Such single crossover mutants are unstable. As an alternative, recombination-independent strategies have been developed that are reliant on retargeted group II introns. One element in particular, the ClosTron, has been devised which provides the facility for the positive selection of mutants. ClosTron-mediated mutant generation is extremely rapid, highly efficient and reproducible. Moreover the mutants made are extremely stable. Its deployment considerably expands current options for functional genomic studies in clostridia.

Introduction

The genus *Clostridium* has recently assumed greater prominence to the endeavours of humankind. Thus, human mortality and morbidity due to *C. difficile* is increasing as a consequence of the emergence of a highly virulent clone of *C. difficile*, necrotic enteritis in fowl has dramatically increased due to *C. perfringens* and there are heightened public concerns over the use of *C. botulinum* by bioterrorists. On the other hand, industrial strains such as *C. acetobutylicum* (the organism at the heart of the acetone–butanol–ethanol fermentation process) have now assumed even greater importance in the drive to reduce carbon emissions through the generation of biofuels such as butanol. Moreover, experimental advances have now demonstrated that clostridia have a significant role to play in treating cancer, by capitalizing on the unique ability of spores to selectively germinate in tumours (e.g. *C. sporogenes* and *C. novyi*). As a consequence, there is an urgent need to better understand the basic biology of these important bacteria as a prerequisite for both countering the diseases they cause, and exploiting their beneficial properties in industry and medicine.

Insight into clostridial physiology has been considerably enhanced in recent years by an explosion in the number of available genome sequences. However, maximum exploitation of the data generated is being precluded by an absence of effective genetic tools for the directed inactivation of the hypothetical genes identified. It is only through mutation that the functional identity of any particular gene can be definitively ascribed. In this respect, functional genomic studies within the genus lag significantly behind those being undertaken in other genera.

Generation of random mutants

In many other bacterial genera, random mutational tools have played a pivotal role in ascribing gene function. Principal amongst these tools are transposons, such as Tn5 and Tn*10*. Disappointingly, equivalent tools have yet to be developed for any clostridial species. Rather, success to date has been limited to the use of conjugative transposons.

Pathogenic clostridia

From a very early stage in the development of genetic systems in *C. difficile* it was apparent that transferable antibiotic resistance was mediated in the absence of plasmid transfer by conjugative elements (Ionesco, 1980; Smith *et al.*, 1981; Wüst and Hardegger, 1983; Mullany *et al.*, 1990). This was in contrast to *C. perfringens*, where the predominant form of antibiotic resistance transfer was via conjugative plasmids (Rood and Cole, 1991). It was soon apparent that at least some of the elements involved shared homology with the frequently employed transposon, Tn*916* of *Enterococcus faecalis* (Hächler *et al.*, 1987). Indeed, the element Tn5397 of strain 630 was subsequently thoroughly investigated (Roberts *et al.*, 2001; Sebaihia *et al.*, 2006) and found to share extensive similarity to Tn*916*, essentially differing only in their respective integration and excision systems. These obvious similarities led, in part, to studies which first investigated the ability of Tn*916* to transfer to *C. difficile* (Mullany *et al.*, 1991) and other clostridial species (Volk *et al.*, 1988; Lyristis *et al.*, 1994) from a variety of bacterial donors, and thence to characterization of this element as a potential mutagenic tool (Lin and Johnson, 1991; Hussain *et al.*, 2005).

In *C. botulinum*, Tn*916* was shown to transfer from *E. faecalis* donors at frequencies ranging from 10^{-8} to 10^{-5} transconjugants per recipient, depending upon the donor strain (Lin and Johnson, 1991). Although the analysis of relatively few transconjugants was described, the site of insertion appeared different in all transconjugants on the basis of Southern blot analysis, suggesting that the element may be inserting randomly. However, a high percentage of the clones obtained contained more than one independent Tn*916* insertion. Thus, only 6 of 13 transconjugants appeared to carry a single copy of the transposon, six carried two copies while one contained three copies. The predilection of Tn*916* for multiple insertion compromises its utility as a mutagenic tool. Nonetheless, 20 of 200 randomly selected transconjugants were identified that did not grow or grew poorly on minimal medium. Growth of 11 of these could be restored by the supplementation of the medium with casein acid hydrolysate. In four cases, the requirement for a specific amino (alanine, serine or lysine) acid was established. In a later study, mutants were reported in which the entire neurotoxin locus (ca. 70 kb DNA segment) appeared to be deleted (Lin and Johnson, 1995), although any relationship between this event and the acquisition of Tn*916* by the organism was not established.

Initial results with Tn*916* in *C. difficile* using the non-toxinogenic recipient strain, CD37 were somewhat disappointing, as the element appeared to insert at a preferred site, or 'hot-spot' (Mullany *et al.*, 1991; Wang *et al.*, 2000), thereby precluding its use as a random mutagen. Subsequently, however, it transpired that Tn*916*, and its derivative Tn*916*ΔE, integrates at random in other *C. difficile* strains (Roberts *et al.*, 2003; Hussain *et al.*, 2005), including the genome sequence (Sebaihia *et al.*, 2006) strain, CD630. However, although it appeared to integrate randomly, 6 out of 10 transconjugants analysed contained more than one copy of the element.

Tn*916* has found significant use in *C. perfringens* as a mutagenic tool. Its first reported use was in the isolation of a pleiotropic mutant of *C. perfringens* that produced reduced levels of phospholipase C, protease and sialidase, and did not produce any detectable perfringolysin O activity (Lyristis *et al.*, 1994). Subsequent analysis revealed that the transposon had inserted into a gene encoding the sensor kinase *virS*, part of a two-component system (VirRS) now known to play a central role in the regulation of virulence factor gene expression (Banu *et al.*, 2000). Tn*916* was subsequently used to isolate several auxotrophs of *C. perfringens* (Kaufmann *et al.*, 1996) and a range of mutants defective in their ability to produce either α-toxin or θ-toxin. Nine independently derived mutants were obtained, several of which were derived from large, apparently Tn*916*-mediated, deletion events. In both studies, the results were again complicated by observation that multiple insertions of the transposon invariably occurred. Indeed, in the study of Awad and Rood (1997), only one of the mutants contained a single insertion of Tn*916*, while in the study of Kaufmann *et al.* (1996) multiple insertions accounted for between 65% and 75% of mutants isolated. In a more recent study, only 12 of 42 independently isolated mutants had single copy insertions of Tn*916* (Briolat and Reysset, 2002).

Non-pathogenic clostridia

The use of Tn*916* has also been pursued as a mutagenic tool for investigation of primary metabolism in the solventogenic clostridia of interest to industry (Table 10.1). Transfer of the transposon to *C. acetobutylicum* DSM 792 from *E. faecalis* was first demonstrated by Bertram and Dürre (1989) at frequencies of $<10^{-8}$ per recipient. Insertion appeared random, but only approx. 50% of the transconjugants obtained contained a single copy of Tn*916*. In a subsequent study, 100 isolates from the 7000 transconjugants obtained were shown to be susceptible to allyl alcohol and bromobutyrate, and therefore likely to be defective in the production of solvents (Betram *et al.*, 1990). Three different classes of mutant were isolated, including a class unable to form any solvents indicative of an inactivation of a regulatory gene. Subsequent analysis of one of these mutants (Sauer and Dürre, 1992) revealed that Tn*916* had inserted into *thrA* (encoding $tRNA^{Thr}$), suggesting that the use of the codon ACG may play some regulatory role. Transfer of Tn*916* to another strain of *C. acetobutylicum*, ATCC 824, was also reported by Mattsson and Rogers (1994). Over 50% of the transconjugants isolated had more than one insertion of the transposon. Various classes of mutants were isolated, defective in the production of one or more solvents and/or spore formation. Mutants deficient in solvent production were also isolated from *C. saccharobutylicum* P262 (formerly *C. acetobutylicum*), as well as sporulation-deficient and metronidazole-resistant mutants using Tn*916* (Babb *et al.*, 1993). The subsequent analysis of one of the metronidazole resistance mutants led to the identification of a *sum* gene, apparently essential for clostridial differentiation and sporulation in this particular clostridial species (Collett *et al.*, 1997).

In contrast to the other solventogenic species, *C. beijerinckii* NCIMB 8052 (formerly *C. acetobutylicum*) has a hot-spot for Tn*916*. Consequently, this transposon could not be used for mutant isolation. In contrast, Tn*1545* apparently inserted randomly into the genome. Transfer from either *B. subtilis* or *E. faecalis* donors was demonstrated at appreciable frequencies, typically at 10^{-5} per recipient (Woolley *et al.*, 1989). The mutagenic properties of Tn*1545* were subsequently exploited (Kashket and Cao, 1993: 1995) to isolate a mutant of this strain deficient in 'degeneration' (the loss in prolonged culture of the ability to produce solvents). It transpired that the transposon had inserted into a gene encoding peptide deformylase (*fms*) and that the observed resistance to degeneration was merely a consequence of a reduction in growth rate due to only a partial inactivation of the gene (Evans *et al.*, 1998). In a later study, a Tn*1545*-derived mutant which displayed an increase in its tolerance to butanol was further investigated (Liyanage *et al.*, 2000). Intriguingly, this study demonstrated that the transposon had in fact inserted into an intergenic region and it was hypothesized to be causing a reduction in the activity of the adjacent gene, identified as *gldA*, through production of anti-sense from a Tn*1545* located promoter. Why the demonstrable reduction in the encoded glycerol dehydrogenase activity should result in a butanol tolerant phenotype is unclear.

Directed mutants using host-mediated recombination

The directed inactivation of genes through homologous recombination has played a central role in the assignment of function in the physiological investigations of all manner of bacterial genera and species, as well as in many eukaryotes. With the advent of the genomics era, the need for such technologies has accelerated considerably.

The basic concept of such strategies is rather simple (summarized in Fig. 10.1). An inactivated copy of the gene in question is introduced into the target organism on an extrachromosomal element, where relatively rare recombination events between the wild-type copy of the gene and the inactivated form occur. Should a single recombi-

Table 10.1 Published mutants obtained in different *Clostridium* spp.

Organism	Gene	Single/double crossover	Knock-out marker	Plasmid marker	Number of mutants isolated[a]	Reference
Recombination mutants						
C. perfringens						
Strain 13	*virR*	Double	*tet*	None	Not given	Shimizu et al. (1994)
Strain 13	*pfoA*, *plc*	Double	*ermBP*	None	Not given	Awad *et al.* (1995)
Strain 13	*colA*	Double	*ermBP*	*catQ*	Two from 30 independent experiments	Awad *et al.* (2000)
Strain 13	*pfoA*	Single	None	*tetM*	Not given	Awad *et al.* (2001)
Strain 13	*pfoR*	Single	None	*tetM*	Not given	Awad *et al.* (2002)
Strain 13	VR-RNA	Double	*ermBP*	None	Not given	Shimizu *et al.* (2002a)
Strain 13	*luxS*	Single	None	*ermBP*	Not given	Ohtani *et al.* (2002)
Strain 13	*virX*	Single	None	*ermBP*	Not given	Ohtani *et al.* (2002)
Strain 13	*cadA*	Single	None	*ermBP*	Not given	Okumura et al. (2005)
Strain 13	*pfoA*, *plc*	Single	None	*ermBP*	Three to several	O'Brien and Melville (2004)
Strain 13	*pilT*, *pilC*	Single	None	*ermBP*	Not given	Varga *et al.* (2004)
Strain 13	*clpC*	Single	None	*catP*	Not given	Tanaka *et al.* 2007)
Strain SM101	*cpe*	Double	*catP*	*ermBP*	Not given	Sarker *et al.* (1999)
Strain 13	*spo0A*	Double	*catP*	*ermBP*	One mutant	Huang *et al.* (2004)
Strain 13	*ccpA*	Double	*catP*	*ermBP*	One mutant	Varga *et al.* (2004)
Strain 13	*ssp3*	Double	*catP*	*ermBP*	One mutant	Raju *et al.* (2006)
Strain FM4969	*cpe*	Double	*catP*	*ermBP*	Not given	Sarker *et al.* (1999)
Strain EHE-NE18	*plc*	Double	*catP*	*ermBP*	Two mutants	Keyburn *et al.* (2006)
C. difficile						
Strain 37	*gldA*	Single	None	*ermB*	Not given	Liyanage *et al.* (2001)
Strain 630	*rgaR*, *rgbR*,	Single	None	*ermBP*	Not given	O'Connor *et al.* (2006)
Strain 630	*codY*, CD1276	Single	None	*catQ*	Not given	Dineen *et al.* (2007)
C. septicum						
strain BX96	*csa*	Double	*ermB*	*catQ*	Not given	Kennedy *et al.* (2005)

C. acetobutylicum						
Strain ATCC 824	*buk*, *pta*	Single	None	*ermB*	0.8 to 0.9 'colonies' per μg	Green *et al.* (1996)
Strain ATCC 824	*aad/adhE1*	Single	None	*ermB*	0.8 to 0.9 'colonies' per μg	Green and Bennett (1996)
Strain ATCC 824	*solR/orf5*	Single	None	*ermB*	Not given	Nair *et al.* (1999)
Strain ATCC 824	*spo0A*	Double (?)	*ermB*	*cat*	> 24	Harris *et al.* (2002)
C. beijerinckii						
Strain NCIMB 8052	*gutD*, *spo0A*	Single	None	*ermB*	10^{-6} to 10^{-7} per recipient	Wilkinson and Young (1994)
Strain NCIMB 8052	*scrR*,*scrB*	Single	None	*ermB*	Not given	Reid *et al.* (1999)
C. tyrobutyricum						
Strain ATCC 25755	*pta*	Single	None	*ermB*	10 per experiment (1 per μg DNA)	Zhu *et al.* (2005)
Strain ATCC 25755	*ack*	Single	None	*ermB*	10 per experiment (1 per μg DNA)	Liu *et al.* (2006)
Intron retargeted mutants						
C. perfringens						
Strain 13	*plc*	NAP	NAP	*catQ*	Mutants present in at a frequency of 10% in 2 of 36 colonies analysed	Chen *et al.* (2005)
Strain 13	*pfoA*	NAP	NAP	*catQ*	Mutants present at an unreported frequency in 42 of 44 colonies analysed	Chen *et al.* (2007)
C. difficile strain 630	*pyrF*, *spo0A*, CD0153, CD0552, CD3563	NAP	*ermB*	*catP*	≥ 1000 Em^R integrants per experiment, of which an average of 82% were mutants	Heap *et al.* (2007)
C. acetobutylicum strain ATCC 824	*pyrF*, *spo0A*, CAC0081, CAC0080, CAC0078, CAC2208	NAP	*ermB*	*catP*	≥ 1000 Em^R integrants per experiment, of which an average of 75% were mutants	Heap *et al.* (2007)
C. sporogenes strain NCTC 10696	*pyrF*, *spo0A*, CBO2641, CBO3558	NAP	*ermB*	*catP*	≥ 100 Em^R integrants per experiment, of which an average of 55% were mutants[b]	Heap *et al.* (2007)
C. botulinum strain ATCC 3502	CBO0332A, CBO0339, CBO0336, CBO0340	NAP	*ermB*	*catP*	≥ 100 Em^R integrants per experiment, of which an average of 60% were mutants[b]	Heap *et al.* (2007)

[a]Data, where given, provide an indication of the ease with which mutants were obtained.
[b]The total number of erythromycin-resistant (Em^R) colonies obtained was in these cases increased 10-fold (i.e. equivalent to the result with *C. difficile* and *C. acetobutylicum*) when expression from *fac* was made constitutive by inactivation of the pMTL007 *lacI* repressor gene (Heap *et al.*, 2007).

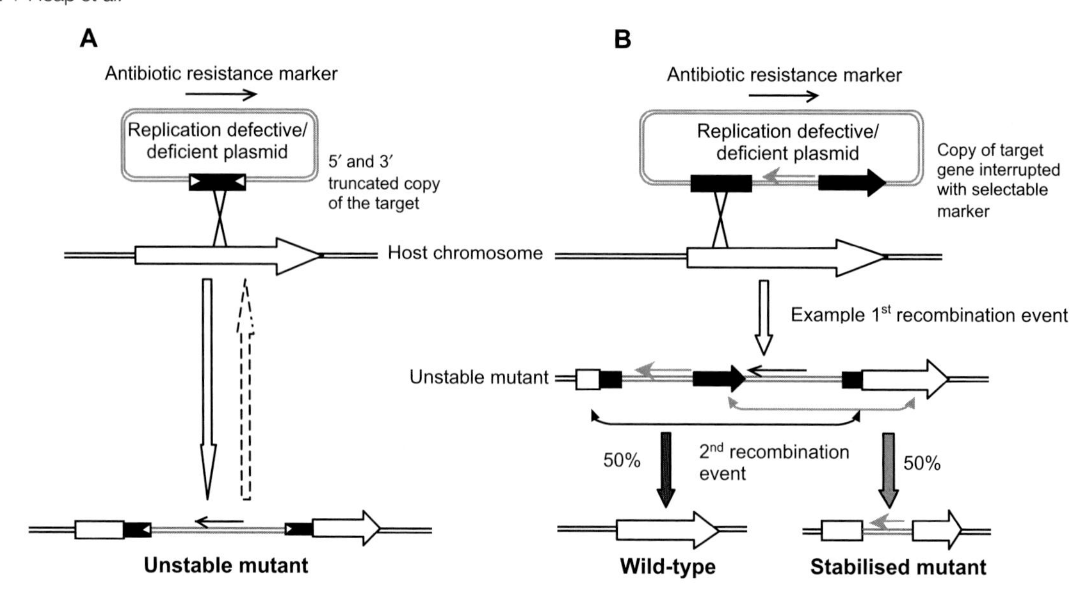

Figure 10.1 Schematic representation of single (A) and double (B) crossover recombination. (A) Due to the inherent instability of mutants generated by single crossover recombination, there is the potential for the mutant to revert to wild-type. (B) During double crossover recombination, following the first recombination event which generates an unstable mutant with a fully functional gene of interest, a second recombination event occurs leading to 50% of the population reverting to the wild-type gene of interest, the other 50% generating a stable mutant.

nation event take place (single crossover), then the entire element will integrate into the genome by a Campbell-like mechanism. If recombination occurs across two regions of homology (double crossover), then the intervening segment of DNA will be inserted into the genome, replacing the equivalent wild-type region by reciprocal exchange. For a single crossover event to be mutagenic, the introduced DNA needs to encompass a central portion of the gene, such that upon integration, the two duplicated copies of the targeted gene that ensue lack either the 5′-end or the 3′-end of the gene, and that the two distinct copies of the incomplete gene are non-functional. For double crossover, the fragment that replaces the wild-type region needs to be inactivated, either through insertion or deletion of DNA (Fig. 10.1).

As the desired recombination events are relatively rare, mechanisms are put in place to allow their selection. This essentially involves using plasmids that are either defective or deficient in their ability to replicate in the host and which carry a marker that will only be maintained if the desired recombination event takes place. Most commonly plasmids are used which are incapable of replication (so called suicide vectors) and which carry an antibiotic resistance gene. By selecting for antibiotic resistance, clonal cells may only arise if the marker gene integrates. For single crossover approaches, the marker gene can reside on the vector backbone, as the entire vehicle becomes integrated. In the case of double crossover strategies, the marker gene is most conveniently placed centrally within the region to be exchanged, where it is responsible for inactivation of the gene. One consequence of this approach is that, for convenience, a second antibiotic resistance gene needs to be placed elsewhere on the vector in order that loss of the vehicle can be determined.

Over and above this basic strategy all manner of levels of sophistication may be introduced, as and when the needs arises. In particular, non-polar deletion mutants may in principle be constructed by omitting an antibiotic resistance marker from the plasmid-borne non-functional allele, and instead including a counter-selectable marker in addition to the antibiotic resistance marker on the vector backbone. In this configuration, integration of the plasmid into the target locus by a first Campbell-like event is positively selected by antibiotic-resistance, and a subsequent second recombination event causing the excision of the plasmid can be selected by virtue of loss of the counter-selection marker, yielding wild-type revertants and the desired deletion mutants in similar proportions. Prominent amongst such strategies has been the use of the *B. subtilis sacB* gene, the presence of which confers lethality on the host in the presence of exogenous sucrose (Kaniga *et al.*, 1991).

Mutants in *C. perfringens*

The first reported examples of directed gene knock-out in the genus were achieved in the species *C. perfringens* (Table 10.1), and in particular in strain 13 (Shimizu *et al.*, 1994; Awad *et al.*, 1995). This strain is highly competent to DNA transfer by electroporation, a factor that may be related to the absence of any type II restriction systems in its genome (Shimizu *et al.*, 2002b). In these initial experiments, replication-deficient suicide vectors were employed carrying a copy of the gene to be inactivated into which was centrally inserted either a *tet* (Shimizu *et al.*, 1994) or an *ermBP* (Awad *et al.*, 1995) selectable marker. In both studies, the vectors were designed with the intention of achieving reciprocal exchange, by double crossover. However, as the *tet* and *ermBP* were the only selectable clostridial marker on the plasmid, those antibiotic resistance transformants obtained in *C. perfringens* were composed of both single crossover (non-mutagenic) and double crossover integrants (mutagenic). The authors were able to distinguish between the two because simple phenotypic plate test were available for measuring the activities of both toxins, e.g. haemolysis on horse blood agar in the case of *virR* (Shimizu *et al.*, 1994) and *pfoA* (Awad *et al.*, 1995).

In the study of Awad *et al.* (1995), some evidence was presented to indicate that the extent of homologous DNA in the knock-out cassette is important, as integrants with the initial *pfoA* suicide vector could not be obtained until the homologous DNA flanking the *ermBP* gene was extended to 3.0 kb. In their experiments, neither the number of erythromycin colonies (integrants) nor the number of mutants obtained is given. Suffice to say, the number of erythromycin-resistant transformants was apparently some 10^5- to 10^6-fold lower than when a replication competent shuttle plasmid was used, while the number of θ-toxin mutants isolated represented between 1.9% and 2.8% of the total erythromycin-resistant colonies obtained in three of four experiments. In the case of *plc* an unstated number of mutants were obtained in only 2 of 10 independent transformation experiments. These early experiments served to illustrate that following transformation with suicide vector constructs, single crossover, non-mutagenic events are in the majority, but that rarer double crossover, mutant integrants may be detected through appropriate screening.

Although inactivation of genes such as *plc* and *pfoA* can be detected via a simple plate assay, inactivation of most genes will not result in an easily discernible phenotypic change. It is, therefore, generally advisable to endow the plasmid backbone with a second antibiotic resistance marker, in addition to the antibiotic resistance gene inserted into the plasmid-borne copy of the gene being targeted. Thus, in a subsequent study which sought to inactivate the *colA* gene of strain 13, Awad *et al.* (2000) adopted the strategy of Sarker *et al.* (1999), and placed a second marker (*catQ*) on the plasmid, in addition to the *ermBP* gene inserted in the plasmid-borne copy of *colA*. By screening the 360 erythromycin-resistant transformants obtained for colonies that were sensitive to chloramphenicol, they were able to distinguish single crossover integrants from mutagenic, double reciprocal crossovers. In total, just two mutants were obtained from a total of 360 erythromycin-resistant transformants, which were themselves derived from 30 independent transformations. Thus, double crossover integrants represented approximately 0.5% of the total number of integrants obtained.

As intimated above, the use of two antibiotic genes was first deployed in a *Clostridium* species by Sarker *et al.* (1999) for mutant isolation in the more disease relevant, *cpe*-bearing strains SM101 and FM4969. Here, *catQ* was used to inactivate *cpe* and a second marker (*ermBP*) was placed on the plasmid. Most significantly, this allowed the initial selection and isolation of rare single crossover integrants on the basis of erythromycin (*ermBP*) resistance, and then the subsequent screening of such a clone for the presence of chloramphenicol-resistant and erythromycin-sensitive derivatives within the population in which the plasmid, along with *ermBP*, had excised to yield the desired reciprocal exchange of *cpe* for *cpe::catQ*. This two-step approach has become the method of choice in most subsequent attempt to generate double crossover mutants in the genus (see Table 10.1), and is the only practical approach where an initial transformation experiment, may yield just one or two transformants, e.g. a *plc* mutant of the chicken pathogen strain EHE-NE18 (Keyburn *et al.*, 2006) and a *csa* mutant in *C. septicum* (Kennedy *et al.*, 2005).

In perhaps a reflection of the low frequency of occurrence of directly selected double crossover mutants, a significant number of *C. perfringens* mutants in strain 13 have also been generated through the use of suicide plasmids designed to specifically generate single crossover mutants, e.g. in *pfoA* (Awad *et al.*, 2001), in *pfoR* (Awad and Rood, 2002), *luxS* and *virX* (Ohtani *et al.*, 2002). In a variation of this approach the Melville laboratory used multimeric forms (by preparation of the plasmid in a *recA+ E. coli* host) of a suicide plasmid carrying either *ermBP* or *catQ* to generate a double *pfoA, plc* mutant of strain 13 through the sequential generation of single crossover mutants (O'Brien and Melville, 2004). The same approach was later used to generate *pilT* and *pilC* mutants (Varga *et al.*, 2006), where six to seven copies of the plasmid were found to have integrated. The practical advantages of this method are unclear.

Mutants in *C. difficile*

To date, only a handful of mutants have been generated in the increasingly important pathogen *C. difficile* (Table 10.1). The first reported example was described by Liyanage *et al.* (2001), who used well established methods successfully exploited in the solventogenic strain *C. beijerinckii* NCIMB 8052 (Wilkinson and Young, 1994) to inactivate a gene identified as *gldA*. In essence, internal fragments of the gene were cloned into the *oriT*-based suicide vector pMTL31 which carries an *ermB* gene, and the resultant plasmids introduced into a toxin-minus strain of *C. difficile* (CD37) by conjugative transfer from an *E. coli* donor. Integrants were selected on the basis of acquisition of resistance to erythromycin. However, disruption of *gldA* apparently proved lethal, and the transconjugants obtained developed no farther than pinpoint colonies. Nevertheless, convincing evidence was obtained that the plasmids had indeed integrated, through the generation of DNA fragments by PCR of a size consistent with integration. Moreover, integrants were also generated in which inactivation of *gldA* was deliberately not sought, allowing the subsequent development of 'normal' sized colonies. DNA fragments consistent with integration of the plasmid by single crossover were also obtained using appropriate primers in a PCR assay and chromosomal DNA from these clones. However, in neither case was the frequency of transfer quantified.

In an alternative approach, Haraldsen and Sonenshein (2003) conjugated a Tn*916* derivative carrying a copy of the *C. difficile sigK* gene from *B. subtilis* to the toxinogenic *C. difficile* strain CD196. Unexpectedly, the modified transposon did not enter the chromosome at the preferred hotspot; rather it integrated through homologous recombination into the *sigK* gene. To date, however, this system has not proven to be generally applicable to the generation of gene knock-outs in other clostridial genes. More promising is an approach recently described by O'Connor *et al.* (2006), in which a replication-defective plasmid has been used to generate single crossover knockouts in a variant of *C. difficile* strain 630, cured of its *ermB* gene. The plasmid utilized (pJIR1456) was based on the replicon of pIP404, which had previously been shown to replicate extremely inefficiently in *C. difficile* (Purdy *et al.*, 2002). Two genes were inactivated, *rgaR* (CD3255) and *rgbR* (CD1089), through single crossover integration of the plasmid. Southern blotting of the *C. difficile rgaR* mutant revealed the presence of 'looped out', independently replicating plasmid in some cells in the population (O'Connor *et al.*, 2006). The potentially unstable nature of this type of gene knock-out was further emphasized in the study of Dineen *et al.* (2007), who used the same system to isolate a single crossover mutant in the *codY* gene of strain 630. In this instance, the mutagenic clone obtained grew very slowly and could only be maintained in the presence of thiamphenicol. In the absence of antibiotic selection, the culture became dominated by wild-type cells in which excision of the integrated plasmid apparently restored the wild-type *codY* gene.

Solventogenic clostridia

There has been a longstanding desire to bring about gene inactivation in the industrially important solventogenic bacteria, both as a route to the better understanding fermentative pathways and as a means of bringing about rational improvements to product formation through metabolic engineering. Directed gene inactivation was first achieved in the organism *C. beijerinckii* NCIMB 8052, at a time when it was designated *C. acetobutylicum*. Thus, mutants in *gutD* (glucitol dehydrogenase) and *spo0A* were generated by cloning internal fragments of both genes into a mobilizable (*oriT*) suicide plasmid, pMTL30, which carried an *ermB* gene for selective purposes. The plasmids generated were introduced into NCIMB 8052 by conjugative transfer from an *E. coli* donor. Integrants arose at frequencies of 10^{-6} to 10^{-7} per recipient, which represented some two orders of magnitude lower than the transfer frequency observed (10^{-4} to 10^{-5}) with replication proficient plasmids (Wilkinson and Young, 1994). Characterization of the mutants obtained indicated that a single copy of the plasmid had integrated by a Campbell-like mechanism. In a later study, Reid *et al.* (1999) used essentially the same system to generate mutants in the *scrB* (sucrose hydrolase) and *scrR* (transcriptional regulator). Inexplicably, the mutants that arose in this study were the result of the insertion of multiple copies of the plasmid.

To date, essentially five published mutations have been made by homologous recombination in *C. acetobutylicum*. Four (*butK*, CAC3075; *pta*, CAC1742; *aad*, CACP0162, and; *solR*, CACP061) were made by single crossover integration of a replication deficient plasmid introduced by electroporation (Green *et al.*, 1996; Green and Bennett, 1996; Nair *et al.*, 1999), and generally arose at frequencies of 0.8 to 0.9 'colonies' per μg of DNA (Green *et al.*, 1996). In these integrants, plasmid sequences at the target site were flanked by two directly repeated copies of the DNA segment directing integration. As a consequence, they were segregationally unstable, e.g. losses per 30 generations of between 1.8 and 3.0×10^{-3} for *buk* and *pta* in *C. acetobutylicum* (Green *et al.*, 1996). This compares with between 0.37 and 1.3×10^{-3} for *C. beijerinckii* (Wilkinson and Young, 1994).

In a more recent study, after several attempts to generate a spo0A knockout with suicide plasmids, Harris *et al.* (2002) adopted an alternative strategy, similar to the double crossover strategy of Sarker *et al.* (1999), in which they used a plasmid incorporating the pIM13 replicon commonly used in cloning studies in this organism. The plasmid made (pETSPO) carried a chloramphenciol resistance gene and a knock-out cassette in which *ermB* interrupted the *spo0A* gene. To obtain a mutant, cells transformed with pETSPO were transferred onto solidified reinforced clostridial medium supplemented with erythromycin (40 mg/l) every 24 h for 5 consecutive days. Colonies were replica plated onto media containing either erythromycin or thiamphenicol (25 mg/l) to identify those that were erythromycin resistant but thiamphenicol sensitive. A single clone was isolated. However, the expected double crossover event had not occurred. Instead, a crossover event occurred between two 10-nt homologous sequences (5′-ACGACCAAAA-3′) that were present in the 3′ end of the repL structural gene and upstream of *ermB*. Loss of the 3-kb fragment between these 10-nt-long homologous sequences resulted in inactivation of *spo0A* through insertion of a 2.1-kb fragment containing *ermB*.

Subsequent to the study of Harris *et al.* (2002), a similar approach was reported at Clostridium IX (Rice University, Houston, TX, USA, 18–21 May 2006), in which a pIM13-based plasmid was employed to generate a double knock-out of the gene encoding the Cac8241 type II restriction enzyme and in *ctfAB* of the megaplasmid pSOL1 (Saint-Prix *et al.*, 2006). In this approach the knock-out cassette was constructed such that the entire target gene was replaced with an *ermB* gene. Furthermore, the *ermB* gene itself was flanked by FRT sites. Thus, by subsequent introduction of a copy of the yeast FLP recombinanse gene, they were able to 'flip-out' the *ermB* fragment, reducing the potential for polar effects and making the marker available for possible future cloning or mutagenesis in these strains.

Directed mutants using a re-targeted group II intron

To date the isolation of mutants by recombination has been an infrequent occurrence. In order to detect such rare events the DNA to be integrated needs to be introduced at high frequencies. Thus, the clostridial strain in which gene integration is most easily achieved (*C. perfringens* strain 13) has the highest transformation frequency. In *C. acetobutylicum* the relatively lower frequencies obtained are compensated for by using 1–2 mg of DNA per transformation. The frequencies of transfer in toxinogenic strains of *C. difficile*, and in particular the genome strain CD630, are relatively poor (between 10^{-6} and 10^{-7} transconjugants per donor cell), and most likely provide a rational explanation as to why homologous recombination is difficult to demonstrate using suicide vectors. The most obvious solutions to this bottleneck are to either (1) maximize the effective frequency of DNA transfer (most obviously by developing a plasmid conditional or defective for replication), thereby increasing the likelihood of the detection of rare recombination events, or (2) integrate the DNA by a process which is largely independent of host recombination factors. In the respect of the latter, the use of group II introns represents a particularly attractive option.

Background

Group II introns are elements which interrupt protein-coding or RNA genes, and are widespread in the chromosomes of both eukaryotic organelles and bacterial genomes. They are catalytically active RNA molecules characterized by their ability to self-splice out of the RNA transcript of their host gene via a mechanism essentially identical to that of eukaryotic nuclear introns, to which they are therefore thought to be ancestral. Some group II introns are also mobile, containing an ORF encoding a multidomain intron-encoded protein (IEP). The IEP contains a number of enzymatic activities which allow the intron lariat (the spliced-out intron RNA) to be used as a template for insertion of the intron into intron-less copies of the host gene within the cell, with high specificity and efficiency.

The mechanistic details of the replication process of mobile group II introns, exemplified by the intron from the *ltrB* gene of *Lactococcus lactis* (the '*L1.LtrB* intron') have been revealed by the findings of numerous studies, especially those of the Lambowitz laboratory. The *ltrB* gene is found on conjugative plasmids and can therefore encounter intron-less copies of the gene after conjugal transfer of the plasmid. The specificity of the integration process, known as retro-homing, was found to be largely determined by basepairing between certain sequences within the intron RNA and the target site DNA. 'Rules' governing target recognition by the intron have been established and used to demonstrate that by changing the intron sequence, its target specificity can be altered (Mohr *et al.*, 2000). This process was refined and automated such that *Ll.LtrB*-derived introns could be designed to target to almost any gene of interest, limited only by a minimal recognition sequence consensus determined by the IEP (Perutka *et al.*, 2004). The IEP ORF can be removed from within the intron and provided *in cis* or *in trans* without affecting intron mobility, in a manner akin to the construction of mini-transposons. Such introns can then be denied splicing and further mobility after the desired insertion is achieved by removing the IEP-encoding plasmid. The insertional mutation will therefore be unconditional, and secondary mutations prevented. These re-targeted introns have often been referred to as 'Targetrons'.

Targetron-mediated inactivation of clostridial genes

The use of Targetron technology within the genus *Clostridium* was first exemplified in *C. perfringens* by Chen *et al.* (2005). The basic Targetron vector pACD3 was modified both to include a clostridial replicon and replace the Gram-negative selectable marker with a clostridial equivalent (derived from plasmid pJIR750) and to substitute the T7 promoter responsible for transcription of the *Ll.LtrB* group II intron with that of the promoter region of the beta-2 toxin gene (*cpb2*) from a *C. perfringens* type A isolate. The group II intron of the final vector (pJIR750ai) was modified such that the target specificity of the intron was consistent with a site within the *plc* gene of *C. perfringens*. Following introduction of the plasmid into strain 13 by electroporation, 38 individual colonies were screened using PCR and primers that flanked the predicted insertion site. The use of template DNA derived from at least two colonies resulted in PCR products consistent with both the wild type gene and a mutant form in which the *Ll.LtrB* element had inserted.

Bacteria from these two colonies were plated on egg yolk BHI plates, and approximately 10% of the colonies formed were not surrounded by halos characteristic of alpha-toxin production. PCR analysis of these colonies, and subsequent sequencing of the product, confirmed that the element had indeed inserted at the predicted site (position 50/51) of the *plc* gene. Western blotting confirmed that alpha-toxin was not produced, and the mutation was shown to be stably maintained.

Positive selection of gene inactivation

Intron integration frequencies vary widely between target sites, and can make the screening effort required to isolate a mutant prohibitively laborious, particularly if no simple phenotypic screen for gene inactivation is available. Although most of the intron sequence (or more generally the encoded structure) cannot be altered without deleterious effects, the region normally encoding the IEP (domain IV) is non-structural, allowing the insertion of 'cargo' sequence. The Lambowitz laboratory constructed an artificial 'twintron' (observed in nature where one intron is nested within another) such that the *Ll.LtrB* intron domain IV contained an antibiotic resistance marker, itself interrupted by a group I intron which abolished antibiotic resistance. These three elements were orientated relative to one another in such a way that the group I intron would self-splice out and be lost from the RNA of the parental *Ll.LtrB* intron, restoring an antisense copy of the antibiotic resistance marker. Subsequent integration of the *Ll.LtrB* intron results in interruption of the target gene and insertion of a functional antibiotic resistance marker, allowing positive selection for gene knock-out by acquisition of resistance to the appropriate antibiotic (Karberg *et al.*, 2001; Zhong *et al.*, 2003). Such a marker is described as a retrotransposition-activated selectable marker (RAM) to denote that the restoration of the marker ORF is diagnostic of its passing through the RNA intermediate. A plasmid incorporating this feature, pACD4K-C, is now available from Sigma Aldrich, where the RAM is based on a modified *kan* gene which, when activated, confers resistance to kanamycin upon cells in which the intron has integrated into the chromosome, mutating the target gene (Fig. 10.2).

The ClosTron: a universal gene knock-out system for clostridia

The generation of the *plc* mutant of *C. perfringens* by Chen *et al.* (2005) did not make use of a RAM element, but instead relied upon a combination of PCR screening and a simple phenotypic plate assay. Such convenient assays are available only for a small minority of genes. Moreover, the *plc* mutant was constructed in a species and strain (*C. perfringens* strain 13) in which the majority of clostridial directed mutants have thus far been generated. Of greater utility would be a more fully fledged system analogous to pACD4K-C, incorporating a clostridial RAM which would allow positive selection for integration of group II intron elements into any gene, without recourse to a phenotypic screen, and in clostridial species in which, unlike *C. perfringens*, mutants have seldom or never previously been obtained.

ClosTron development

As the majority of the features present on plasmid pACD4K-C would not function in clostridia, an extensive reconstruction was necessary. To date, two distinct RAM elements have been described. The Sigma Aldrich RAM of pACD4K-C based on kanamycin selection (http://www.sigma-genosys.com/targetron/) and an earlier version based on trimethroprim selection (Zhong *et al.*, 2003). Since neither antibiotic is generally selectable in clostridial species, a central requirement was to develop an alternative clostridial RAM. The *ermB* gene from the *Enterococcus faecalis* plasmid pAMβ1 was chosen as a starting point, as this gene has been widely used in the construction of *E. coli*/*Clostridium* shuttle vectors (Mauchline *et al.*, 1999). The incorporation of the group I intron of the *td* gene of phage T4 requires that the element is immediately flanked by specific exon sequences necessary for self-catalytic splicing. As the introduction of the required sequences could not be made within the *ermB* coding region without alteration to the primary sequence of the encoded enzyme, a linker region was inserted between the *ermB* ORF and its promoter (Heap *et al.*, 2007). The presence of the *td* intron prevents expression of the gene, and therefore plasmids carrying this modified element do not confer erythromycin resistance upon the host cell. Following retrotransposition and splicing-out of *td*, the *ermB* ORF is extended

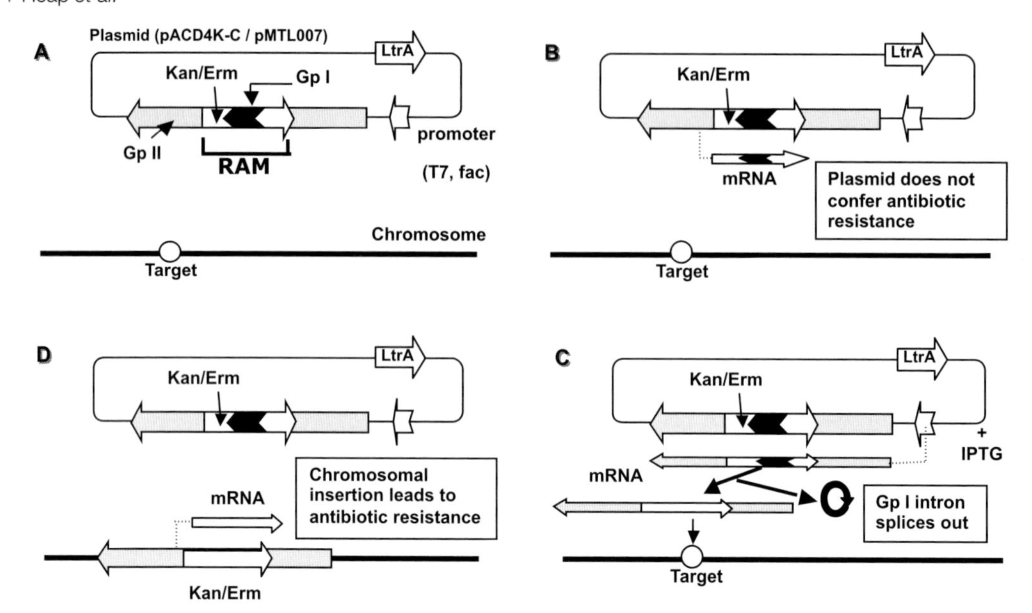

Figure 10.2 Positive selection for mutagenesis using a retrotransposition-activated selectable marker (RAM). (A) An appropriate TargeTron or ClosTron plasmid is first transferred into the host organism. The relative orientations of the Gp II intron and RAM element (comprising a Kan or Erm antibiotic resistance marker and Gp I intron) are crucial. (B) The plasmid-borne antibiotic resistance gene does not confer antibiotic resistance, since it is interrupted by a Gp I intron. The Gp I intron cannot self-splice from the antibiotic resistance gene transcript, as it is in the reverse orientation. (C) The Gp II intron RNA is transcribed from a strong promoter on the plasmid. In this transcript, the Gp I intron is in the forward orientation, and self-splices. However, antibiotic resistance is still not conferred, because the antibiotic resistance gene is in the reverse orientation. The mature Gp II intron RNA undergoes retrotransposition into the target gene in some cells in the population. (D) After completion of the retrotransposition process, an intron containing a functional antibiotic resistance gene (lacking the Gp I intron) is localized to the chromosome. Only these integrant cells acquire antibiotic resistance, and can be readily selected on solid media supplemented with the appropriate antibiotic.

by 12 codons. The addition of the encoded amino acids to the ErmB protein did not interfere with function, and cells (*E. coli* or clostridial) carrying this gene become resistant to erythromycin (Heap *et al.*, 2007).

Crucial to the success of the strategy was to replace the natural promoter of *ermB* with a more active transcription signals, in the form of the promoter of the thiolase gene (*thlA*) derived from *C. acetobutylicum* ATCC 824 (Girbal *et al.*, 2003). This is because the natural promoter of the pAMβ1 *ermB* gene is relatively inefficient due to an unusually large, suboptimal distance (21 bp) between its -35 and -10 regions. The vegetative promoter consensus is 17 bp. Indeed, until the *ermB* promoter was replaced with *thlA*, no erythromycin-resistant colonies could be obtained using the prototype vector, indicating that expression of *ermB* in a single copy chromosome insertion was insufficient to confer erythromycin resistance on the host cells (Heap *et al.*, 2007).

In keeping with the strategy adopted in the Sigma Aldrich vector pACD4K-C, the final clostridial vector, pMTL007, carried a clostridial promoter (*fac*) from which transcription could be induced by addition of exogenous IPTG in much the same way as the promoter it replaced, T7. However, by introducing a frameshift into the *lacI* gene on pMTL007, it was shown that constitutive expression of *fac* was at least equally effective in the generation of mutants. Indeed, in the organism in which *fac* was under the tightest regulatory control, *C. sporogenes*, the total number of mutants generated was higher, by an order of magnitude, when *fac* was made constitutive rather than when transcription was induced by IPTG. As a consequence, more recent derivatives of this vector no longer incorporate the *lacI* gene, relying on constitutive expression of the group II intron. An example is illustrated in Fig. 10.3.

ClosTron validation

Following the construction of pMTL007, its effectiveness was initially tested in 3 different clostridial species, namely *C. difficile*, *C. acetobutylicum* and *C. sporogenes* (Heap *et al.*, 2007). Two genes were targeted in each case. These were *pyrF* and *spo0A*. The former generated strains with a growth requirement for the supply of exogenous uracil, while the latter resulted in strains that were no longer able to efficiently sporulate. The changes necessary to re-target individual plasmids were identified using a web-based computer algorithm, use of which is provided as part of the Sigma Aldrich TargeTron Gene Knockout System kit. All genes were inactivated at the first attempt with high efficiency. Thereafter, the procedure was further validated by targeting several further genes of the three clostridial species, as well as four genes in *C. botulinum*. Every gene targeted was successfully inactivated (Heap *et al.*, 2007).

In typical experiments with *C. acetobutylicum* and *C. difficile* (Heap *et al.*, 2007), 1000s of erythromycin-resistant colonies were obtained in each individual experiment. The numbers obtained in *C. sporogenes* were an order of magnitude lower, unless a constitutive version of pMTL007 was employed (carrying a frameshift in *lacI*) in which case equivalent numbers were obtained to experiments with the other two clostridial species. Every erythromycin-resistant colony obtained may be assumed to have been a consequence of insertion of the group II intron and *ermB* into the chromosome. In most cases, all of these integration events analysed, or at least a majority, took place at the intended target site. Even at the lowest frequency obtained (2.5%), the correct mutant was readily identified by screening pools of erythromycin-resistant candidate clones. The insertion point of a representative clone of every mutant isolated was sequenced, and shown to be precisely as intended. In keeping with the previously described properties of this system, no evidence was obtained for multiple insertion of the element, and the mutations generated were extremely stable (Heap *et al.*, 2007). Thus, in the case of the three *pyrF* mutants, no evidence of reversion to wild-type uracil prototrophy was apparent when late exponential cultures of each were plated out without dilution onto minimal media lacking uracil.

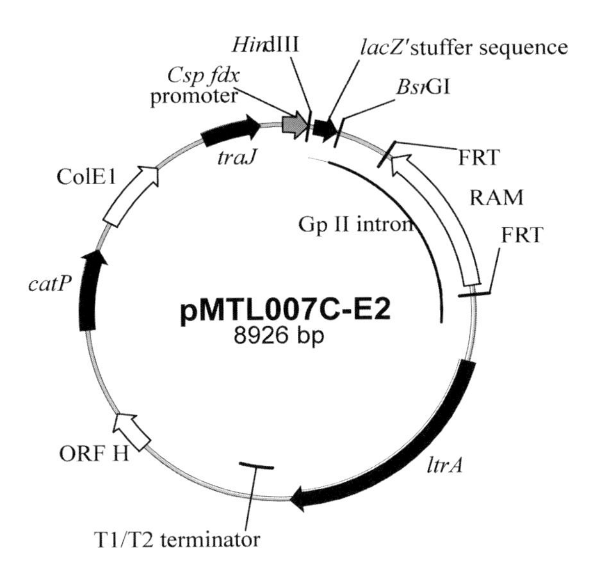

Figure 10.3 The ClosTron plasmid pMTL007-CE2. The second generation ClosTron plasmid pMTL007C-E2 uses the strong *fdx* promoter from *C. sporogenes* to direct expression of the Gp II intron, and contains FRT sites flanking the RAM to facilitate FLP-mediated marker removal. The plasmid also features a *lacZ'* stuffer sequence that is replaced during intron re-targeting, allowing clones containing successfully re-targeted plasmids to be identified by blue/white screening, restriction analysis or PCR.

ClosTron procedure

Inactivation of genes is remarkably straightforward. Retargeting of the group II intron is achieved by subcloning a ~350 bp fragment into the plasmid (e.g. pMTL007-CE2, Fig. 10.3) which changes the target specificity of the group II intron. This fragment is generated by SOE PCR using primers designed by inputting the sequence of the target gene into the web-based programme at http://www.sigma-genosys.com/targetron/, in combination with template DNA supplied in the Sigma Aldrich TargeTron Gene Knockout System kit. The modified plasmid is then introduced into the intended clostridial host either by conjugation (e.g. C. *difficile* 630Δerm or C. *sporogenes*) or by electroporation (e.g. C. *acetobutylicum*, C. *perfringens* or C. *botulinum*). In the case of pMTL007, production of group II intron RNA is induced by the addition of IPTG. The induced cells are then plated onto media supplemented with erythromycin (Em). No induction is necessary using plasmids such as pMTL007-CE2 (Fig. 10.3). The nature of the system is such that a functional *ermB* gene can only arise if the group II intron element has integrated into the chromosome.

To ensure that the insertion is mutagenic in those instances where the insertion of the intron is in the sense strand, it is important that LtrA is not present within the cell, as it will bring about splicing of the intron from the mRNA. This is ensured by screening colonies for loss of the plasmid carrying *ltrA*, measured by absence of resistance to thiamphenicol. The replicon employed shows a gradation in its segregational stability in different clostridial species, but in all cases to date plasmid-free colonies are readily isolated. The replicon is most unstable in the case of C. *difficile*, where it replicates very inefficiently (Purdy *et al.*, 2002) and plasmid loss is ensured by simply restreaking the integrant clone.

PCR screening is used to easily identify the desired knock-outs, but can represent the step where workers unfamiliar with the method can experience some difficulty. Screening is best accomplished by designing three primers (Fig. 10.4). Pair P1 and P2 flank the site of insertion of the group II inton + *ermB* fragment. In the wildtype strain the use of these primers in a PCR will generate a relatively small fragment, compared to the mutant, which will generate a fragment which is some 1800 bp larger. Primers P1 and P3 (EBS Universal, specific to the element itself, and known to be an effective primer) may be used to demonstrate that the element has inserted into the correct gene. Most importantly, primers P1 and P2, besides distinguishing between the mutant and wild type, can be employed to ensure that the chromosomal DNA preparations being made are of a sufficient quality for efficient PCR. Moreover, once P1 and P2 have been shown to generate an appropriate DNA fragment, then one can be reasonably confident that P1 and P3 (a tried and tested primer) are going to work.

Routinely, it is only necessary to screen four erythromycin colonies for insertion of the element into the correct locus. Generally at least 25% of the clones, if not all, are found to contain the desired insertion. If none of the clones prove positive, it is advisable to flood an agar plate carrying at least 100 colonies and harvest the mixture of cells. DNA can then be prepared from this pool, and screened as normal with primers P1 and P3. If a PCR product of the expected size is not evident, then the particular re-targeting region being deployed should be abandoned and a second construct with a different retargeting region constructed. If a band of the right size is evident, then the clone is present, and screening of further erythromycin-resistant colonies is worthwhile. To speed up identification a grid system may be used in which pools of ten colonies can be screened at once.

Issues and refinements

As with any insertional mutagen, there are practical issues to the use of the ClosTron which must be borne in mind by the operator. Care should be taken in selecting the site of insertion. The introduction of the intron-element at a distal location in the gene may not result in a mutagenic phenotype, if the truncated protein produced still retains activity. Similarly, insertion at a proximal site may, depending on the reading frame, generate a novel ORF encoding a fore-shortened protein with potential activity. However, the most important consideration is one of polar effects. Insertion of DNA into any gene with one or more transcriptionally linked

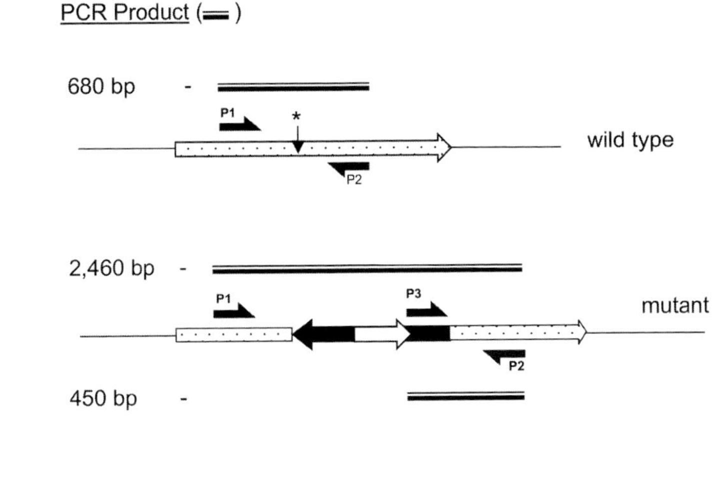

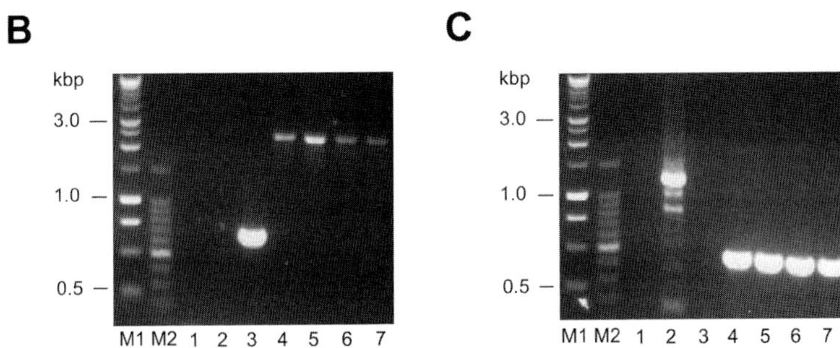

Figure 10.4 Screening for successful integration of the ClosTron into the targeted gene. (A) PCR primers P1 and P2 are designed to anneal to chromosomal sequence either side of the ClosTron insertion site (*). Primer P3 (i.e. EBS universal primer) anneals specifically to the ClosTron itself. PCR screening with primers P1 and P2 (B) or with primers P3 and P2 (C). Lane 1, H_2O no template control; lane 2, pMTL007; lane 3, genomic DNA from wild-type *C. difficile* parental strain; lanes 4–7, genomic DNA from four independent erythromycin-resistant clones. M1, 1-kb ladder (Promega); M2, 100-bp ladder (Promega).

genes downstream can interfere with the expression of the downstream gene(s). In the case of the ClosTron, the situation is exacerbated by the presence of the powerful *thlA* promoter, which, depending upon the orientation of insertion of the intron, can lead to increased expression of downstream genes, or to increased expression of upstream genes via antisense and transcriptional interference effects. Such effects can occur regardless of whether the adjacent genes are in an operon or not.

To tackle this issue, we have constructed second generation versions of pMTL007, in which the *ermB*-based RAM, and consequently the integrated *ermB* gene, is flanked by FRT sites. An example of such a plasmid, pMTL007-CE2, is shown in Fig. 10.3. In preliminary studies, this derivative plasmid has been used to generate a number of mutants in the Agr quorum sensing system of *C. acetobutylicum*. A segregationally unstable plasmid carrying the *flp* gene has then been introduced into these mutants, and following growth in the absence of antibiotic selection, cells isolated that are sensitive to erythromycin (J. Scott, unpublished data). In all cases (*agrA*, *agrC* and *agrB*), direct genome sequencing has shown that precise excision of the *erm* gene occurred to leave a single FRT site. Screening of cells from this population for thiamphenicol-sensitivity has enabled the isolation of clones that have additionally lost the FLP recombinase gene-bearing plasmid.

'Flipping-out' the *ermB* gene should considerably reduce any potential polar effects, most obviously because it eliminates the transcriptional effects of the strong *thlA* promoter. The remaining insertion comprises the group II intron itself. This 940 bp region contains stop codons in all 6 reading frames, thus preventing translational read-through. Any polar effects will therefore be confined to interference of mRNA transcription. However, as the intron is normally efficiently transcribed (into the pre-mRNA prior to splicing) it should not cause premature termination of transcripts. It follows that the effect of insertion is expected to be confined to the physical extension of the length of the mRNA transcript produced. It is therefore likely that polar effects, if any, will be minimal.

Beyond minimization of polar effects, a major consequence of the facility to remove the *ermB* gene is the potential to make multiple mutations. Thus, following the removal of *ermB* from a ClosTron-generated mutant, a second gene can be targeted, and its inactivation selected on the basis of acquisition of erythromycin resistance.

Conclusions and future prospects

Despite the best efforts of numerous laboratories around the globe, in the preceding decades the generation of mutants in the genus *Clostridium* has not been immensely productive. Of particular disappointment are the absence of effective transposons by which random mutations may be generated. The use of conjugative transposons is disadvantaged by poor efficiency, their large size and most unfavourably, their predilection to insert in multiple copies. In other bacterial species, the lack of such elements has been solved by either creating transposons through the addition of an antibiotic resistance marker to an endogenous IS element, or through the adaptation of a well-used transposon to function in the intended host. One would predict that such strategies will in the not too distant future be applied to *Clostridium*. Such elements are essential if hitherto unknown functions are to be determined, and would prove to be particularly valuable in unravelling the molecular pathogenesis of clostridial pathogens.

The most mutants generated to date have been in the *C. perfringens* strain 13, in part a reflection of its intrinsic high frequency of transformation. In those organisms that have less efficient routes to the introduction of vector DNA, the most productive strategy has been to design vectors that are capable of generating a double crossover, through insertion of an antibiotic resistance marker into the plasmid-borne copy of the gene being targeted, and initially selecting an appropriate antibiotic-resistant integrant that has arisen by single crossover insertion of the plasmid. Once obtained, screening of this clonal population for cells in which the plasmid excises to give a double crossover is then possible. In the absence of a simple phenotypic screen to detect such a mutagenic event, it is advisable to use a vehicle which carries a second antibiotic resistance gene through which loss of the vector component of the integrated plasmid can be monitored. This type of approach would be considerably strengthened by the availability of a negative selection marker, which would also open the way to the construction of non-polar deletion mutants, which have not been reported to-date. Indeed the first tentative approach in this direction has been made by Philippe Soucaille's laboratory, through the adaptation of the *B. subtilis* method described by Fabret *et al.* (2002) for use in *C. acetobutylicum* (pers. communication). It relies on sensitivity of cells to 5-fluorouracil, a consequence of its conversion to 5-fluoro-dUMP by uracil-phosphoribosyltransferase, encoded by the *upp* gene. Fabret *et al.* (2002) deleted *upp* from the genome, enabling them to use *upp* on the knock-out vector as a negative selection marker. Thus, loss of the *upp* gene following excision of the plasmid can be selected by the isolation of 5-fluorouracil-resistant colonies. An identical strategy has been applied to *C. acetobutylicum* ATCC 824. Whilst extremely powerful, the advisability of using a strain that is mutated in *upp* for virulence studies in a pathogen is questionable.

In certain instances, the efficiency of DNA transfer may be so low that the frequency of occurrence of the desired single crossover integration of the introduced suicide vector is such that they are seldom, or never, detected. In these instances, it would be preferable to introduce a plasmid that is 'replication-deficient' as opposed to a 'replication-defective', suicide vector. This enables the isolation of a clonal population carrying the constructed 'knock-out' cassette on an autonomous plasmid, prior to the process of screening for integrants. Ideally the plasmid should be conditional for replication, e.g. have a temperature-sensitive phenotype. Once obtained, cells may be transferred to culture under the non-permissive condition, and the desired integrant selected for. As every cell carries a copy of the plasmid at the outset, it would be equivalent to transforming every cell with a suicide plasmid. Therefore, it is likely that rare double crossover mutants could be directly selected. Unfortunately, at present no such conditional vector has been described in the genus. A search for such a vector would clearly pay dividends. It is intriguing that in at least two clostridia (*C. acetobutylicum* and *C. difficile*, Saint-Prix *et al.*, 2006; O'Connor *et al.*, 2006) the suicide vector route has been abandoned in favour of using plasmids which appear generally defective in their ability to propagate. This provides similar advantages to the use of a conditional vector, in that just a single transformant/transconjugant is required, and that the entire cell population carries a copy of the plasmid at the outset.

In contrast to plasmid-based, recombination dependent methods, strategies that make use of elements such as the ClosTron provide a new impetus to functional genomic studies in the genus. In a few short months, the ClosTron system has been used at Nottingham and in European collaborators' laboratories to generate over 60 *stable* mutants in *C. difficile*, *C. acetobutylicum*, *C. sporogenes* and *C. botulinum*. Prior to this the total published number of mutants in these clostridia was 10 (Table 10.1). The procedure is extremely rapid (taking as few as 10–14 days per mutant), highly efficient (hundreds of retargeted clones are obtained per experiment) and reproducible (100% success rate among non-essential genes), and is revolutionizing functional genomic studies in clostridia. It will facilitate virulence factor discovery in pathogenic species, such as *C. difficile*, and the generation of improved solvent-producing strains of *C. acetobutylicum* by metabolic engineering. However, its use as an integrational tool should be tempered with some degree of caution, particularly with respect to the possibility of polar effects – a problem in common with previous clostridial mutants constructed by inserting marker genes via allelic exchange. That being said, since its publication, a number of unpublished refinements to pMTL007 have been made. Most notably, derivatives have been made in which the RAM element is flanked by FRT sites. This allows subsequent excision of the retargeted *ermB* gene, through *in vivo* expression of the yeast FLP recombinase. The addition of this facility reduces polar effects and paves the way for the creation of multiple mutations using the same RAM element. Further refinements are envisaged which should eventually lead to the creation of non-polar deletion mutants.

Acknowledgements

The authors wish to acknowledge the financial support of the BBSRC (grants BB/D522797/1, BD/D522289/1 BB/F003390/1), the MRC (G0601176) and Morvus Technology Ltd.

References

Awad, M.M., Bryant, A.E., Stevens, D.L., and Rood, J.I. (1995). Virulence studies on chromosomal alpha-toxin and theta-toxin mutants constructed by allelic exchange provide genetic evidence for the essential role of alpha-toxin in *Clostridium perfringens*-mediated gas gangrene. Mol. Microbiol. *15*, 191–202.

Awad, M.M., and Rood, J.I. (1997). Isolation of alpha-toxin, theta-toxin and kappa-toxin mutants of *Clostridium perfringens* by Tn*916* mutagenesis. Microb. Pathog. *22*, 275–284.

Awad, M.M., Ellemor, D.M., Bryant, A.E., Matsushita, O., Boyd, R.L., Stevens, D.L., Emmins, J.J., and Rood, J.I. (2000). Construction and virulence testing of a collagenase mutant of *Clostridium perfringens*. Microb. Pathog. *28*, 107–117.

Awad, M.M., Ellemor, D.M., Boyd, R.L., Emmins, J.J., and Rood J.I. (2001). Synergistic effects of alpha-toxin and perfringolysin O in *Clostridium perfringens*-mediated gas gangrene. Infect. Immun. *69*, 7904–7910.

Awad, M.M., and Rood J.I. (2002). Perfringolysin O expression in *Clostridium perfringens* is independent of the upstream pfoR gene. J. Bacteriol. *184*, 2034–2038.

Babb, B.L., Collett, H.J, Reid, S.J., and Woods, D.R. (1993). Transposon mutagenesis of *Clostridium acetobutylicum* P262: isolation and characterization of solvent deficient and metronidazole resistant mutants. FEMS Microbiol. Lett. *14*(3), 343–8.

Banu, S., Ohtani, K., Yaguchi, H., Swe, T., Cole, S.T., Hayashi, H., and Shimizu, T. (2000). Identification of novel VirR/VirS-regulated genes in *Clostridium perfringens*. Mol. Microbiol. 35, 854–864.

Bertram, J., and P. Dürre. (1989). Conjugal transfer and expression of streptococcal transposons in *Clostridium acetobutylicum*. Arch. Microbiol. *151*, 551–557.

Bertram, J., Kuhn, A., and Dürre, P. (1990). Tn*916*-induced mutants of *Clostridium acetobutylicum* defective in regulation of solvent formation. Arch. Microbiol. *153*, 373–377.

Briolat, V., and Reysset, G. (2002). Identification of the *Clostridium perfringens* genes involved in the adaptive response to oxidative stress. J. Bacteriol. *184*, 2333–43.

Chen, Y., McClane, B.A., Fisher, D.J., Rood, J.I., and Gupta, P. (2005). Construction of an alpha toxin gene knockout mutant of *Clostridium perfringens* type A by use of a mobile group II intron. Appl. Environ. Microbiol. *71*, 7542–7.

Chen, Y., Caruso, L., McClane B., Fisher, D., and Gupta, P. (2007). Disruption of a toxin gene by introduction of a foreign gene into the chromosome of *Clostridium perfringens* using targetron-induced mutagenesis. Plasmid. *58*, 182–189.

Collett, H.J., Woods, D.R., and Reid, S.J. (1997). Characterisation of a transposon-induced pleiotropic mutant of *Clostridium acetobutylicum* P262. Anaerobe *3*, 411–421.

Dineen, S.S., Villapakkam, A.C., Nordman, J.T., and Sonenshein, A.L. (2007). Repression of *Clostridium difficile* toxin gene expression by CodY. Mol. Microbiol. *6*, 206–219.

Evans, V.J., Liyanage, H., Ravagnani, A., Young, M., and Kashket, E.R. (1998). Truncation of peptide deformylase reduces the growth rate and stabilizes solvent production in *Clostridium beijerinckii* NCIMB 8052. Appl. Environ. Microbiol. *64*, 1780–1785.

Fabret, C., Ehrlich, S.D., Noirot, P. (2002) A new mutation delivery system for genome-scale approaches in *Bacillus subtilis*. Mol. Microbiol. *46*, 25–36.

Girbal, L., Mortier-Barriere, I., Raynaud, F., Rouanet, C., Croux, C., and Soucaille, P. (2003). Development of a sensitive gene expression reporter system and an inducible promoter–repressor system for *Clostridium acetobutylicum*. Appl. Environ. Microbiol. *69*, 4985–4988.

Green, E.M., and Bennett, G.N. (1996). Inactivation of an aldehyde/alcohol dehydrogenase gene from *Clostridium acetobutylicum* ATCC 824. Appl. Biochem. Biotechnol. *57/58*, 213–221.

Green, E.M., Boynton, Z.L., Harris, L.M., Rudolph, F.B., Papoutsakis, E.T., and Bennett, G.N. (1996). Genetic manipulation of acid formation pathways by gene inactivation in *Clostridium acetobutylicum* ATCC 824. Microbiology.*142*, 2079–2086.

Hächler, H., Kayser, F.H., Berger-Bachi, B., (1987). Homology of a transferable tetracycline resistance determinant of *Clostridium difficile* with *Streptococcus* (*Enterococcus*) *faecalis* transposon Tn*916*. Antimicrob. Agents Chemother. *31*, 1033–1038.

Haraldsen, J.D., and Sonenshein A.L. (2003) Efficient sporulation in *Clostridium difficile* requires disruption of the σ^k gene. J. Mol. Microbiol. *48*, 811–821.

Harris, L.M., Welker, N.E., and Papoutsakis, E.T. (2002). Northern, morphological, and fermentation analysis of *spo0A* inactivation and overexpression in *Clostridium acetobutylicum* ATCC 824. J. Bacteriol. *184*, 3586–3597.

Heap, J.T., Pennington, O.J., Cartman, S.T., Carter, G.P., and Minton, N.P. (2007). The ClosTron: a universal gene knock-out system for the genus *Clostridium*. J. Microbiol. Methods *70*, 452–464.

Huang, I.H., Waters, M., Grau, R.R., and Sarker, M.R. (2004). Disruption of the gene (*spo0A*) encoding sporulation transcription factor blocks endospore formation and enterotoxin production in enterotoxigenic *Clostridium perfringens* type A. FEMS Microbiol. Lett. *233*, 233–240.

Huang, I.H., and Sarker, M.R. (2006). Complementation of a *Clostridium perfringens spo0A* mutant with wild-type *spo0A* from other *Clostridium* species. Appl. Environ. Microbiol. *72*, 6388–6393.

Hussain, H.A., Roberts, A.P., and Mullany, P. (2005). Generation of an erythromycin-sensitive derivative of *Clostridium difficile* strain 630 (630Deltaerm) and demonstration that the conjugative transposon Tn*916*DeltaE enters the genome of this strain at multiple sites. J. Med. Microbiol. *54*, 137–141.

Ionesco, H. (1980). Transferable tetracycline resistance in *Clostridium difficile*. (author's transl.). Ann. Microbiol. (Paris) *131A*, 171–9. French.

Kaniga K, Delor I, Cornelis GR (1991) A wide-host-range suicide vector for improving reverse genetics in gram-negative bacteria: inactivation of the *blaA* gene of *Yersinia enterocolitica*. Gene *109*, 137–141.

Karberg, M., Guo, H., Zhong, J., Coon, R., Perutka, J. and Lambowitz, A.M. Group II introns as controllable gene targeting vectors for genetic manipulation of bacteria. (2001). Nat. Biotechnol. *19*, 1162–1167.

Kashket, E.R., and Cao, Z.Y. (1993). Isolation of a degeneration-resistant mutant of *Clostridium acetobutylicum* NCIMB 8052. Appl. Environ. Microbiol. *59*, 4198–4202.

Kashket, E.R. and Cao, Z.Y. (1995). Clostridial strain degeneration. FEMS Microbiol. Rev. *17*, 307–315.

Kaufmann, P., Lehmann, Y., and Meile, L. (1996). Conjugative transposition of Tn*916* *Enterococcus faecalis*, *Escherichia coli*, *Clostridium perfringens*. Syst. Appl. Microbiol. *19*, 35–39.

Kennedy, C.L., Krejany, E.O., Young, L.F., O'Connor, J.R., Awad, M.M., Boyd, R.L., Emmins, J.J., Lyras, D., and Rood, J.I. (2005). The alpha-toxin of *Clostridium septicum* is essential for virulence. Mol. Microbiol. *57*, 1357–1366.

Keyburn, A.L., Sheedy, S.A., Ford, M.E., Williamson M.M., Awad, M.M., Rood, J.I., and Moore, R.J. (2006). Alpha-toxin of *Clostridium perfringens* is not an essential virulence factor in necrotic enteritis in chickens. Infect. Immun. *74*, 6496–6500.

Lin, W.J., and Johnson, E.A. (1991). Transposon Tn*916* Mutagenesis in *Clostridium botulinum*. Appl. Env. Microbiol. *57*, 2946–2950.

Lin W.J., and Johnson, E.A. (1995). Genome analysis of *Clostridium botulinum* type A by pulsed-field gel electrophoresis. Appl. Environ. Microbiol. *61*, 4441–4447.

Liu, X., Zhu, Y., and Yang, S.H. (2006). Construction and characterization of *ack* deleted mutant of *Clostridium tyrobutyricum* for enhanced butyric acid and hydrogen production. Biotechnol. Prog. 22, 1265–1275. doi10.1021/bp.060082.

Liyanage, H., Young, M., and Kashket, E.R. (2000). Butanol tolerance of *Clostridium beijerinckii* NCIMB 8052 associated with down-regulation of *gldA* by antisense RNA. J. Mol. Microbiol. Biotechnol. *2*, 87–93.

Liyanage, H., Kashket, S., Young, M., and Kashket, E.R. (2001). *Clostridium beijerinckii* and *Clostridium difficile* detoxify methylglyoxal by a novel mechanism involving glycerol dehydrogenase. Appl. Environ. Microbiol. *67*, 2004–2010.

Lyristis, M., Bryant, A.E., Sloan, J., Awad, M.M., Nisbet, I.T., Stevens, D.L., and Rood, J.I. (1994). Identification and molecular analysis of a locus that regulates extracellular toxin production in *Clostridium perfringens*. Mol. Microbiol. *12*, 761–777.

Mattsson, D.M., and Rogers, P. (1994). Analysis of Tn*916*-induced mutants of *Clostridium acetobutylicum* altered in solventogenesis and sporulation. J. Ind. Microbiol. *13*, 258–268.

Mauchline, M.L., Davis, T.O. and Minton, N.P. (1999). In: Clostridia: Manual of Industrial Microbiology and Biotechnology, A.L. Demain and J.E. Davies, ed. (ASM Press), pp. 475–492.

Mermelstein, L.D., and Papoutsakis, E.T. (1993). *In vivo* methylation in *Escherichia coli* by the *Bacillus subtilis* phage phi 3T I methyltransferase to protect plasmids from restriction upon transformation of *Clostridium acetobutylicum* ATCC 824. Appl. Environ. Microbiol. *59*, 1077–1081.

Mohr, G., Smith, D., Belfort, M. and Lambowitz, A.M. (2000). Rules for DNA target-site recognition by a lactococcal group II intron enable retargeting of the intron to specific DNA sequences. Genes Dev. *14*, 559–573.

Mullany, P., Wilks, M., Lamb, I., Clayton, C., Wren, B., and Tabaqchali S. (1990). Genetic analysis of a tetracycline resistance element from *Clostridium difficile* and its conjugal transfer to and from *Bacillus subtilis*. J. Gen. Microbiol. *136*, 1343–1349.

Mullany, P., Wilks, M., and Tabaqchali, S. (1991). Transfer of Tn*916* and Tn*916* delta E into *Clostridium difficile*: demonstration of a hot-spot for these elements in the *C. difficile* genome. FEMS Microbiol. Lett. *79*, 191–194.

Nair, R.V., Green, E.M., Watson, D.E., Bennett, G.N., and Papoutsakis, E.T. (1999). Regulation of the sol locus genes for butanol and acetone formation in *Clostridium acetobutylicum* ATCC 824 by a putative transcriptional repressor. J. Bacteriol. *181*, 319–330.

O'Brien, D.K., and Melville, S.B. (2004). Effects of *Clostridium perfringens* alpha-toxin (PLC) and perfringolysin O (PFO) on cytotoxicity to macrophages, on escape from the phagosomes of macrophages, and on persistence of *C. perfringens* in host tissues. Infect Immun. *72*, 5204–5215.

O'Connor, J.R., Lyras, D., Farrow, K.A., Adams, V., Powell, D.R., Hinds, J., Cheung, J.K., and Rood, J.I. (2006). Construction and analysis of chromosomal *Clostridium difficile* mutants.Mol. Microbiol. *61*, 1335–1351.

Ohtani, K., Bhowmik, S.K., Hayashi, H., and Shimizu, T. (2002). Identification of a novel locus that regulates expression of toxin genes in *Clostridium perfringens*. FEMS Microbiol. Lett. *209*, 113–118.

Okumura, K., Kawsar, H.I., Shimizu, T., Ohta, T., Hayashi, H., and Shimizu T. (2005). Identification and characterization of a cell-wall anchored DNase gene in *Clostridium perfringens*. FEMS Microbiol. Lett. *242*, 281–285.

Perutka, J., Wang, W., Goerlitz, D. and Lambowitz, A.M. (2004). Use of computer-designed group II introns to disrupt *Escherichia coli* DExH/D-box protein and DNA helicase genes. J. Mol. Biol. *336*, 421–439.

Purdy, D., O'Keeffe, T.A., Elmore, M., Herbert, M., McLeod, A., Bokori-Brown, M., Ostrowski, A. and Minton, N.P. (2002). Conjugative transfer of clostridial shuttle vectors from *Escherichia coli* to *Clostridium difficile* through circumvention of the restriction barrier. Mol. Microbiol. *46*, 439–452.

Raju, D., Waters, M., Setlow, P., and Sarker M.R. (2006). Investigating the role of small, acid-soluble spore proteins (SASPs) in the resistance of *Clostridium perfringens* spores to heat. BMC Microbiol. *6*:*50*, doi10.1186/1471-2180-6-50.

Reid, S.J., Rafudeen, M.S., and Leat, N.G. (1999). The genes controlling sucrose utilization in *Clostridium beijerinckii* NCIMB 8052 constitute an operon. Microbiology *145*, 1461–1472.

Roberts, A.P., Hennequin, C., Elmore, M., Collignon, A., Karjalainen, T., Minton, N., Mullany, P. (2003). Development of an integrative vector for the expression of antisense RNA in *Clostridium difficile*. J. Microbiol. Methods 55, 617–624.

Roberts, A.P., Johanesen, P.A., Lyras, D., Mullany, P., and Rood, J.I. (2001). Comparison of Tn5397 from *Clostridium difficile*, Tn*916* from Enterococcus faecalis and the CW459tet(M) element from *Clostridium perfringens* shows that they have similar conjugation regions but different insertion and excision modules. Microbiology *147*, 1243–1451.

Rood, J.I., and Cole, S.T. (1991). Molecular genetics and pathogenesis of *Clostridium perfringens*. Microbiol. Rev. 55, 621–48.

Saint-Prix, F. (2006) Development of a system for chromosomal integration and gene deletion in *Clostridium acetobutylicum* ATCC 824. In: Abstracts of the 9th International Workshop on the Regulation of Metabolism, Genetics and Development of the Solvent- and Acid-producing Clostridia, Houston, USA,

Sarker, M.R., Carman, R.J., and McClane, B.A. (1999). Inactivation of the gene (cpe) encoding *Clostridium perfringens* enterotoxin eliminates the ability of two *cpe*-positive *C. perfringens* type A human gastrointestinal disease isolates to affect rabbit ileal loops. Mol. Microbiol. 33, 946–958. Erratum in Mol. Microbiol. (2000) 35, 249.

Sauer, U., and Dürre, P. (1993). Possible function of tRNA(Thr)ACG in regulation of solvent formation in *Clostridium acetobutylicum*. FEMS Microbiol. Lett. *11*), 343–3348.

Sebaihia, M., Wren, B.W., Mullany, P., Fairweather, N.F., Minton, N., Stabler, R., Thomson, N.R., Roberts, A.P., Cerdeno-Tarraga, A.M., Wang, H., Holden, M.T., Wright, A., Churcher, C., Quail, M.A., Baker,

S., Bason, N., Brooks, K., Chillingworth, T., Cronin, A., Davis, P., Dowd, L., Fraser, A., Feltwell, T., Hance, Z., Holroyd, S., Jagels K., Moule, S., Mungall, K., Price, C., Rabbinowitsch, E., Sharp, S., Simmonds, M., Stevens, K., Unwin, L., Whithead, S., Dupuy, B., Dougan, G., Barrell, B., and Parkhill, J. (2006). The multidrug-resistant human pathogen *Clostridium difficile* has a highly mobile, mosaic genome. Nat. Genet. *38*, 779–786.

Shimizu, T., Ba-Thein, W., Tamaki, M. and Hayashi, H. (1994). The virR gene, a member of a class of two-component response regulators, regulates the production of perfringolysin O, collagenase, and hemagglutinin in *Clostridium perfringens*. J. Bacteriol. *176*, 1616–1623.

Shimizu, T., Yaguchi, H., Ohtani, K., Banu, S., and Hayashi, H. (2002a). Clostridial VirR/VirS regulon involves a regulatory RNA molecule for expression of toxins. Mol. Microbiol. *43*, 257–265.

Shimizu, T., Ohtani, K., Hirakawa, H., Ohshima, K., Yamashita, A., Shiba, T., Ogasawara, N., Hattori, M., Kuhara, S., and Hayashi, H.(2002b). Complete genome sequence of *Clostridium perfringens*, an anaerobic flesh-eater. Proc. Natl. Acad. Sci. USA *99*, 996–1001.

Smith, C.J., Markowitz, S.M., and Macrina, F.L. (1981). Transferable tetracycline resistance in *Clostridium difficile*. Antimicrob. Agents Chemother. *19*, 997–1003.

Varga, J., Stirewalt, V.L., and Melville S.B. (2004). The CcpA protein is necessary for efficient sporulation and enterotoxin gene (*cpe*) regulation in *Clostridium perfringens*. J. Bacteriol. *186*, 5221–5229.

Varga, J.J., Nguyen, V., O'Brien, D.K., Rodgers, K., Walker, R.A., and Melville, S.B. (2006). Type IV pili-dependent gliding motility in the Gram-positive pathogen *Clostridium perfringens* and other Clostridia. Mol. Microbiol. *62*, 680–694.

Volk, W.A., Bizzini, B., Jones, K.R., and Macrina, F.L. (1988). Inter- and intrageneric transfer of Tn*916* between *Streptococcus faecalis* and *Clostridium tetani*. Plasmid *19*, 255–259.

Wang, H., Roberts, A.P., and Mullany, P. (2000). DNA sequence of the insertional hot spot of Tn*916* in the *Clostridium difficile* genome and discovery of a Tn*916*-like element in an environmental isolate integrated in the same hot spot. FEMS Microbiol. Lett. *192*, 15–20.

Wilkinson, S.R., and Young, M. (1994). Targeted gene integration of genes into the *Clostridium acetobutylicum* chromosome. Microbiol. *140*, 89–95.

Woolley, R.C., Pennock, A., Ashton, R.J., Davies A., Young, M. (1989). Transfer of Tn*1545* and Tn*916* to *Clostridium acetobutylicum*. Plasmid. *22*, 169–174.

Wüst, J., and Hardegger, U. (1983). Transferable resistance to clindamycin, erythromycin, and tetracycline in *Clostridium difficile*. Antimicrob. Agents Chemother. *23*, 784–786.

Zhong, J., Karberg, M. and Lambowitz, A.M. (2003). Targeted and random bacterial gene disruption using a group II intron (targetron) vector containing a retrotransposition-activated selectable marker. Nucleic Acids Res. *31*, 1656–1664.

Zhu, Y., Liu, X., and Yang, S.H. (2005). Construction and characterization of *pta* gene-deleted mutant of *Clostridium tyrobutyricum* for enhanced butyric acid fermentation. Wiley InterScience. doi10.1002/bit.20354

Clostridia in Anti-tumour Therapy

11

Asferd Mengesha, Ludwig Dubois, Kim Paesmans, Brad Wouters, Philippe Lambin and Jan Theys

Abstract

Although traditional anticancer therapies are effective in the management of many patients, there are a variety of factors that limit their effectiveness in controlling some tumours. These observations have led to interest in alternative strategies to selectively target and destroy cancer cells. In that context, *Clostridium*-based tumour targeted therapy holds promise for the treatment of solid tumours. Upon systemic administration, various strains of non-pathogenic clostridia have been shown to infiltrate and selectively replicate within solid tumours. This specificity is based upon the unique physiology of solid tumours, which is often characterized by regions of hypoxia and necrosis. Clostridial vectors can be safely administered and their potential to deliver therapeutic proteins has been demonstrated in a variety of preclinical models. However, there are several issues that are still unknown and remain major challenges. In this chapter, we will review the potential use of *Clostridium* in cancer treatment and discuss the major advantages, challenges and shortcomings of bacterial systems for tumour-specific therapy. In addition, we will highlight the requirements needed to advance the approach into clinical trials.

Clostridia as part of anti-cancer therapy?

Cancer is a disease with a high incidence in the western world. The use of conventional treatment modalities results in ~50% cure of cancer patients, whereas failure to control the tumour locally and metastasis result in treatment failure in the other 50%. Thus, for many patients, current treatments are ineffective and overall survival is poor. Major advances have been made in the understanding of the genetic basis of cancer, driving intensive activity worldwide to develop alternative treatment approaches and substantial resources are currently being devoted to discovery and development of new anti-cancer therapies. Included amongst these new approaches is gene therapy, with a wide variety of proposed therapeutic genes. One of the most important problems in the development and use of gene therapy is the safe and specific delivery of genes to the tumour. A variety of viral vector delivery systems and a number of non-viral mechanisms have been developed. Many different approaches have been conceived to produce more selective vector systems (Waehler *et al.*, 2007). Delivery can be targeted to tumour-specific and tissue-specific antigens, and targeted gene expression has been analysed using tissue-specific, disease-specific and/or inducible promoter systems. In addition, progress has been made in recent years towards the development of more efficient delivery methods (e.g. replicative viral vectors, combination of components of several vectors into hybrids with beneficial properties). Without doubt, the progress made in these areas will help to exploit the potential power of gene therapy for cancer, most probably as a part of combinatorial therapy with more established disciplines.

An alternative to viral delivery that has been proposed is the use of clostridia as gene delivery vehicles. Several strains have been genetically engineered to deliver a therapeutic gene into tumours. As such, it can be considered a gene therapy approach. If on the other hand, gene therapy is to be defined as the introduction of a gene, or part of a gene, into the cancer cells (or normal cells) then using recombinant bacteria as anticancer vectors is not gene therapy. Bacteria are not vectors for the introduction of genes into mammalian cells. However, they can and do concentrate in tumours and can carry a gene of interest to produce a protein of choice in tumours. This can be a powerful adjunct to cancer therapy.

We and others have been investigating the potential use of non-pathogenic strictly anaerobic bacteria such as *Clostridium* (Bettegowda *et al.*, 2003; Bettegowda *et al.*, 2006; Cheong *et al.*, 2006; Liu *et al.*, 2002; Minton, 2003; Pawelek *et al.*, 2003; Theys *et al.*, 2003; Theys *et al.*, 2006). Their tumour specificity is based upon the unique physiology of solid tumours, which is often characterized by regions of hypoxia and necrosis. As a group, *Clostridium* has probably achieved greatest prominence as a consequence of their more notorious representatives, such as *C. botulinum* or *C. tetani*. Most members of the genus are, however, non-pathogenic. Although clostridia are strictly anaerobic, most of them can form spores allowing survival but not growth in oxic conditions. It is the very ability to form spores that presents this genus with its potential for treating cancer.

In this chapter, the potential of using clostridia in anti-cancer therapy will be reviewed and discussed. We will outline the importance of hypoxia/necrosis in solid tumours, give an historical overview of clostridial strains that have been investigated as anti-cancer agents until now and highlight the limitations and hence areas that need further investigation in order to move on to extended clinical evaluation.

Importance of hypoxia and necrosis in tumours

Cancer cells are characterized by an aberrantly accelerated metabolism and an exponential proliferation. This results in an imbalance of 'oxygen-supply' and 'oxygen-consumption' leading to heterogeneity in regional oxygenation. This disequilibrium is a major causative factor of tumour hypoxia, a unique microenvironment in locally advanced solid tumours (Vaupel and Mayer, 2007). Similarly, necrotic regions are a common, if not a universal, feature of human solid tumours. These necrotic regions typically occur at a distance from functioning blood vessels beyond the diffusion distance of oxygen. Because these necrotic regions typically develop because of prolonged lack of oxygen, they are usually intimately associated with hypoxic but viable cells.

The importance of hypoxia in solid tumours is linked to the fact that hypoxic cells are intrinsically more resistant to current cancer therapies. Tumour hypoxia is a major factor contributing to the failure of radiotherapy. This is largely because DNA damage produced by ionizing radiation, which would otherwise become fixed and lethal to cells by reacting with O_2 in well-oxygenated conditions, can be restored to its undamaged form under hypoxic conditions (Brown and Wilson, 2004). Tumour hypoxia may also compromise the outcome of conventional chemotherapy. Since the hypoxic tumour cells are distant from functional blood vessels they receive much lowered concentrations of anticancer drugs than the target concentrations (Durand, 1994). Furthermore, hypoxic tumour cells may demonstrate an inhibition of cell cycle progression and proliferation, and hence may be relatively resistant to many anticancer drugs that target rapidly dividing cells (Kizaka-Kondoh *et al.*, 2003).

Several clinical studies in patients with soft tissue (Nordsmark *et al.*, 2001), uterine cervix (Nordsmark *et al.*, 2006) and head and neck (Janssen *et al.*, 2005) carcinomas confirmed that the presence of hypoxia negatively affects the loco-regional tumour control and/or disease free survival after primary radiotherapy. The classical 'oxygen effect' as described above is unlikely to be the only explanation for treatment resistance, since tumour oxygenation has been shown to be the most important prognostic factor for therapy outcome, irrespective of the therapeutic modality (Hockel *et al.*, 1999). Hypoxia-induced modifications of gene expression may also contribute to this poor prognostic outcome, giving rise to a more aggressive locoregional disease and an enhanced invasive capacity (Graeber *et al.*, 1996).

Thus, there is a lot of clinical evidence that hypoxia interferes with the therapy efficacy of solid tumours and also contributes to a more malignant phenotype. On the other hand, hypoxia and necrosis may also be exploited to give a therapeutic advantage. Since it is virtually a unique characteristic of tumour cells, therapies that target hypoxic/necrotic regions may have the potential to kill malignant tumour cells while leaving non-malignant cells relatively unaffected.

There are two potential approaches for hypoxia-targeted tumour therapy. One involves improvement of the oxygenation of the hypoxic regions in solid tumours, combined with classical cancer therapies like radiotherapy or chemotherapy. Examples of this approach include the use of recombinant erythropoietin (EPO) to increase haemoglobin levels (Henke and Guttenberger, 2000), accelerated radiotherapy in combination with carbogen breathing and nicotinamide administration (ARCON) (Kaanders *et al.*, 2002) or the use of radiation-modifying drugs like nimorazoles (Brown, 2000).

The other approach involves the exploitation of the microenvironment of hypoxic tumour cells by targeting the important transcription factor 'hypoxia-inducible factor-1' (HIF-1), use of bioreductive drugs or application of gene therapy. HIF-1 plays pivotal roles in the cellular adaptive response to an hypoxic microenvironment (Semenza, 2001). It is a heterodimeric transcription factor composed of a hypoxia responsive α- (HIF-1α) and a constitutive expressed β-subunit (HIF-1β). Its activity is mainly dependent on the stability of the complex. Under normoxic conditions, HIF-1α is hydroxylated and therefore ubiquitinated by the E3 ubiquitin ligase von Hippel–Lindau (VHL) protein, targeting this complex for degradation. On the other hand, HIF-1α is stabilized under hypoxic conditions and interacts with HIF-1β (Jaakkola *et al.*, 2001). The resulting complex promotes the expression of numerous genes by interacting with their hypoxia-responsive elements (HRE). These genes are involved in crucial aspects of cancer biology, including angiogenesis, cell survival, glucose metabolism and invasion. The importance of HIF-1 as a transcription factor, suggests that HIF-1 and its regulators could be tumour-specific targets for anticancer therapy. Several reports have indeed demonstrated that a variety of approaches can be used to block HIF-1 function (Harris, 2002).

An alternative strategy to overcome the problem of hypoxic tumour cells is to selectively kill them using bioreductive prodrugs. These are typically non-toxic compounds which are reduced by biological enzymes to toxic, active metabolites. They are designed in such way that this reduction occurs only or preferentially in the absence of oxygen. Tirapazamine (TPZ) is the prototype compound in this class of agents. Under hypoxic conditions, it is reduced to a radical that leads to DNA double- and single-strand breaks and base damage. TPZ has a unique O_2 concentration dependency such that its cytotoxicity does not level off at high concentrations, but gradually decreases as the O_2 concentration increases (Koch, 1993). TPZ promotes cell killing with fractionated irradiation and is particularly effective in combination with cisplatin (Rischin *et al.*, 2005). The newest bioreductive prodrug PR-104 is a phosphate pre-prodrug of the dinitrobenzamide mustard which is activated in hypoxia to become a potent bifunctional alkylating agent producing DNA interstrand cross-links. Moreover, PR-104 has demonstrated a substantial bystander effect, killing aerobic as well as hypoxic cells in solid tumours, and producing significant antitumour activity as a single agent alone (Patterson *et al.*, 2007). It is also superior to tirapazamine in combination with irradiation in a SiHa xenograft model (Hicks *et al.*, 2007). More focus on bioreductive prodrugs is reviewed in (Ahn and Brown, 2007).

A third possibility to exploit the microenvironment of hypoxic/necrotic tumour cells, is the use of gene therapy. Successful gene therapy requires not only the identification of an appropriate therapeutic gene for treatment, but also a selective vector system by which the gene can be delivered to the desired site both efficiently and accurately (Lambin *et al.*, 1998). A fundamental requirement of these strategies is that only tumour cells should be exposed as much as possible to the toxic agent, while normal healthy tissues are not affected. Over the past decade, many strategies using both viral and non-viral methods to deliver therapeutic genes have been explored (Seth, 2005; Wolff *et al.*, 2005). Vehicles such as retro- and adenoviruses, liposomes and naked DNA injection or electroporation are currently being evaluated in clinical trials and new delivery systems like gene-attenuated adenoviruses, lentiviruses, polylysine constructs and leukocytes are being developed (Greco *et al.*, 2000). Although so far, trials have met with only little success, they did serve to highlight some deficiencies of the applied approaches and as such, have led to a reassessment of the field. Most importantly, preclinical and clinical data suggest that strategies need to focus less on the choice of the therapeutic gene than on the means of delivering it. Therefore, when making choices regarding suitable vectors for gene therapy for cancer, it is important to recognize both the factors that distinguish a tumour from its surrounding normal tissue as well as the factors that limit successful therapy with available treatments. Tumour hypoxia is a good example. Since this micro-environmental condition arises due to the chaotic organization and irregularity of blood vessels, sufficient delivery of oxygen, nutrients and accordingly also therapeutic agents (i.e. gene delivery vectors) to all cells within the tumour is prevented (Dreher *et al.*, 2006). On the other hand, it also represents a unique environment not found elsewhere in the body. This environment encourages the growth of strictly anaerobic bacteria (Dang *et al.*, 2001b; Liu *et al.*, 2002; Minton *et al.*, 1995a; Theys, 2001). Numerous publications from our group and others provide convincing evidence that although systemically injected clostridial spores are dispersed throughout the body, only those that encounter the hypoxic environment will germinate and multiply (see below). This predisposition of clostridial spores to germinate specifically in hypoxic/necrotic regions of solid tumours makes them an ideal delivery vehicle for anti-cancer agents (Fig. 11.1). The application of *Clostridium* as a gene therapy delivery vehicle offers several advantages:

1 The system exhibits a high degree of tumour selectivity, since hypoxic/necrotic regions are not present in normal tissues and thus represent a unique feature of tumours. Most solid tumours investigated have been shown to contain hypoxic/necrotic regions, even in tumours as small as $1\,mm^3$.
2 There is convincing clinical evidence demonstrating that hypoxia is a primary cause of conventional treatment failure and that it contributes to a more malignant tumour phenotype. Thus, *Clostridium* preferentially targets the areas within tumours that are most difficult to treat with standard agents (Vaupel and Mayer, 2007).
3 The approach is safe. We and others have shown that the bacteria can be eliminated by using antibiotics. This not only allows control over bacterial presence in the tumour but also to maintain control over therapeutic gene expression, since the gene is expressed in the bacterial host. This represents a significant advantage over most other gene therapy approaches, where therapeutic gene expression occurs within the target cells (Theys *et al.*, 2001a).
4 The *Clostridium* approach is feasible and easy to implement. Early clinical experience in various centres has already demonstrated that administration of non-pathogenic clostridial spores is well tolerated by cancer patients. The production of these spores is straightforward, and since the therapeutic spore formulation is oxygen insensitive, they can subsequently be easily handled, stored and administrated (Minton, 2003).

The past: early experience with clostridial oncolysis

The concept of using bacteria as tumour vectors has been vigorously pursued using several *Clostridium* species (Table 11.1). The genus *Clostridium* comprises a large and heterogeneous group of anaerobic, Gram-positive, spore-forming bacteria which develop into metabolically active vegetative cells in the absence (or at very low levels) of oxygen. *Clostridium* species naturally inhabit the intestinal tracts of animals and humans. The pathogenic species *C. tetani*, *C. botulinum* and *C. perfringens* germinate in necrotic tissue and produce toxins that cause tetanus, botulism and gas gangrene, respectively (Ryan *et al.*, 2006). Except for those *Clostridium* species, most of the members are however non-pathogenic.

Clostridium was first associated with cancer in 1813, when Vautier reported tumour regression in patients who suffered gas gangrene following infection of *C. perfringens* (Minton, 2003). More than a century later, in 1927, the first report of tumour lysis of a Flexner–Jobling rat carcinoma

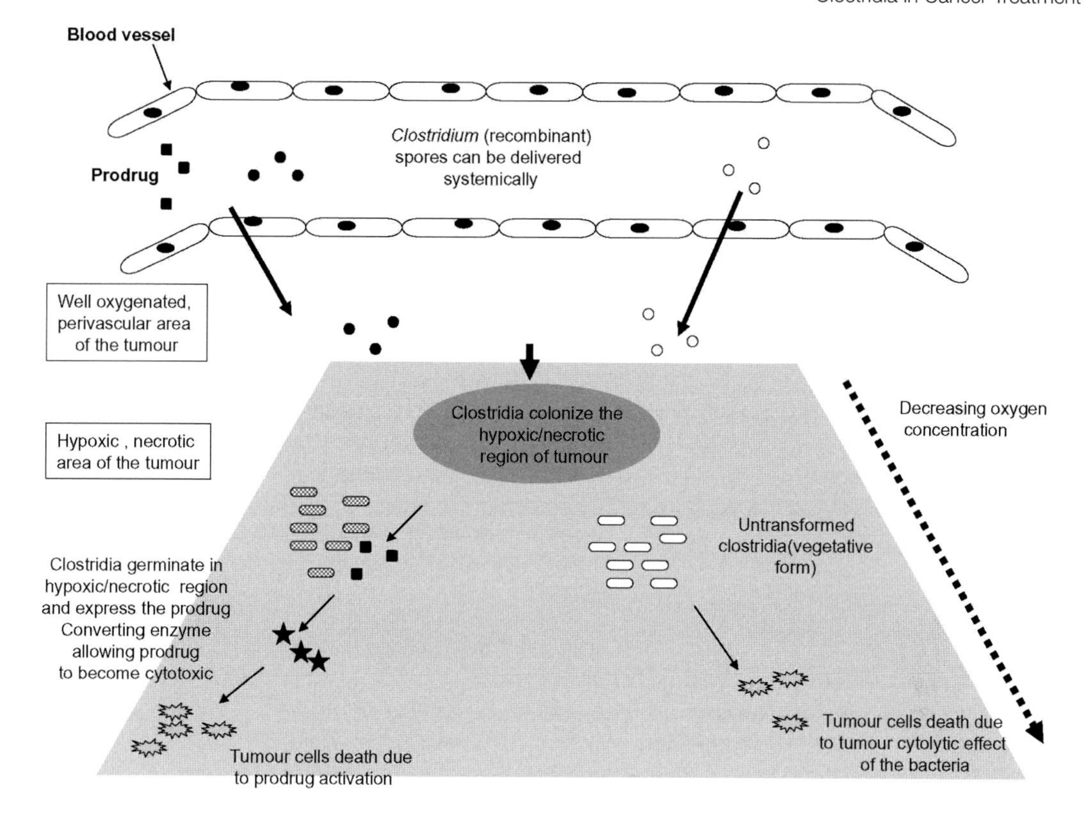

Figure 11.1 Use of clostridia to target the hypoxic/necrotic regions of solid tumours. Wild-type or recombinant clostridia carrying a therapeutic gene (e.g. encoding prodrug converting enzyme such as CD or NTR, or a cytotoxic drug such as TNF-α or IL-2) can be administered systemically. Upon arrival to the hypoxic/necrotic part of the tumour, clostridia germinate and proliferate thereby expressing the transgene, leading to tumour cell death.

by selectively growing *Clostridia* in the tumour was given (Torrey, 1927). A few years later, Connell investigated the therapeutic potential of *C. histolyticum* using sterile filtrates to treat advanced cancers. Tumour regression was observed and attributed to the production of proteolytic enzymes (Connell, 1935). Parker and colleagues tested these effects further in 1947 by injecting a spore suspension of *C. histolyticum* into transplantable mouse sarcomas. Not only a tumour regression and oncolysis (liquification), but also an extended survival of the tumour-bearing mice was seen (Parker, 1947). However, few animals survived this treatment, since *Clostridium*-mediated oncolysis was accompanied by toxic effects and death. In summary, the studies above indicate the potential of using clostridial spores for the induction of tumour lysis.

Malmgren and Flanigan were in 1955 the first to demonstrate the exquisite tumour selectivity of the system using intravenous injection of *C. tetani* spores. They administered up to 1.5×10^5 spores to mammary, fibrosarcoma and hepatoma tumour-bearing CSH/He mice and observed death from tetanus poisoning within 48 hours. However, mice without tumours cleared the bacterium and suffered no side effects (Malmgren and Flanigan, 1955). These results indicated that the spores only germinated within (the anoxic regions of) the tumours causing release of the tetanus toxins, which resulted in death. This study also demonstrated that direct intratumoral spore injection is unnecessary and systemic administration of spores is sufficient for effective tumour colonization. Based on these observations, it was reasoned that *Clostridium*-mediated treatment might of course only be successful if tumour regression could be induced without causing death of the host. Obviously, the use of non-pathogenic *Clostridium* was the alternative. Möse and Möse reported in 1964 that *C. butyricum* M-55 localized and germinated in

Table 11.1. Clostridial strains tested for anti-tumour studies.

Clostridial strain	Tumour model	Strategy/result	Reference
C. histolyticium	Mouse transplanted tumour	Intravenous injection of spores resulted in tumour liquefaction/lysis	Parker RC *et al.* (1947)
C. tetani	Mouse transplanted tumour	Intravenous injection of spores resulted in death of mouse bearing tumour	Malmgren and Flanigan (1955)
C. oncolyticum	Mouse Ehrlich ascites tumour	Intravenous or intra-tumour injection of 10^{6-10} cfu of spores	Möse (1979)
C. oncolyticum (sporogenes)	*In vitro* study	Recombinant for *E. coli* colicin E3 for cancerostatic properties	Schlechte and Elbe (1988)
C. beijerinckii (acetobutylicum)	EMT6 mouse	Recombinant for NTR CDEPT strategy with CB1954, nitroreductase activity detected in tumour lysate	Minton *et al.* (1995b)
C. acetobutylicum	Rhabdomyosarcoma	Recombinant for TNF-α. Recombinant protein detected in tumour, but no control of tumour growth	Theys *et al.* (2001a)
C. acetobutylicum	Rhabdomyosarcoma	Recombinant for *E. coli* CD, CDEPT strategy, CDase activity detected in tumour lysate	Theys *et al.* (2001b)
C. novyi-NT	B16 melanoma, HCT116 xenographs	Combination bacteriolytic therapy (COBALT), tumour regression and complete cure in a considerable proportion of mice that survived	Dang *et al.* (2001b)
C. acetobutylicum	Rhabdomyosarcoma	Recombinant for IL-2, Enhanced anti-tumour effect	Barbe *et al.* (2005)
C. sporogenes NCIMB 10696	Human colorectal carcinoma (HCT116)	Recombinant for NTR, CDEPT strategy with CB1954, high level of colonization 10^8–10^9 cfu/g tumour, repeated treatment cycle, significant tumour growth delay	Theys *et al.* (2006)
C. novyi-NT	HCT116 and CT26 colorectal tumour	Wild-type *C.novyi*-NT plus Doxil, A bacterial protein enhances the release and efficacy of liposomal cancer drugs	Cheong *et al.* (2006)

solid Erhlich ascites tumours causing extensive lysis without any concomitant effect on normal tissues. The procedure resulted in softening of tumours followed by spontaneous discharge of a brownish liquid necrotic mass. However, not all animals did survive this stage of extensive oncolysis, but the animals that survived demonstrated tumour regrowth from the remaining outer rim of viable cells (Mose, 1964). They also investigated whether other strains of clostridia produced oncolysis to a smaller or larger extent than the standard *C. butyricum* M-55. *C. paraputrificum* was the only strain which was clearly superior to the control strain as far as its oncolytic effect was concerned. However, its great disadvantage was the fact that it was not clearly non-pathogenic (Mose, 1964). Similar observations were soon made and extended by a number of investigators using tumours in rats (Thiele *et al.*, 1964b) and hamsters (Engelbart and Gericke, 1964). Anti-tumour effects were also obtained using a new *Clostridium* strain originating from pig skin extracts in various animal-tumour combinations (Mohr *et al.*, 1972). Overall, these early studies indicated that germination of wild-type *Clostridium* was well tolerated in animal systems and frequently resulted in destruction of a significant portion of the tumour.

Based on all the results and knowledge obtained from animal studies, clinical trials were initiated with *C. butyricum* M-55. Möse and Möse demonstrated the absence of pathogenicity in humans of *Clostridium* by injecting themselves with spore suspensions of *C. sporogenes* ATCC 13732, as cited in (Carey *et al.*, 1967). Carey and colleagues were in 1967 the first to initiate a clinical trial. Five patients with neoplastic disease were intravenously injected with 10^{10} *C. butyricum* M55 spores. They reported oncolysis by *C. butyricum* M-55 in patients with the largest tumours, not in the surrounding normal tissues and (smaller) metastases (Carey *et al.*, 1967). Heppner and colleagues administered in 1978 *C. butyricum* M-55 spores to patients with vascular glioblastomas by intracarotid injection. One week after injection, they observed a complete oncolysis with conversion of the glioblastoma into a brain abscess which was then operated. Patients developed fever and positive blood cultures for approximately 1 week and occasionally required supportive care involving treatment with antibiotics. Most patients, however, had no evidence of objective regression or change in tumours and the rate of recurrence remained uninfluenced (Heppner and Mose, 1978). Despite these observations, the clinical reports established the safety of this *Clostridium*-mediated approach as well as the fact that colonization of human tumours occurs following intravenous injection of spores. Although patients tolerated the *Clostridium* treatment well, clinical trials were discontinued because tumour regrowth occurred from a rim of viable tumour cells, leading to an overall lack of clinical benefits. In order to attack this rim of viable tumour cells, combinations of spore administration and irradiation, local hyperthermia and various chemotherapeutic agents, such as 5-fluorodeoxyuridine and cyclophosphamide, were tested. These combined treatments generally resulted in a larger reduction of tumour weight in animal tumour models (Gericke *et al.*, 1979; Schlechte *et al.*, 1982; Thiele *et al.*, 1964a). A significant enhancement of oncolysis could also be obtained by decreasing the oxygen level to 11–12% in the respiratory air of tumour-bearing rats (Mose, 1979). Although these combination therapies in general increased tumour lysis, they did not result in complete tumour responses.

In agreement with the observations mentioned above, clostridia were also tested as a tool for cancer diagnosis. A microbiological tumour test has been developed based on the detection of antibodies directed against vegetative clostridial cells in the serum following intravenous spore injection. This method has been evaluated in various animals with neoplastic disease (Fabricius *et al.*, 1987; Wittmann *et al.*, 1990).

Clostridium as a tumour-specific gene delivery system

The above studies showing the specific colonization of tumours by non-pathogenic clostridial species together with the development of recombinant DNA technology that allowed transformation of (certain strains of) *Clostridium*, prompted the suggestion first made by Brown and colleagues that *Clostridium* could be used as a tumour specific gene delivery system (Lemmon *et al.*, 1994). The proposed tumour-targeted strategy is to inject spores of non-pathogenic clostridia, genetically engineered to produce a specific protein potentially harmful to the tumour when the spores germinate (Fig. 11.1). Because germination to the vegetative form will occur only in the hypoxic/necrotic regions, the protein should be produced only in the tumour. As such, use of these genetically engineered clostridia may provide a potential solution to overcome the described deficiencies of using wild-type strains for cancer treatment.

To date, the introduction of two classes of recombinant protein has been explored:

1. proteins with a direct cytotoxic effect (e.g. TNF alpha, IL-2) and
2. enzymes which convert a non-toxic prodrug into a toxic agent. In this concept of directed-enzyme-prodrug-therapy (DEPT), active drug will be generated only in the vicinity of the tumour, resulting in high concentrations of active drug specifically at the tumour site while at the same time maintaining harmless levels of drugs elsewhere in the body. Since a single 'enzyme' can convert large quantities of prodrug into anti-cancer therapeutic drugs, a limited amount of therapeutic enzyme may be sufficient to eradicate many tumour cells, a phenomenon known as the 'bystand-

er' effect. Furthermore, as the prodrug can diffuse in and out the bacterial cell, there is no need for secretion of the therapeutic protein, thereby avoiding problems with enzymatic breakdown in the extracellular environment (Mengesha *et al.*, 2007). Targeting of appropriate prodrug converting enzymes has been achieved using viral vectors (gene directed enzyme prodrug therapy, GDEPT) (Kerr *et al.*, 1997) or using a fusion of the enzyme component to a monoclonal antibody directed against a tumour specific antigen (antibody directed enzyme prodrug therapy, ADEPT) (Bagshawe, 1987). However, these approaches are limited due to low tumour specificity and often inefficient distribution of the vector throughout the tumour mass. Because germination of clostridial spores is only possible in the unique hypoxic/necrotic areas within the tumour, these limitations do not apply to an approach based on clostridia (clostridia directed enzyme prodrug therapy, CDEPT). Moreover, as opposed to ADEPT which requires either tumour-specific or possibly patient-specific enzyme/antibody conjugates, the clostridial approach, because it depends only on tumour necrosis, will have universal applicability.

The principal focus of CDEPT efforts so far has been the delivery of the bacterial enzymes cytosine deaminase (CD) and nitroreductase (NTR). Both enzymes essentially have no human equivalent.

CD converts 5-fluorocytosine (5-FC) into 5-fluorouracil (5-FU), which is further metabolized into inhibitors of RNA and DNA biosynthesis. There are several reasons why the CD/5-FC combination has received most attention:

1 Because of vascular insufficiency, the distribution of chemotherapeutic drugs in tumours is not uniform. This has been elegantly shown in the clinic for 5-FU in studies where patients could be imaged following drug administration. Levels of 5-FU were variable amongst tumours and correlated with response (Moehler *et al.*, 1998). This suggests that a local intra-tumour production of 5-FU, as achieved with CDEPT, might increase the anti-tumour effect;.
2 5-FU production can be measured directly using non-invasive imaging modalities such as magnetic resonance spectroscopy (MRS).
3 Both 5-FU and 5-FC are well understood, clinically applied drugs.
4 The action of CD is associated with a so-called 'bystander effect'. Its therapeutic efficacy has been observed in a wide variety of solid tumours (Ge *et al.*, 1997; Ichikawa *et al.*, 2000).
5 Since in a curative setting, *Clostridium* is anticipated to be used in combination with other treatment modalities, it is interesting that 5-FU can potentate the effects of ionizing radiation. We have already reported that a conversion efficiency of 5-FC into 5-FU of 1–3% would be sufficient to obtain radiation sensitization enhancement ratios of 1.1–1.2 when combined with a standard 2 Gray (Gy) daily radiation scheme (Lambin *et al.*, 2000).

NTR converts the potent alkylating agent CB1954 to an active cytotoxic species which causes inter-strand cross-links in DNA (Knox *et al.*, 1988). Similar to CD/5-FC, there is a good rationale to choose the NTR/CB1954 combination:

1 Cell killing by activated CB1954 is cell cycle independent. Both proliferating and non-proliferating cells, as they are often present in tumour areas with hypoxic gradients are killed by the converted prodrug. Indeed, the NTR/CB1954 is effective under hypoxia and anoxia (for reviews see Greco and Dachs, 2001; Patterson *et al.*, 2003).
2 Cell killing occurs in tumours which have acquired resistance to other agents such as e.g. cisplatin (Bridgewater *et al.*, 1995).
3 There is no human homologue of NTR, so the prodrug will not be converted to the activated species elsewhere in the body. In addition, conversion of CB1954 to an active species relies on a co-factor present in *Clostridium*, further increasing safety, as no extracellular activation can occur.

4 The action of NTR is associated with a so-called bystander effect via a gap-junction independent mechanism (Bridgewater *et al.*, 1997; Djeha *et al.*, 2000).
5 CB1954 can modify the effects of radiation, and can thus complement radiotherapy treatment (Patterson *et al.*, 2003);.
6 CB1954 has already been extensively tested in a clinical setting (Chung-Faye *et al.*, 2001).

Use of (saccharolytic) strains with suboptimal tumour colonization capacities

From the observations described in the previous paragraphs on the tumour colonization and oncolytic capacities of different clostridial strains, it is clear that different strains can have highly diverse characteristics. It is therefore not surprising that different strains exhibit varying colonization properties and that they may not all be susceptible for genetic engineering. The latter is very important since, if *Clostridium* is to be used as a gene delivery vector, reliable transformation methods allowing the construction of stable recombinant clostridial strains need to be available. However, a generally applicable and reproducible transformation method for gene transfer into *Clostridium* does not exist and transformation protocols described are typically characterized by a narrow, strain-specific character. Initially, protocols for transformation were only available for saccharolytic strains, and thus *C. acetobutylicum* and *C. beijerinckii* were the first to be tested as protein delivery vectors.

C. acetobutylicum DSM792 has been genetically engineered to express and secrete cytotoxic agents like murine tumour necrosis factor-alpha (TNF-α) and rat interleukin 2 (IL-2), two cytokines with known anti-tumour activity. A main advantage of in situ cytokine production is the avoidance of severe side effects that occur when these cytokines are administered systemically. Both TNF-α and IL-2 have been shown to be efficiently secreted and also biologically active as measured in bacterial lysates and supernatants (Barbe *et al.*, 2005; Theys *et al.*, 1999). So far, injection of TNF- α recombinant *C. acetobutylicum* failed to result in growth delay, probably because recombinant protein levels in the tumour were too low. On the other hand, preliminary results with IL-2 producing *C. acetobutylicum*, locally administered to rhabdomyosarcoma-bearing rats, show some promise, as tumour growth delay was observed, especially in combination with fractionated radiotherapy (S. Barbé, PhD thesis, University of Leuven, 2005).

In the mid-nineties Fox and co-workers were the first to clone the *Escherichia coli* CD gene into *C. beijerinckii* by use of a *Clostridium* expression vector. High levels of CD enzyme were found in the bacterial medium thereby sensitizing mouse EMT6 carcinoma cells to 5-FU up to 500-fold (Fox *et al.*, 1996). A few years later, our group demonstrated biologically active CD protein in lysates and supernatants of genetically engineered *C. acetobutylicum* (Theys *et al.*, 2001a). Additionally, tumour specific expression of CD was detected in WAG/Rij rats bearing rhabdomyosarcoma tumours following intravenous injection of spores of recombinant *C. acetobutylicum* (Theys *et al.*, 2001b). Similar gene therapy experiments were performed using the *Escherichia coli* NTR gene again in *C. beijerinckii*. Expression of the NTR gene by *Clostridium in vitro*, enhanced tumour cell death by CB1954 and NTR activity was detected in tumours following intravenous administration to mice bearing EMT6 tumours (Minton *et al.*, 1995b). Despite these encouraging data, animal studies with recombinant *C. beijerinckii* or *C. acetobutylicum* failed to produce anti-tumour activity when the prodrug (5-fluorocytosine for CD-containing tumours or CB1954 for NTR-containing tumours) was injected systemically. The most probable explanation for the observed lack of anti-tumour activity almost certainly resulted from insufficient levels of prodrug-activating enzymes in the tumours due to the low levels of viable recombinant *Clostridium* in the tumours. It has indeed been shown that *C. beijerinckii* or *C. acetobutylicum* produced only 10^5 to 10^6 bacteria per gram tumour. Although these recombinant saccharolytic strains demonstrate suboptimal tumour colonization properties, they provided invaluable information regarding the heterologous gene expression in *Clostridium*. After all, the use of these strains may even be beneficial if only secretion of a desired therapeutic gene is required, since degradation of the protein might be enhanced using proteolytic strains (Van Mellaert *et al.*, 2006).

In an attempt to overcome the limited colonization levels of the saccharolytic strains, combination treatments with vascular targeting agents have been put in place. These agents specifically target the dividing endothelial cells of tumour blood vessels and cause rapid vascular shutdown in tumours, thereby promoting hypoxia, anoxia and necrosis (Brekken *et al.*, 2002). Our group was the first to demonstrate the benefit of such vascular targeting agent as combretastatin A-4 phosphate (CombreAp) to improve *Clostridium* tumour colonization. Systemic administration of (CombreAp) to WAG/Rij rats bearing rhabdomyosarcomas resulted in a vascular shutdown (3–6 hours) leading to extensive necrosis (Landuyt *et al.*, 2000). This phenomenon led to consistent and high levels of clostridial colonization of (very) small tumours that were very inefficiently colonized in the absence of CombreAp. In animals treated with CD-recombinant *C. acetobutylicum* spores, the incidence of CD-positive tumours increased from 58% to 100% when treated concomitantly with CombreAp (Theys *et al.*, 2001b). Recently, notable results using *C. novyi-NT* in a comparable regimen (anti-vascular agent dolastatin in combination with other chemotherapeutics and *Clostridium*) were reported (see below) (Dang *et al.*, 2001b). These data indicate for sure the potential benefit of this kind of combination strategies and merits therefore further investigation.

Use of (proteolytic) strains with optimal tumour colonization capacities

As already mentioned, the lack of satisfactory anti-tumour effects when using recombinant saccharolytic strains may be explained by insufficient levels of therapeutics due to low number of recombinant clostridia at the tumour site. Strikingly, upon systemic administration of spores, colonization levels of the saccharolytic clostridia were shown to be 1000-fold lower compared to proteolytic strains such as *C. sporogenes* (Lambin *et al.*, 1998). The superior tumour colonization efficiency of the latter is probably a consequence of its proteolytic character. On the other hand, the superior proteolytic tumour colonizing *Clostridium* strains have been refractory for a very long time to genetic engineering. A protocol for their transformation was described in 2002 (Liu *et al.*, 2002), but the efficiency has been shown to be extremely low ($1–2 \times 10^2$ transformants per μg of plasmid DNA) and the method has proven not to be reproducible in other labs. Fortunately, a recent gene transfer protocol based on conjugation allows the construction of these superior recombinant strains. Using this methodology, it became possible to transfer plasmid vectors at frequencies of 1.0×10^{-5} transconjugants per donor cell. This can be considered one of the major breakthroughs of the research involving recombinant clostridia, as it is now possible to use *Clostridium* strains with the highest tumour colonization and thus the highest therapeutic gene expression levels.

The first experimental evidence of the improved anti-tumour effects when using strains with high colonization capacities came from studies with CD-recombinant *C. sporogenes* NCIMB10696 as a host. Injection of recombinant spores to tumour-bearing mice resulted in tumour-specific expression of CD (Liu *et al.*, 2002). More importantly, a significant tumour growth delay was achieved when combined with systemic administration of 5-FC. This anti-tumour effect of CD recombinant clostridia/5-FC combination following a single intravenous spore injection was even greater than that produced by maximum tolerated dose of 5-FU given by the same schedule.

More recently, we have tested NTR-recombinant *C. sporogenes* for its antitumour activity following systemic injection of spores in combination with CB1954 prodrug administration. Because in a clinical setting anticancer agents are typically given in multiple cycles to allow patients to recover, it was investigated whether anti-tumour effects could be achieved in multiple treatment cycles (Theys *et al.*, 2006). Our results demonstrated that significant tumour growth delay could be obtained. Intriguingly, repeated treatment cycles of NTR-recombinant *C. sporogenes* resulted in prolonged anti-tumour efficacy.

Similar to strategies used with saccharolytic strains, the benefit of combining enzyme–prodrug therapy using engineered *C. sporogenes* with vascular targeting has been shown. In these experiments, NTR- or CD-expressing recombinant C. *sporogenes* was combined with administration of the vascular-targeting agent DMXAA (5,6-dimethylxantheone-4-acetic acid), 4 hours after spore injection. Administration of DMXAA was estimated to increase tumour colonization fourfold, and complete tumour regression was achieved with both murine SCCVII and human HT29 carcinomas transplanted subcutaneously into C3H and nude mice, respectively (referred in to (Minton, 2003)).

Next to C. *sporogenes*, one of the organisms that have emerged as a potential candidate for anti-cancer treatment is *Clostridium novyi*. C. *novyi* is a pathogenic anaerobic spore-forming strain, which secretes a large number of virulence factors, the most potent of which is α toxin, carried on a prophage. Vogelstein *et al.* cured C. *novyi* of the prophage, resulting in a non-toxin-producing strain attenuated in virulence. This strain was therefore named C. *novyi*-NT ('non-toxic'). Systemic administration of C. *novyi*-NT spores resulted in extensive spreading of vegetative cells throughout the necrotic tumour regions. Similar to the early observations initially made with C. *butyricum* M55, intravenous injection of C. *novyi*-NT spores led to enlargement of the necrotic regions and to tumour growth delay. Despite the observed cytolysis and tumour growth retardation, however, C. *novyi-NT* treatment did also not result in total tumour control. In order to increase the therapeutic effect of C. *novyi*-NT, Dang *et al.* combined the bacteriolytic therapy with low molecular weight conventional chemotherapy (mitomycin C and cytoxan) in a so-called combination bacteriolytic therapy (COBALT) (Dang *et al.*, 2001a). The rationale behind COBALT was that the bacteria would lyse the tumours from the inside out, and low molecular weight chemotherapeutic agents would attack cancer cells in the well-perfused, non-necrotic region. To enhance the effect of the chemotherapy agents and bacteria, the vascular targeting agent dolastatin (D-10), was added in these combination experiments. The benefit of this addition to COBALT is that (1) vascular stasis increases the extent of hypoxia thereby increasing the size of the region affected by C. *novyi-NT* and (2) that vascular shutdown results in trapping of extravagated molecules in tumours, thereby enhancing exposure to therapeutic agents in combination therapy. COBALT therapy did produce impressive results with two different tumour cell lines grown subcutaneously in mice. It should, however, also be noted that the observations made were tumour-type dependent and that some combinations led to severe toxicity as a consequence of the so-called 'tumour lysis' syndrome (Dang *et al.*, 2004; Dang *et al.*, 2001a).

The low level of oxygenation within tumours is a major cause of radiation treatment failures. To overcome tumour hypoxia related resistance to radiation therapy, combination with *Clostridium novyi*-NT treatment appeared feasible since these intratumoral multiplying clostridia eliminated those tumour cells that are most resistant to radiation. The combination of C. *novyi*–NT and two radiotherapy treatment modalities has also been investigated. Treatment of athymic nu/nu mice bearing HCT116 xenographs with C. *novyi-NT* spores and fractionated RT (2 Gy/day for 5 days) resulted in significant tumour shrinkage as compared to either treatments alone or to control tumours (Bettegowda *et al.*, 2003). However, this combination therapy was tumour type specific and, in all cases, the small number of residual cells that remained subsequently proliferated causing tumour recurrence. C. *novyi-NT* was also combined with brachytherapy as it can deliver higher doses of radiation in a more specific and less toxic fashion. Using this setting, complete cure of two tumour types was achieved (Bettegowda *et al.*, 2003). The mechanism through which these bacteria enhance the therapeutic effect is that C. *novyi-NT* targets those components of tumours that are least sensitive to radiation, because they are poorly oxygenated. Recent experiments have suggested that damage to microvascular endothelial cells might be an important component of the radiation effects. Such microvascular damage is predicted to increase the niche for C. *novyi-NT* growth by creating more hypoxic areas within tumours, thereby increasing tumour colonization.

Because the mechanisms through which radiation sensitizers and C. *novyi-NT* function are distinct, it is also possible that radiation therapy could be further improved by combining *C. novyi-NT* with one of the several radiosensitizing agents. Interestingly, the efficacy of the combination treatment was shown to be independent of the tumour volumes suggesting that a wide range of tumour volumes may be treated with this strategy. In traditional radiation therapy tumour sensitivity is inversely correlated to tumour size where large tumour volumes are associated with decreased tumour oxygen tension, providing a basis for radio resistance. This decrease in oxygen tension, however, allows for greater germination of *C. novyi-NT.* Accordingly, responses were observed in tumours of diverse sizes, with complete regressions and cures observed in tumours ranging from 50 to 1000 mm^3 in volume (Bettegowda *et al.*, 2003). Furthermore, the efficacy of combined bacteriolytic radiation therapy was independent of the immune status as both syngeneic mouse B16 melanoma and xenogeneic human tumour cell lines with immunocompromised nude mice showed significantly improved results of radiotherapy through selective clostridial destruction of hypoxic tumour districts (Bettegowda *et al.*, 2003). This in line with our observations in rats, showing that treatments involving clostridia are well tolerated and do not seem to elicit an immune response (Theys *et al.*, 2001b).

The genome sequence has recently been determined and exploited for transcriptome analysis (Bettegowda *et al.*, 2006; Dang *et al.*, 2001a). The genome of *C. novyi* NT is smaller in size and coding capacity than those of the four previously sequenced *Clostridium* species. The genome analysis also revealed that *C. novyi* NT has significant cytolytic capability, as manifested by the presence of several genes encoding secreted proteins and lipid-degrading enzymes. Interestingly, transcriptome analysis revealed that these extracellular enzymes are highly expressed in *C. novyi* NT-infected tumours. This is of particular relevance to anti-tumour therapeutics because these extracellular enzymes might contribute to the destruction of the tumour. Furthermore, one of these extracellular enzymes, phospholipase C, is known to trigger host inflammatory responses and induce anti-tumour immunity. The cytolytic effects of the proteases and lipases, coupled with the phospholipase-C-induced host inflammatory response, could contribute to the destruction of the tumour tissue (Cheong *et al.*, 2006). Indeed, although the mechanisms of the elicited anti-tumour effects by C. novyi-NT are not yet fully understood, it could be partially explained as follows. Once germinated, the bacteria destroy adjacent cancer cells through the secretion of lipases, proteases, and other degradative enzymes. At the same time, the host reacts to this localized infection, producing cytokines such as IL-6, MIP-2, G-CSF andTIMP-1 that attract a massive influx of inflammatory cells, initiated largely by neutrophils and followed within a few days by monocyte and lymphocyte infiltration (Agrawal *et al.*, 2004). The inflammation may also directly contribute to the destruction of tumour cells through the production of reactive oxygen species, proteases, and other degradative enzymes. Moreover, it stimulates a potent cellular immune response that can subsequently destroy residual tumour cells not lysed by the bacteria. It was particularly notable that the induced immune response, when combined with the bacteriolytic effects of *C. novyi-NT*, could eradicate large established tumours.

A recent study from the lab of Dr. Vogelstein also provides a new insight on tumour-targeted drugs (Cheong *et al.*, 2006). In this study the authors use *C. novyi*-NT to release liposomal drugs within the tumour. Since *C. novyi*-NT lyses red blood cells, Cheong *et al.* hypothesized that its membrane-disrupting properties could be exploited to enhance the release of liposome-encapsulated drugs within tumours and used a form of doxorubicin that is incorporated within a liposome (Doxil). These liposomes can gradually penetrate the fenestrated vascular endothelium found in some types of tumours and preferentially accumulate in the tumour. Using two tumour models, it was shown that systemic injection of *C. novyi-NT* spores in combination with Doxil resulted in complete eradication of the tumours and enhanced survival as compared with treatment with either agent alone or with free doxorubicin. This finding opens a novel possibility for cancer therapy as many types of drugs can be incorporated in liposomes and their preferential release within solid tumours would constitute a powerful approach for drug targeting.

The authors suspected that the enhanced antitumour effect of Doxil might be due to secretion of a lipase. Interestingly, analysis of protein fractions from *C. novyi-NT* culture medium revealed a single band with strong lipase activity. Subsequent Mass spectrometry analysis identified the lipase as the protein product of the *NT01CX2047* gene of the *C. novyi*-NT.

Is there a future for *Clostridium*-mediated anti-cancer therapy in a clinical setting?

Despite significant progresses achieved so far, there are still several issues that need to be addressed before prokaryote-based therapy becomes a standard practice in the clinic.

The vast majority of preclinical studies with clostridial vectors were conducted in only a limited number of murine experimental models. Consequently, detailed toxicological, pharmacologic and pharmacokinetic information with clostridial vectors in a wider range of animal models is still missing and extensive pharmaceutical studies need to be put in place before a clostridial vector can be routinely used in a clinical setting. One of the most important outstanding issues for recombinant *Clostridium* vectors is undoubtedly the development of bacterial strains that carry no antibiotic resistance genes along with the therapeutic gene. Furthermore, the development of a clinically relevant delivery vector requires the construction of a strain in which the introduced genes will be integrated into the *Clostridium* chromosome so that consistent gene expression is maintained over time and to prevent the occurrence of horizontal gene transfer. Recent advances in the field of clostridial engineering have suggested that these challenging issues can be resolved in the near future. The latter tasks are not relevant for wild-type *C. novyi*-NT, at least as long as this strain does not carry a therapeutic gene. Most importantly, since use of *C. novyi*-NT has been shown to induce the so-called 'tumour lysis' syndrome, a means to counteract this issue must be achieved. This might be a general concern when using proteolytic strains: although they are far more effective in tumour colonization, their more aggressive properties might induce toxicity problems. There is no doubt that these issues need to be solved before clinical trials with cancer patients can be initiated.

Although for therapeutic efficacy of clostridial vectors, the nature of the best therapeutic gene can be argued, the levels of biologically active compound at the tumour site should be as high as possible. Optimizing the production of the therapeutic agent may require interference of the protein production process at different levels (transcription, translation). In addition, benefit may be anticipated from the use of more effective enzyme/prodrug combinations, more soluble prodrug derivatives and enzymes with improved catalytic properties. Whether secretion of the therapeutic compound from the prokaryotic cell is advantageous, remains unclear. Although more extensive spreading of the compound within the tumour consequently increasing the therapeutic effect is expected, an increase in the likelihood of therapeutic protein leakage out of the tumour, and subsequent specific immune response against the protein may be seen.

Concluding remarks

Notwithstanding the lessons that yet have to be learned and despite some barriers that need to be torn down, the application of live prokaryote vectors hold considerable promise for treating solid tumours. Recent progress made in the field of genetic engineering of *Clostridium* will certainly allow the required progress to develop a relevant vector for clinical testing. The arguments nonetheless remain as to which strain would be the best to use, which therapeutic gene should be integrated or which treatment combination would be most preferred.

Currently, the most pressing need to establish the potential of prokaryote-mediated therapy in humans is for well-designed clinical trials. These critical evaluations will facilitate the transfer of bacterial-based treatment to routine clinical practice by relieving the fears and concerns that some clinicians may have to deliberately infect patients with live bacteria.

References

Agrawal, N., Bettegowda, C., Cheong, I., Geschwind, J.F., Drake, C.G., Hipkiss, E.L., Tatsumi, M., Dang, L.H., Diaz, L.A., Jr., and Pomper, M. (2004). Bacteriolytic therapy can generate a potent immune response against experimental tumors. Proc. Natl. Acad. Sci. USA *101*, 15172–15177.

Ahn, G.O., and Brown, M. (2007). Targeting tumors with hypoxia-activated cytotoxins. Front. Biosci. *12*, 3483–3501.

Bagshawe, K.D. (1987). Antibody directed enzymes revive anti-cancer prodrugs concept. Br. J. Cancer 56, 531–532.

Barbe, S., Van Mellaert, L., Theys, J., Geukens, N., Lammertyn, E., Lambin, P., and Anne, J. (2005). Secretory production of biologically active rat interleukin-2 by Clostridium acetobutylicum DSM792 as a tool for anti-tumor treatment. FEMS Microbiol. Lett. *246*, 67–73.

Bettegowda, C., Dang, L.H., Abrams, R., Huso, D.L., Dillehay, L., Cheong, I., Agrawal, N., Borzillary, S., McCaffery, J.M., Watson, E. L. (2003). Overcoming the hypoxic barrier to radiation therapy with anaerobic bacteria. Proc. Natl. Acad. Sci. USA *100*, 15083–15088.

Bettegowda, C., Huang, X., Lin, J., Cheong, I., Kohli, M., Szabo, S. A., Zhang, X., Diaz, L. A., Jr., Velculescu, V. E., Parmigiani, G. (2006). The genome and transcriptomes of the anti-tumor agent *Clostridium novyi*-NT. Nat. Biotechnol. *24*, 1573–1580.

Brekken, R.A., Li, C., and Kumar, S. (2002). Strategies for vascular targeting in tumors. Int. J. Cancer *100*, 123–130.

Bridgewater, J.A., Knox, R.J., Pitts, J.D., Collins, M.K., and Springer, C.J. (1997). The bystander effect of the nitroreductase/CB1954 enzyme/prodrug system is due to a cell-permeable metabolite. Hum. Gene Ther. *8*, 709–717.

Bridgewater, J.A., Springer, C.J., Knox, R.J., Minton, N.P., Michael, N.P., and Collins, M. K. (1995). Expression of the bacterial nitroreductase enzyme in mammalian cells renders them selectively sensitive to killing by the prodrug CB1954. Eur. J. Cancer *31A*, 2362–2370.

Brown, J.M. (2000). Exploiting the hypoxic cancer cell: mechanisms and therapeutic strategies. Mol. Med. Today 6, 157–162.

Brown, J.M., and Wilson, W.R. (2004). Exploiting tumor hypoxia in cancer treatment. Nat. Rev. Cancer 4, 437–447.

Carey, R.W., Holland, J.F., Whang, H.Y., Neter, E., and Bryant, B. (1967). Clostridial oncolysis in man. Eur. J. Cancer 3, 37–46.

Cheong, I., Huang, X., Bettegowda, C., Diaz, L. A., Jr., Kinzler, K.W., Zhou, S., and Vogelstein, B. (2006). A bacterial protein enhances the release and efficacy of liposomal cancer drugs. Science *314*, 1308–1311.

Chung-Faye, G., Palmer, D., Anderson, D., Clark, J., Downes, M., Baddeley, J., Hussain, S., Murray, P.I., Searle, P., Seymour,L. (2001). Virus-directed, enzyme prodrug therapy with nitroimidazole reductase: a phase I and pharmacokinetic study of its prodrug, CB1954. Clin. Cancer Res. 7, 2662–2668.

Connell, H. (1935). The study and treatment of cancer by proteolytic enzymes. A preliminary report. Can. Med. Ass J 33, 364–370.

Dang, L.H., Bettegowda, C., Agrawal, N., Cheong, I., Huso, D., Frost, P., Loganzo, F., Greenberger, L., Barkoczy, J., Pettit, G.R. (2004). Targeting vascular and avascular compartments of tumors with C. *novyi*-NT and anti-microtubule agents. Cancer Biol. Ther. 3, 326–337.

Dang, L.H., Bettegowda, C., Huso, D.L., Kinzler, K.W., and Vogelstein, B. (2001a). Combination bacteriolytic therapy for the treatment of experimental tumors. Proc. Natl. Acad. Sci. USA *98*, 15155–15160.

Dang, L.H., Bettegowda, C., Huso, D.L., Kinzler, K.W., and Vogelstein, B. (2001b). Combination bacteriolytic therapy for the treatment of experimental tumors. Proc. Natl. Acad. Sci. USA *98*, 15155–15160.

Djeha, A.H., Hulme, A., Dexter, M.T., Mountain, A., Young, L.S., Searle, P.F., Kerr, D.J., and Wrighton, C.J. (2000). Expression of Escherichia coli B nitroreductase in established human tumor xenografts in mice results in potent antitumoral and bystander effects upon systemic administration of the prodrug CB1954. Cancer Gene Ther *7*, 721–731.

Dreher, M.R., Liu, W., Michelich, C.R., Dewhirst, M.W., Yuan, F., and Chilkoti, A. (2006). Tumor vascular permeability, accumulation, and penetration of macromolecular drug carriers. J. Natl. Cancer Inst. *98*, 335–344.

Durand, R.E. (1994). The influence of microenvironmental factors during cancer therapy. In Vivo *8*, 691–702.

Engelbart, K., and Gericke, D. (1964). Oncolysis by Clostridia. V. Transplanted tumors of the hamster. Cancer Res. *24*, 239–242.

Fabricius, E.M., Schneeweiss, U., and Schmidt, W. (1987). Methodological aspects of a serodiagnostic *Clostridium* tumor test – experience with spontaneous canine tumors. Zentralbl. Bakteriol. Mikrobiol. Hyg. [A] *265*, 99–112.

Fox, M.E., Lemmon, M.J., Mauchline, M.L., Davis, T.O., Giaccia, A.J., Minton, N.P., and Brown, J.M. (1996). Anaerobic bacteria as a delivery system for cancer gene therapy: *in vitro* activation of 5-fluorocytosine by genetically engineered clostridia. Gene Ther. *3*, 173–178.

Ge, K., Xu, L., Zheng, Z., Xu, D., Sun, L., and Liu, X. (1997). Transduction of cytosine deaminase gene makes rat glioma cells highly sensitive to 5-fluorocytosine. Int. J. Cancer *71*, 675–679.

Gericke, D., Dietzel, F., and Ruster, I. (1979). Further progress with oncolysis due to local high frequency hyperthermia, local x-irradiation and apathogenic clostridia. J. Microw. Power *14*, 163–166.

Graeber, T.G., Osmanian, C., Jacks, T., Housman, D.E., Koch, C.J., Lowe, S.W., and Giaccia, A. J. (1996). Hypoxia-mediated selection of cells with diminished apoptotic potential in solid tumors. Nature *379*, 88–91.

Greco, O., and Dachs, G.U. (2001). Gene directed enzyme/prodrug therapy of cancer: historical appraisal and future prospectives. J. Cell. Physiol. *187*, 22–36.

Greco, O., Patterson, A.V., and Dachs, G.U. (2000). Can gene therapy overcome the problem of hypoxia in radiotherapy? J. Radiat. Res. (Tokyo) *41*, 201–212.

Harris, A.L. (2002). Hypoxia – a key regulatory factor in tumor growth. Nat Rev Cancer 2, 38–47.

Henke, M., and Guttenberger, R. (2000). Erythropoietin in radiation oncology – A review. 1st international conference, Freiburg, June 11–12, 1999. Oncology 58, 175–182.

Heppner, F., and Mose, J.R. (1978). The liquefaction (oncolysis) of malignant gliomas by a non pathogenic *Clostridium*. Acta Neurochir. 42, 123–125.

Hicks, K.O., Myint, H., Patterson, A.V., Pruijn, F.B., Siim, B.G., Patel, K., and Wilson, W.R. (2007). Oxygen dependence and extravascular transport of hypoxia-activated prodrugs: comparison of the dinitrobenzamide mustard PR-104A and tirapazamine. Int. J. Radiat. Oncol. Biol. Phys. 69, 560–571.

Hockel, M., Schlenger, K., Hockel, S., and Vaupel, P. (1999). Hypoxic cervical cancers with low apoptotic index are highly aggressive. Cancer Res. 59, 4525–4528.

Ichikawa, T., Tamiya, T., Adachi, Y., Ono, Y., Matsumoto, K., Furuta, T., Yoshida, Y., Hamada, H., and Ohmoto, T. (2000). In vivo efficacy and toxicity of 5-fluorocytosine/cytosine deaminase gene therapy for malignant gliomas mediated by adenovirus. Cancer Gene Ther. 7, 74–82.

Jaakkola, P., Mole, D.R., Tian, Y.M., Wilson, M.I., Gielbert, J., Gaskell, S.J., Kriegsheim, A., Hebestreit, H.F., Mukherji, M., Schofield, C.J. (2001). Targeting of HIF-alpha to the von Hippel-Lindau ubiquitylation complex by O2-regulated prolyl hydroxylation. Science 292, 468–472.

Janssen, H.L., Haustermans, K.M., Balm, A.J., and Begg, A.C. (2005). Hypoxia in head and neck cancer: how much, how important? Head Neck 27, 622–638.

Kaanders, J.H., Bussink, J., and van der Kogel, A.J. (2002). ARCON: a novel biology-based approach in radiotherapy. Lancet Oncol. 3, 728–737.

Kerr, D.J., Young, L.S., Searle, P.F., and McNeish, I.A. (1997). Gene directed enzyme prodrug therapy for cancer. Adv. Drug Deliv. Rev. 26, 173–184.

Kizaka-Kondoh, S., Inoue, M., Harada, H., and Hiraoka, M. (2003). Tumor hypoxia: A target for selective cancer therapy. Cancer Sci. 94, 1021–1028.

Knox, R.J., Friedlos, F., Jarman, M., and Roberts, J.J. (1988). A new cytotoxic, DNA interstrand crosslinking agent, 5-(aziridin-1-yl)-4-hydroxylamino-2-nitrobenzamide, is formed from 5-(aziridin-1-yl)-2,4-dinitrobenzamide (CB 1954) by a nitroreductase enzyme in Walker carcinoma cells. Biochem. Pharmacol. 37, 4661–4669.

Koch, C.J. (1993). Unusual oxygen concentration dependence of toxicity of SR-4233, a hypoxic cell toxin. Cancer Res. 53, 3992–3997.

Lambin, P., Nuyts, S., Landuyt, W., Theys, J., De Bruijn, E., Anne, J., Van Mellaert, L., and Fowler, J. (2000). The potential therapeutic gain of radiation-associated gene therapy with the suicide gene cytosine deaminase. Int J Radiat Biol 76, 285–293.

Lambin, P., Theys, J., Landuyt, W., Rijken, P., van der Kogel, A., van der Schueren, E., Hodgkiss, R., Fowler, J., Nuyts, S., de Bruijn, E. (1998). Colonisation of *Clostridium* in the body is restricted to hypoxic and necrotic areas of tumors. Anaerobe 4, 183–188.

Landuyt, W., Verdoes, O., Darius, D. O., Drijkoningen, M., Nuyts, S., Theys, J., Stockx, L., Wynendaele, W., Fowler, J.F., Maleux, G. (2000). Vascular targeting of solid tumors: a major 'inverse' volume-response relationship following combretastatin A-4 phosphate treatment of rat rhabdomyosarcomas. Eur. J. Cancer 36, 1833–1843.

Lemmon, M.J., Elwell, J.H., Brehm, J.K., Mauchline, M.L., Minton, N.P., Giaccia, A.J., and Brown, J. M. (1994). Anaerobic bacteria as a gene delivery system to tumors. Proc Am Assoc Cancer Rese. 35, 374.

Liu, S.C., Minton, N.P., Giaccia, A.J., and Brown, J.M. (2002). Anticancer efficacy of systemically delivered anaerobic bacteria as gene therapy vectors targeting tumor hypoxia/necrosis. Gene Ther. 9, 291–296.

Malmgren, R.A., and Flanigan, C.C. (1955). Localization of the vegetative form of *Clostridium tetani* in mouse tumors following intravenous spore administration. Cancer Res 15, 473–478.

Mengesha, A., Dubois, L., Chiu, R.K., Paesmans, K., Wouters, B.G., Lambin, P., and Theys, J. (2007). Potential and limitations of bacterial-mediated cancer therapy. Front Biosc.i 12, 3880–3891.

Minton, N.P. (2003). Clostridia in cancer therapy. Nat. Rev. Microbiol. 1, 237–242.

Minton, N.P., Mauchline, M.L., Lemmon, M.J., Brehm, J.K., Fox, M., Michael, N.P., Giaccia, A., and Brown, J.M. (1995a). Chemotherapeutic tumor targeting using clostridial spores. FEMS Microbiol. Rev. 17, 357–364.

Moehler, M., Dimitrakopoulou-Strauss, A., Gutzler, F., Raeth, U., Strauss, L.G., and Stremmel, W. (1998). ^{18}F-labeled fluorouracil positron emission tomography and the prognoses of colorectal carcinoma patients with metastases to the liver treated with 5-fluorouracil. Cancer 83, 245–253.

Mohr, U., Hondius Boldingh, W., Emminger, A., and Behagel, H.A. (1972). Oncolysis by a new strain of Clostridium. Cancer Res. 32, 1122–1128.

Mose, J., and Mose, G. (1964). Oncolysis by Clostridia. I. Activity of *Clostridia butyricum* (M-55) and other nonpathogenic Clostridia against the Ehrlich carcinoma. Cancer Res. 24, 212–216.

Mose, J.R. (1979). [Experiments to improve the oncolysis – effect of clostridial-strain M55 (author's transl)]. Zentralbl. Bakteriol. [Orig A] 244, 541–545.

Nordsmark, M., Alsner, J., Keller, J., Nielsen, O.S., Jensen, O.M., Horsman, M.R., and Overgaard, J. (2001). Hypoxia in human soft tissue sarcomas: adverse impact on survival and no association with p53 mutations. Br. J. Cancer 84, 1070–1075.

Nordsmark, M., Loncaster, J., Aquino-Parsons, C., Chou, S.C., Gebski, V., West, C., Lindegaard, J.C., Havsteen, H., Davidson, S.E., and Hunter, R. (2006). The prognostic value of pimonidazole and tumor pO2 in human cervix carcinomas after radiation therapy: a prospective international multi-center study. Radiother. Oncol. 80, 123–131.

Parker, R., Plummber, H., Siebenmann, C., and Chapman, M. (1947). Effect of histolyticus infection and toxin on transplantable mouse tumors. Proc. Soc. Exp. Biol. Med. 66, 461–465.

Patterson, A.V., Ferry, D.M., Edmunds, S.J., Gu, Y., Singleton, R.S., Patel, K., Pullen, S.M., Hicks, K. O., Syddall, S.P., Atwell, G.J. (2007). Mechanism of action and preclinical antitumor activity of the novel hypoxia-activated DNA cross-linking agent PR-104. Clin. Cancer Res. *13*, 3922–3932.

Patterson, A.V., Saunders, M.P., and Greco, O. (2003). Prodrugs in genetic chemoradiotherapy. Curr. Pharm. Des. *9*, 2131–2154.

Pawelek, J.M., Low, K.B., and Bermudes, D. (2003). Bacteria as tumor-targeting vectors. Lancet Oncol. *4*, 548–556.

Rischin, D., Peters, L., Fisher, R., Macann, A., Denham, J., Poulsen, M., Jackson, M., Kenny, L., Penniment, M., Corry, J. (2005). Tirapazamine, cisplatin, and radiation versus fluorouracil, cisplatin, and radiation in patients with locally advanced head and neck cancer: a randomized phase II trial of the Trans-Tasman Radiation Oncology Group (TROG 98.02). J. Clin. Oncol. *23*, 79–87.

Ryan, R.M., Green, J., and Lewis, C.E. (2006). Use of bacteria in anti-cancer therapies. Bioessays *28*, 84–94.

Schlechte, H., and Elbe, B. (1988). Recombinant plasmid DNA variation of *Clostridium oncolyticum*--model experiments of cancerostatic gene transfer. Zentralbl. Bakteriol. Mikrobiol. Hyg. [A] *268*, 347–356.

Schlechte, H., Schwabe, K., Mehnert, W.H., Schulze, B., and Brauniger, H. (1982). Chemotherapy for tumors using clostridial oncolysis, antibiotics and cyclophosphamide: model trial on the UVT 15264 tumor. Arch. Geschwulstforsch. *52*, 41–48.

Semenza, G.L. (2001). HIF-1, O(2), and the 3 PHDs: how animal cells signal hypoxia to the nucleus. Cell *107*, 1–3.

Seth, P. (2005). Vector-Mediated Cancer Gene Therapy: An Overview. Cancer Biol Ther *4*.

Theys, J., Barbe, S., Landuyt, W., Nuyts, S., Van Mellaert, L., Wouters, B., Anne, J., and Lambin, P. (2003). Tumor-specific gene delivery using genetically engineered bacteria. Curr. Gene Ther. *3*, 207–221.

Theys, J., Landuyt, W., Nuyts, S., Van Mellaert, L., Bosmans, E., Rijnders, A., Van Den Bogaert, W., van Oosterom, A., Anne, J., and Lambin, P. (2001a). Improvement of *Clostridium* tumor targeting vectors evaluated in rat rhabdomyosarcomas. FEMS Immunol. Med. Microbiol. *30*, 37–41.

Theys, J., Landuyt, W., Nuyts, S., Van Mellaert, L., van Oosterom, A., Lambin, P., and Anne, J. (2001b). Specific targeting of cytosine deaminase to solid tumors by engineered *Clostridium acetobutylicum*. Cancer Gene Ther *8*, 294–297.

Theys, J., Landuyt, W., Nuyts, S., Van Mellaert, L., Lambin, P. and Anne, J. (2001). Anaerobic bacteria as a tumor-specific delivery system of therapeutic proteins. Cancer Detect. Prevent. *25*, 548–557.

Theys, J., Nuyts, S., Landuyt, W., Van Mellaert, L., Dillen, C., Bohringer, M., Durre, P., Lambin, P., and Anne, J. (1999). Stable *Escherichia coli–Clostridium acetobutylicum* shuttle vector for secretion of murine tumor necrosis factor alpha. Appl. Environ. Microbiol. *65*, 4295–4300.

Theys, J., Pennington, O., Dubois, L., Anlezark, G., Vaughan, T., Mengesha, A., Landuyt, W., Anne, J., Burke, P.J., Durre, P. (2006). Repeated cycles of Clostridium-directed enzyme prodrug therapy result in sustained antitumor effects *in vivo*. Br. J. Cancer 95, 1212–1219.

Thiele, E.H., Arison, R.N., and Boxer, G.E. (1964a). Oncolysis by clostridia. III. Effects of clostridia and chemotherapeutic agents on rodent tumors. Cancer Res *24*, 222–233.

Torrey, J. C., Kahn, M.C. (1927). The treatment of Flexner–Jobling rat carcinomas with bacterial proteolytic ferments. J. Cancer Re.s *11*, 334–376.

Van Mellaert, L., Barbe, S., and Anne, J. (2006). Clostridium spores as anti-tumor agents. Trends Microbiol. *14*, 190–196.

Vaupel, P., and Mayer, A. (2007). Hypoxia in cancer: significance and impact on clinical outcome. Cancer Metastasis Rev. *26*, 225–239.

Waehler, R., Russell, S.J., and Curiel, D.T. (2007). Engineering targeted viral vectors for gene therapy. Nat. Rev. Genet *8*, 573–587.

Wittmann, W., Fabricius, E.M., Schneeweiss, U., Schaepe, C., Benedix, A., Weissbrich, C., and Schwanbeck, U. (1990). Application of microbiological cancer test to cattle infected with bovine leucosis virus. Arch. Exp. Veterinarmed. *44*, 205–212.

Wolff, J., Lewis, D.L., Herweijer, H., Hegge, J., and Hagstrom, J. (2005). Non-viral approaches for gene transfer. Acta Myol. *24*, 202–208.

Metabolic Networks in *Clostridium acetobutylicum*: Interaction of Sporulation, Solventogenesis and Toxin Formation

12

Peter Dürre

Abstract

Clostridia are one of the few bacterial genera able to undergo cell differentiation. They can either grow vegetatively or form endospores, the most resistant survival form of all living organisms. Some species, e.g. *Clostridium acetobutylicum*, link the metabolic network of sporulation to that of solventogenesis (formation of acetone and butanol). This gives them an ecological advantage by preventing toxic effects of acidic end products from the fermentation and allows them to stay longer metabolically active. In other clostridia, even toxin formation is coupled to sporulation. The key component for these links at the molecular level is the response regulator Spo0A in its phosphorylated form. In contrast to bacilli, clostridia do not possess a phosphorelay for Spo0A activation. Instead, phosphorylation is catalysed directly by still unknown kinases or by butyryl phosphate. In addition to Spo0A~P, various other regulators are required to control the different metabolic networks. Systems biology is a new approach to understand these processes and their interaction at the molecular level and to adapt them for biotechnological use.

Introduction

Systems biology is an emerging research field that aims at understanding the complex interactions of metabolic processes and regulatory networks in a living cell. In addition, the quantitative dynamics are an essential parameter that will allow construction of models and predictions of behaviour under altered conditions. The approach might be compared to understanding how a car or a radio is functioning and running, having at hand first only a list of all its parts and some pictures of its exterior and interior (Kitano, 2002; Lazebnik, 2002). Molecular biology has provided valuable tools that help in identifying the players in the game and their role by using the so-called 'omics'-technologies. Genomics provides the gene inventory of a living cell, but does not inform on use under specific conditions or redundancy. Transcriptomics is helping to fill this gap by determining the expression of genes in a given state by comparing to another strain or another condition. However, microarrays only allow determination of mRNA abundance and thus clustering of genes expressed in a specific metabolic situation, but do not provide insight into the causality of regulatory steps in a complex network. Also, posttranscriptional events cannot be detected this way. Proteomics aims at showing the set of proteins required by the cell under specific conditions. In combination with genomics (number of respective genes) it also allows detection of posttranslational reactions (more than one spot of a given protein in a two-dimensional gel). However, this powerful technique is still hampered by the inability of detection in gels of all proteins present in a cell (hydrophilic versus hydrophobic proteins, abundance, varying isoelectric points, etc.). A subdiscipline in this area refers to the detection of glycoproteins, which also are present in prokaryotes (Schmidt *et al.*, 2003; Upreti *et al.*, 2003; Hitchen and Dell, 2006). Metabolomics aims at determining all small metabolites in a cell and their concentration, and fluxomics tries to reveal the dynamic changes of substrate, products, and small metabolites. All these techniques are used by a scientific consortium, formed in 2007, to elucidate the systems biology of *Clostridium acetobutylicum* (project COSMIC within the SysMO program, www.sysmo.net).

C. acetobutylicum is a bacterium with a long-standing history in biotechnology and even politics (Dürre, 2005a; Dürre, 2007a). It was used for industrial production of the solvents acetone and butanol at a large scale until about 1950, when cheap crude oil rendered the petrochemical production more economic. However, due to the increasing oil prices and also the current climate debate, the pendulum is swinging back and the biotechnological solvent synthesis has already been reintroduced in several countries (Dürre, 2007a). Especially butanol is an important bulk product in the chemical industry and has also the potential of becoming a superior biofuel (Dürre, 2007a). The formation of acetone and butanol by *C. acetobutylicum* is a metabolic response to endangering environmental conditions (decreasing pH caused by production of acetic and butyric acids) and is tightly linked to the ultimate survival programme of sporulation (Dürre, 2005a). Meanwhile, a lot of information has been obtained regarding the regulations and interactions of the metabolic networks of solventogenesis and endospore formation. However, there is still a long way to go for a complete understanding of these processes.

Sporulation

Various survival forms are found in microorganisms. They can be grouped into cysts, exospores, and endospores (Lengeler *et al.*, 1999). All these morphological, biochemical, and metabolic changes are examples of bacterial cell differentiation. Cysts form by conversion of the whole cell into a thick-walled structure and are less resistant towards environmental stress than both exo- and endospores. The difference between the latter two is mainly their generation. Exospores are produced outside of the mother cell or by budding from it. They also lack some of the components found in endospores and are less thermoresistant. Finally, endospores are produced within the mother cell, typically starting by asymmetric septation, followed by engulfing of the forespore by the cytoplasmic membrane of the mother cell, and then formation of peptidoglycan and protein layers (Labbé, 2005). Endospores represent the most resistant cell type known, providing shelter against acids, chemical disinfectants, desiccation, heat, and radiation. In case of some *Bacillus* endospores, survival for 250 million years has been claimed (Vreeland *et al.*, 2000). In general, only one endospore is formed per cell. In some bacteria, however, even multiple (up to five) endospores are formed in a single mother cell. Such examples are *Anaerobacter polyendosporus* and *Metabacterium polyspora* (Angert *et al.*, 1996; Siunov *et al.*, 1999). The persistence of endospores also presents a serious problem to the pharmaceutical and the food industry while trying to ensure a longer shelf life of their products. Established preservation methods are addition of preservatives, decrease of pH, heating, radiation, and withdrawal of water (Lund and Peck, 2000). They might be joined in future by combining heat and pressure, high-voltage electric discharges, high-intensity laser, ultrahigh hydrostatic pressure, and UV light.

The ability to form endospores is almost completely restricted to some members of the Gram-positive bacterial phylum (Sonenshein, 2000; Dürre, 2005b). The largest genera of endospore producers are the clostridia and the bacilli. In addition to endospore formation, *Clostridium* is originally characterized by a Gram-positive-type cell wall, an anaerobic metabolism (though some species are more or less oxygen-tolerant), and the inability of dissimilatory sulphate reduction. 16S rDNA analysis proved to be a much better taxonomic parameter and more than 150 species are currently validly described (http://www.bacterio.cict.fr/c/clostridium.html, http://dx.doi.org/10.1007/bergeysoutline200310) (Dürre, 2007b). However, many species have been attributed to other and also newly formed genera, and the genus *Clostridium* still represents a major challenge for taxonomists. Evolutionary, the clostridia can be recognized as a separate class approximately 2.7 billion years ago (Battistuzzi *et al.*, 2004; Paredes *et al.*, 2005). At that time, the atmosphere was still anaerobic. Only with oxygenic photosynthesis, as invented by cyanobacteria or their respective predecessors, oxygen was released and aerobic metabolism became possible. As a consequence, the genus *Bacillus* appeared approximately 2.3 billion years ago. However, it should be kept in mind that its members are also still able to perform fermentation and anaerobic respiration. Phylogenetically, both *Clostridium* and *Bacillus* belong to the order *Firmicutes*. Thus, it is not surprising that the basic process of endospore formation is conserved in both genera. However, despite their fairly close relationship, they also exert significant differences in the regulation of the process, reflecting the need to respond in part to different environmental stimuli (Dürre, 2005b; Paredes *et al.*, 2005).

The major difference between the two genera is signal transduction at the initiation of sporulation (Fig. 12.1). Bacteria in general rely on so-called two-component systems to sense environmental changes and to adapt their metabolic responses accordingly. Typically, one of the components is a sensor kinase, which on receiving a signal changes its conformation, dephosphorylates ATP, and phosphorylates a histidine residue in the C-terminal kinase domain. From there, the phosphoryl group is transferred to an aspartate in the N-terminal receiver domain of the response regulator, the second component of the system. Phosphorylation of the aspartate causes a conformational change and allows DNA binding or interaction with proteins by the N terminus of the regulator. The initiation of endospore formation in the genus *Bacillus* is even more sophisticated. Signal transduction is mediated by a series of His to Asp phosphate-transfers. Five sensor kinases (KinA–KinE) are able to phosphorylate the first receiver Spo0F at an aspartate residue. From there, another transfer to a histidine residue in Spo0B is catalysed. Finally, the phosphate is transduced to an aspartate in Spo0A, which in its phosphorylated form serves as a transcription factor and represents the master regulator of sporulation (Perego and Hoch, 2002). This chain of transmitters and receivers has been designated phosphorelay.

Unlike *Bacillus*, the clostridia do not possess such a phosphorelay (as evident from the completed clostridial genome sequencing projects), but instead directly phosphorylate the master regulator Spo0A, which also acts as a transcription factor. Spo0A~P enables increased transcription of an alternative sigma factor, σ^H, which, in turn, together with Spo0A~P, is used for transcription of the σ^F gene in the forespore, while Spo0A~P (in combination with σ^A) induces transcription of the σ^E gene in the mother cell (Fig. 12.2). Later, a second pair

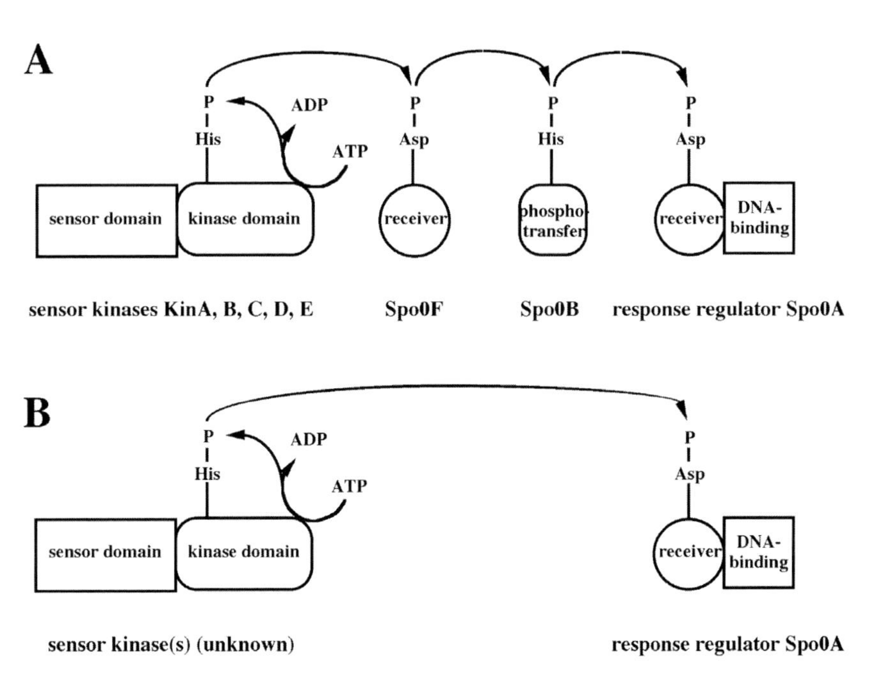

Figure 12.1 Signal transduction for initiation of endospore synthesis. (A) Signalling pathway in bacilli (phosphorelay); (B) signalling pathway in clostridia.

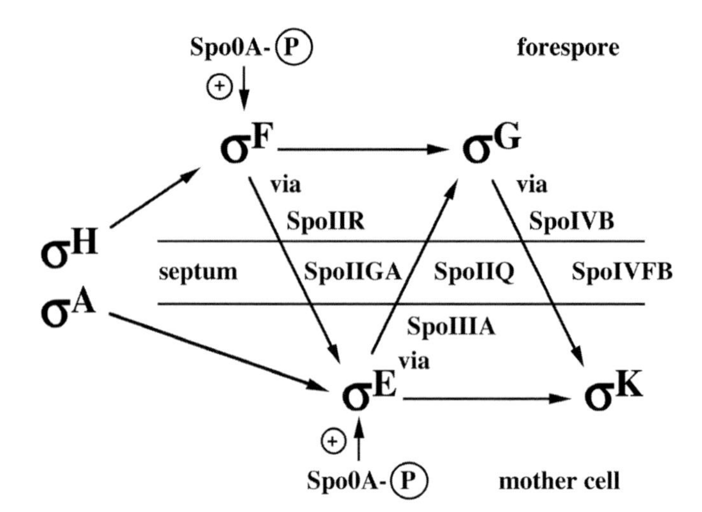

Figure 12.2 'Crisscross' regulation of sporulation by sigma factors in *Bacillus subtilis*. The corresponding genes have also been identified in the sequenced genomes of clostridia, only *spoIVFB* is present only in some *Bacillus* and some *Clostridium* species (Stragier, 2002).

of sigma factors is induced, σ^G in the forespore and σ^K in the mother cell. Regulation of activity of these proteins is achieved in the forespore by action of anti- and anti-anti-sigma factors and in the mother cell by proteolytic activation. There are also signalling pathways between the two compartments, which led to introduction of the term 'crisscross' regulation (Losick and Stragier, 1992). The sigma factor-dependent pathways are obviously conserved in the genera *Bacillus* and *Clostridium*, as demonstrated by their common presence and expression pattern (Sauer *et al.*, 1994; Wong *et al.*, 1995; Sauer *et al.*, 1995; Santangelo *et al.*, 1998; Arcuri *et al.*, 2000). Interrupted *sigK* genes can be found in both genera, but rather represent exceptions. In *B. subtilis* and *C. tetani*, a prophage-like element of 48 and 47 kbp, repectively, represents the *sigK* intervening sequence (*skin*). Orientation of the insert is identical in both cases. In *C. difficile*, the situation is different. The *skin* element is shorter (14.6 kbp), and the *sigK* gene lacks a sequence encoding an N-terminal propeptide, indicating a different activation mechanism (Haraldsen and Sonenshein, 2003). More details are provided in recent reviews (Dürre and Hollergschwandner, 2004; Dürre, 2005b). It should be noted that the formation of exospores in e.g. *Myxococcus xanthus* is regulated by a cascade of various transcription factors rather than sigma factors as in bacilli and clostridia (Kroos, 2007). There is also an extensive involvement of enhancer-binding proteins and σ^{54}-dependent promoters in *M. xanthus*.

Another difference refers to formation of storage material and thus altered morphology. Clostridia start to produce granules of slightly different polysaccharides, collectively called granulose, at the onset of sporulation (Woods and Jones, 1986). Due to this process, the cells swell and then represent the so-called clostridial stage. Under the microscope, typical phase-bright, swollen, cigar-shape cells become visible (Fig. 12.3). Granulose consists of α1,4-linked glucopyranose units, occasionally with 1,6-branching. The material serves as an endogenous carbon and energy source and is consumed during spore synthesis. While granulose is found in most clostridia, poly-β-hydroxybutyrate has been detected in *C. botulinum* and fulfills a similar role (Emeruwa and Hawirko, 1973). Bacilli, on the other hand, do not form such swollen cells prior to sporulation. Another morphological difference is the appearance of pin-like, ribbon-like, or tubular appendages of clostridial spores (Labbé, 2005). Their function has not yet been elucidated.

Solventogenesis

The formation of acetone and butanol in *C. acetobutylicum* and some other clostridia (as well as in addition 2-propanol (isopropanol) in *C. beijerinckii*) occurs during the switch from

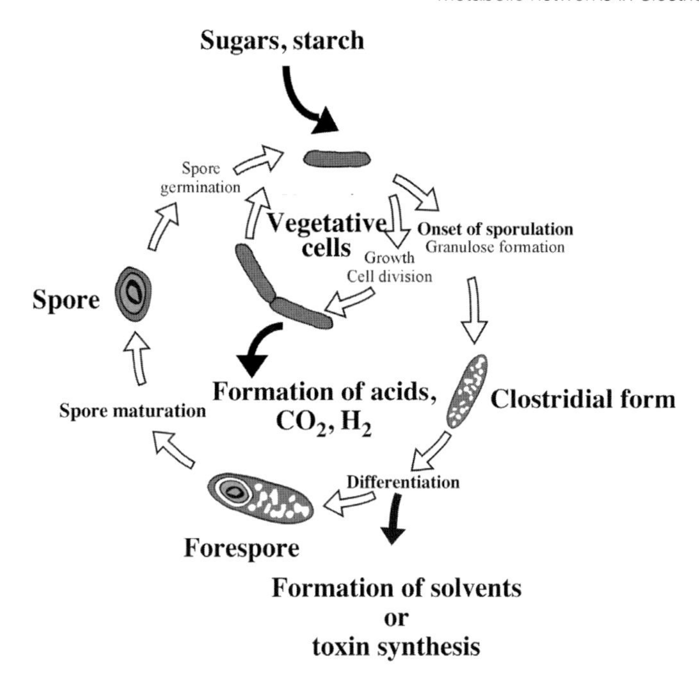

Figure 12.3 Vegetative growth and cell differentiation of clostridia. The inner circle shows vegetative growth, performing a fermentation (e.g. butyrate). The outer circle shows various stages of cell differentiation during sporulation. In parallel, solvents such as acetone and butanol or toxins such as Cpe or C2 are produced.

exponential to stationary growth phase and is coupled to the onset of sporulation (Fig. 12.3). The physiological basis for solventogenesis is the decreasing pH (due to butyrate and acetate formation during logarithmic growth), which jeopardizes cell survival. Anaerobic bacteria are unable to maintain homeostasis at a constant pH. The decreasing external value is paralleled by a decreasing internal value, which is in general 1 pH unit higher (Dürre *et al.*, 1988). When the butyrate fermentation, employed by most clostridia, produced enough acids for lowering the pH to approximately 4–4.5, the acids become undissociated and thus are able to diffuse across the cytoplasmic membrane. Owing to the more alkaline interior, they dissociate there and lead to a collapse of the transmembrane pH gradient and thereby to cell death. Conversion of butyrate and acetate into butanol and acetone allows the cell to raise the external pH and thus provides an ecological advantage over non-solventogenic competitors. However, butanol in higher concentrations is toxic as well (Moreira *et al.*, 1981; Bowles and Ellefson, 1985), and, thus, the solventogenic clostridia just buy time for a longer living and start, in parallel to solvent formation, the synthesis of endospores. This is the physiological link between the two metabolic networks (Dürre, 2005a, 2007a). It is important to note that, in contrast to bacilli, nutrient limitations are not the signal for initiation of sporulation. This might explain the lack of the phosphorelay components in clostridia.

The metabolic reactions leading to acid and solvent formation in *C. acetobutylicum* are summarized in Fig. 12.4. Acetoin, ethanol, and lactate are only formed in minor amounts or under special conditions (Dürre, 2005a). Additional major fermentation products are CO_2 and H_2. *C. acetobutylicum* has originally been isolated on starchy substrates (Weizmann, 1915), whereas many solventogenic clostridia grow better on molasses. Although all these strains had been designated *C. acetobutylicum*, it has meanwhile been shown that the worldwide isolated and used strains are grouped into 4 different species: *C. acetobutylicum*, *C. beijerinckii*, *C. saccharobutylicum*, and *C. saccharoperbutylacetonicum* (Keis *et al.*, 2001). Only few additional enzymes are required for the switch from acidogenesis to solventogenesis. Butyrate and acetate are converted into their respective coenzyme A-derivatives by action of a

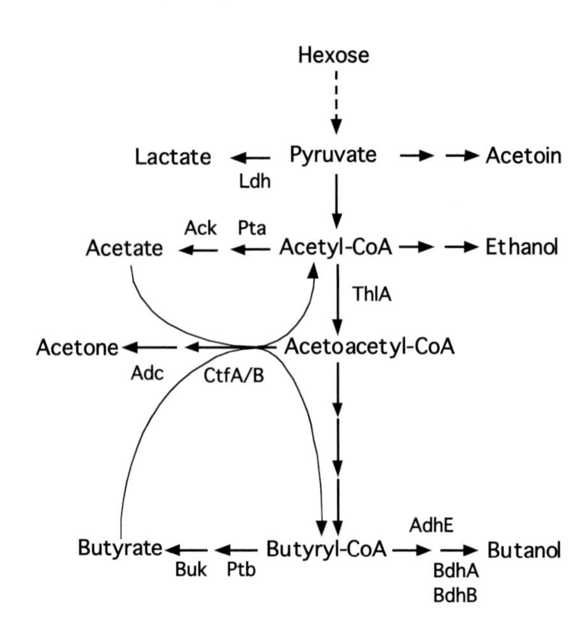

Figure 12.4 Pathways of acidogenesis and solventogenesis in *Clostridium acetobutylicum*. Ack, acetate kinase; Adc, acetoacetate decarboxylase; AdhE, bifunctional butyraldehyde/butanol dehydrogenase; BdhA, butanol dehydrogenase A; BdhB, butanol dehydrogenase B; Buk, butyrate kinase; CtfA/B, acetoacetyl-CoA:butyrate/acetate-coenzyme A transferase; Ldh, lactate dehydrogenase; Pta, phosphotransacetylase; Ptb, phosphotransbutyrylase; ThlA, thiolase.

CoA transferase (consisting of the subunits CtfA and CtfB). Acetone stems from decarboxylation of acetoacetate, catalysed by acetoacetate decarboxylase (Adc). Butanol is formed from butyryl-CoA by a bifunctional butyraldehyde/butanol dehydrogenase (AdhE, sometimes also called Aad). There are also two separate butanol dehydrogenases, BdhA and BdhB, that are able to catalyze the last step of production of this alcohol (Welch *et al.*, 1989). When reduced substrates such as glycerol are fermented, no acetone but butanol and increased amounts of ethanol are formed ('alcohologenic fermentation'). Under these conditions, a second butyraldehyde/butanol dehydrogenase becomes active (AdhE2), which is very homologous to AdhE (Fontaine *et al.*, 2002). C. *acetobutylicum* is the only organism known so far to possess two such closely related enzymes. Species producing 2-propanol, such as C. *beijerinckii*, contain additional primary/secondary alcohol dehydrogenases (Chen, 1995). A detailed summary of the various enzymes and their properties is provided in a recent review (Dürre, 2005a).

The respective genes *adc, bdhA, bdhB, adhE, ctfA/B*, and *adhE2* have all been shown to be induced at the level of transcription prior to the formation of solvents (Gerischer and Dürre, 1992; Walter *et al.*, 1992; Fischer *et al.*, 1993; Fontaine *et al.*, 2002). They are organized in five different operons. *adc, adhE2, bdhA*, and *bdhB* are monocistronic transcription units, the first two being localized on the 192-kbp megaplasmid pSOL1 and the last two on the 3.94-Mbp chromosome (Nölling *et al.*, 2001). *adhE, ctfA*, and *ctfB* form the so-called *sol* operon, which is also located on the megaplasmid, contiguous to *adc*, but with convergent direction of transcription (Fischer *et al.*, 1993). This organization in C. *acetobutylicum* is unique, as the other solventogenic clostridia (as far as known) all contain a *sol* operon combining the genes *ald* or *bld* (encoding a butyraldehyde dehydrogenase), *ctfA, ctfB*, and *adc* (Chen and Blaschek, 1999; Kosaka *et al.*, 2007). The separation of *adc* and *ctfA/B* genes allows C. *acetobutylicum* to vary the butanol/acetone ratio and thus to react to the redox state of the substrate(s) (Bahl *et al.*, 1986; Dürre *et al.*, 1995).

A large number of other genes is also induced during the onset of solventogenesis (and sporulation), which are not directly involved in these processes as far as known. Identification was made possible by generation of C. *acetobutylicum*

DNA-microarrays (reviewed in Tummala *et al.*, 2005; Paredes *et al.*, 2005). As already mentioned, butanol is toxic in higher amounts and thus exerts a massive stress onto the cells. Thus, it is not surprising that stress response genes are also induced at the onset of solventogenesis. *hsp90* seems to be the first upregulated member of this class, followed by *hrcA* and *groESL* (Tomas *et al.*, 2004; Alsaker *et al.*, 2004; Alsaker and Papoutsakis, 2005). However, this order is obviously not fixed, as in a mutant deficient in butyrate kinase the first stress gene upregulated is *hsp18*, followed by *groESL*, *clpC*, and *dnaK*. *hsp90* expression occurs significantly later. Other mutants show again a different pattern (Paredes *et al.*, 2005). It should also be noted that results from Northern blots and DNA-microarrays are sometimes contradictory (fortunately only in rare cases). An example is the order of expression of the alcohol dehydrogenase genes. Northern blots revealed that *adhE* is the first gene being induced, followed by *bdhB* (Sauer and Dürre, 1995), whereas the microarrays indicated a sequential expression of first *bdhB* (plus *adc*, *spo0A*, and *hsp90*) and then *adhE* and the other genes required for solventogenesis (Alsaker and Papoutsakis, 2005). The difference might be caused by different cultivation conditions, indicating the need for standardized procedures. Another puzzling result is that microarray experiments indicated that addition of butanol (the stress factor) leads to enhanced expression of the *sol* operon, encoding the butanol-forming genes (Tomas *et al.*, 2004; Alsaker *et al.*, 2004).

Also related to solventogenesis are genes required for motility and chemotaxis. The respective sets are downregulated close to the beginning of the stationary growth phase (Alsaker and Papoutsakis, 2005). This reflects the situation in *B. subtilis* and also indicates an essential role of glycosylation in the regulation of these gene clusters (Paredes *et al.*, 2005). Support comes from the finding that flagellin of *C. acetobutylicum* is a glycoprotein (Lyristis *et al.*, 2000) and that inactivation of the glycosylase/deglycosylase Orf5 (erroneously sometimes designated SolR, see last paragraph) has significant effects on solvent formation (Thormann and Dürre, 2001; Harris *et al.*, 2001; Thormann *et al.*, 2002, Dürre, 2005a).

An unexpected finding was the induction of an operon (*serCAXS*), encoding presumably enzymes for serine biosynthesis and incorporation into protein (serine aminotransferase, 3-phosphoglycerate dehydrogenase, unknown protein (3-phosphoserine phosphatase?), seryl-tRNA synthetase). The original detection stemmed from proteome comparison under acidogenic and solventogenic conditions and was then confirmed at the level of transcription (Schaffer *et al.*, 2002). As no proteins with high serine content are known to play a role in solventogenesis or sporulation, the physiological function of this induction still remains to be elucidated.

Sporulation-coupled toxin formation

C. acetobutylicum does not produce toxins and is therefore regarded as the model organism of apathogenic clostridia. Some species of this genus (less than 10%), however, form dangerous toxins (see also other chapters of this book; Dürre, 2007b). In several cases, toxin production is linked to sporulation, reflecting the link of solventogenesis to endospore synthesis (Fig. 12.3).

The most intensively studied example is the *C. perfringens* enterotoxin (Cpe) (McClane, 2005). It causes a common form of food-borne disease. It binds to eukaryotic claudins in the host membrane and, together with other membrane proteins, forms various complexes. The large ones interact with tight junction proteins and cause thus massive damage. The Cpe protein represents a significant portion of the bacterial cell in form of a crystal structure and is clearly visible by electron microscopy (Labbé, 2005). Synthesis only starts during sporulation and is dependent on the sporulation-specific sigma factors σ^E and σ^K, which are only active in the mother cell. The *cpe* gene does not contain a signal sequence, so the protein cannot be exported. Release is only achieved upon lysis of the mother cell (Zhao and Melville, 1998; Popoff and Stiles, 2005). One of the physiological responses of the eukaryotic host to toxin release is diarrhoea, thus spreading the concomitantly produced spores and eventually giving them the opportunity to be taken up by a new host.

Another sporulation-linked toxin is the binary toxin C2 of *C. botulinum* (Nakamura *et al.*,

1978). It comprises two proteins, a binding component C2II and the catalytic component C2I, which acts as an ADP-ribosyltransferase. C2II, after proteolytic activation to C2IIa, attaches to the host membrane and forms a complex. C2I binds to it, and the complex is taken up by the host cell by endocytosis. Acidification of the endosome leads finally to the release of C2I into the cytosol. There, it attacks G-actin monomers and prevents their polymerization (Barth and Aktories, 2005). The molecular events linking the toxin synthesis to sporulation have not yet been elucidated in detail.

Finally, the genes encoding the large clostridial cytotoxins ToxA and ToxB of *C. difficile* are expressed only weakly during exponential, but massively during stationary growth phase (Popoff and Stiles, 2005). The proteins act as glycosyltransferases and glycosylate G-proteins of the target host such as Rho, Rac, Cdc42, Rap, and Ral, thereby inactivating them. This causes a blockade of signal cascades important for cell differentiation, enzyme activation, organization of actin cytoskeleton, proliferation, and transcription (Barth and Aktories, 2005). Regulation of *toxA* and *toxB* is mediated by the alternative sigma factor TcdR (formerly known as TxeR). Similar proteins are required for the neurotoxin gene expression in *C. botulinum* and *C. tetani* (Sonenshein *et al.*, 2005). It is not yet clear, whether a link to sporulation exists.

Regulatory interactions

The key component for linking the metabolic networks of sporulation, solventogenesis, and also formation of some toxins is the phosphorylated response regulator Spo0A. The presence of its gene has meanwhile even become the essential taxonomic marker for attributing a bacterium to endospore formers. A recent review compares the features of Spo0A proteins from Gram-positive endospore formers (Stephenson and Lewis, 2005). Spo0A~P binds to a special DNA motif, the 0A box, which is conserved in bacilli and clostridia (TGNCGAA)(Ravagnani *et al.*, 2000). Then, activation of either σ^A- or σ^H-dependent promoters is possible. Despite the fact that the phosphorylated regulator forms a dimer, many Spo0A-controlled operons are preceded by only one 0A box. In *B. subtilis*, the concentration of Spo0A~P must reach a threshold in order to initiate sporulation. Some regulated operons require a low threshold for control, others a high threshold. The phosphorelay, with its additional phosphatases acting on Spo0F~P and Spo0A~P, allows a gradual increase of Spo0A level and activity, thereby mediating successive regulation of low- and high-threshold genes. This gradual increase is critical for triggering sporulation (Fujita and Losick, 2005). However, Spo0A does not only initiate sporulation. Approximately 120 genes are meanwhile known to be under its control in *B. subtilis* (Molle *et al.*, 2003). These genes are essential for cell motility, chemotaxis, competence, expression of extracellular proteases, carbohydrate-degrading enzymes, and antibiotics. Although most of the target genes are activated, some are directly or indirectly repressed (e.g. *abrB* and *sinR*) (Burkholder and Grossman, 2000). Some, but not all clostridial *spo0A* genes can complement respective mutant strains from other species. *C. acetobutylicum*- and *C. tetani*-Spo0A was found to be functional in a *C. perfringens spo0A* mutant, while the corresponding genes from *C. botulinum* and *C. difficile* did not restore sporulation (Huang and Sarker, 2006). Also, *C. botulinum spo0A* could not complement a *B. subtilis* mutant (Wörner *et al.*, 2006).

An interesting question is, how Spo0A is phosphorylated in clostridia. An alternative phosphorelay seems unlikely, leaving two potential possibilities (Paredes *et al.*, 2005). First, one or more of the orphan kinases, found in all clostridia sequenced so far, could directly phosphorylate the protein. The recently developed targeted gene knock-out technology for clostridia (Heap *et al.*, 2007) will certainly help to check out the various candidates. Time-course expression profiles indicated four orphan kinases in *C. acetobutylicum* (out of 6), possibly interacting with Spo0A (Paredes *et al.*, 2005). Second, butyryl phosphate could directly serve as the phosphate donor, similar to the role of acetyl phosphate in *Escherichia coli* and other bacteria (Paredes *et al.*, 2005). Butyryl phosphate concentrations show a strong correlation with solventogenesis (Zhao *et al.*, 2005), not only in the wild type, but also in various mutants (Paredes *et al.*, 2005).

The coupling of toxin formation to sporulation is hardly investigated. *cpe* expression is dependent on σ^E- and σ^K-controlled promoters (Zhao and Melville, 1998). Again, Spo0A is the key link, as without this regulator no sporulation-specific sigma factors can be synthesized (Fig. 12.2). Supporting data come from experiments with a *spo0A*-negative mutant, unable to produce Cpe (Huang *et al.*, 2004).

Spo0A~P is also directly involved in regulation of solventogenesis. All five operons of *C. acetobutylicum* mentioned before contain 0A boxes in the vicinity of their promoters. *sol*, *bdhB*, and *adhE2* contain a single Spo0A-binding sequence upstream of the transcription start points (in case of *sol* the 0A box is reversed), *bdhA* two, and *adc* even three 0A boxes (one of which is reversed) (Ravagnani *et al.*, 2000; Dürre, 2004). However, deletion of the binding motifs upstream of *adc* did reduce, but not completely abolish regulation (Ravagnani *et al.*, 2000; Böhringer, 2002). Also, inactivation of the *C. acetobutylicum spo0A* gene did not lead to a total loss of butanol and acetone formation (Harris *et al.*, 2002). This clearly indicates the presence of additional regulators. They also should be specific, as the ratio of acetone to butanol formation can vary depending on the substrates fermented (Bahl *et al.*, 1986; Girbal and Soucaille, 1998). A report, claiming that the *orf5* gene upstream of the *sol* operon in *C. acetobutylicum* should encode a transcriptional repressor (SolR) (Nair *et al.*, 1999) proved to be wrong. Orf5 represents a gycosylase/deglycosylase, and the effect on solventogenesis observed was due to erroneously subcloning part of the regulatory region of the *sol* operon (Thormann and Dürre, 2001; Thormann *et al.*, 2002). However, DNA affinity chromatography with the control regions of the *adc* and *sol* operons in *C. acetobutylicum* revealed specific binding of several proteins, which are currently under investigation (B. Schiel, N. Nold, and P. Dürre, unpublished). No clues are available as of yet for additional regulators of the *bdhA* and *bdhB* operons. Upstream of *adhE2*, a putative FNR binding site has been identified, which might be responsible for the redox state-dependent regulation (Fontaine *et al.*, 2002).

Using primer extension experiments, single transcription start points have been mapped upstream of *adc*, *bdhA*, and *bdhB*. They allowed to deduce typical σ^A-dependent promoters (Gerischer and Dürre, 1992; Walter *et al.*, 1992). In contrast, 2 transcription start points have been found upstream of *sol* and *adhE2* (Fischer *et al.*, 1993; Nair *et al.*, 1994; Fontaine *et al.*, 2002). For both operons, the distal promoters showed almost perfect homology to σ^A-dependent motifs, whereas the putative proximal ones only displayed very weak homology. A thorough analysis revealed that the proximal transcription start point of the *sol* operon indeed stems from an RNA processing event (Thormann *et al.*, 2002). The similarity of organization of *adhE2* renders it likely that also in this case RNA processing is an additional means of regulation. The processing site of the *sol* transcript is not found in the heterologous host *Escherichia coli* (Fischer *et al.*, 1993), indicating the presence of a specific RNase in *C. acetobutylicum*. The promoter strengths have been compared using reporter genes based on *lucB* from *Photinus pyralis* as well as *lacZ* from *Thermoanaerobacterium thermosulfurigenes* (Feustel *et al.*, 2004). The relative strengths are shown in Table 12.1. As a comparison, a typical transcription unit for acidogenesis has been included (the promoter of the *ptb-buk* operon,

Table 12.1 Promoter strength of operons required for solventogenesis (and acidogenesis) in *Clostridium acetobutylicum*

Promoter	Relative strength (%)[a]
P_{ptb}	100
P_{adc}	100
p_{bdhB}	77
p_{bdhA}	4.5
p_{sol}	1.2

[a]Data have been determined using *lacZ* and *lucB* reporter genes (Feustel *et al.*, 2004).

encoding phosphotransbutyrylase and butyrate kinase). The promoters of *adc* and *bdhB* are in the same range as that of *ptb*. It is not surprising that P*bdhA* shows much lower activity, as the respective gene product probably only serves as a sink for reducing equivalents (Dürre, 2005a). However, the low activity of *sol* is certainly surprising. A possible explanation might be the fact that this operon is transiently expressed and only required for the onset of solventogenesis (Sauer and Dürre, 1995).

A further level of regulation of solventogenesis is protein modification. Adc is co- or post-translationally modified, in an as of yet unknown form (Schaffer *et al.*, 2002). The physiological function of the modification is also not known. As it is the only solventogenic enzyme, which is stable after purification, it is tempting to speculate that the modification helps to stabilize the protein (or to prevent its degradation in the cell, respectively).

Another interesting phenomenon is the regulation of the thiolase gene *thlA*. The gene product is needed for acidogenesis as well as solventogenesis (Dürre, 2007a). A time course profile of transcription revealed strong expression during acid formation, followed by a steep decrease, and followed again by an increase at the onset of butanol and acetone formation (Winzer *et al.*, 2000). Nothing is known about transcription factors that are responsible for this regulation.

Thus, despite the significant progress made during the last 17 years, there are still many open questions with respect to regulation. Answering them will allow full exploitation of the metabolic properties of *C. acetobutylicum*. Especially butanol is not only an important bulk substrate for the chemical industry, but has meanwhile become an attractive biofuel (Dürre, 2007a). In the past, from about 1915 to 1950, commercial butanol fermentation was a dominant biotechnological process worldwide. Cheap oil led to a drastic decline (Dürre, 2007a). Increasing crude oil prices and decreasing reserves clearly dictate now the necessity to look for alternatives or additions of/to gasoline. Thus, *C. acetobutylicum* is bound to play an important economic role in future again.

Acknowledgements

Work in my laboratory was supported by grants from the BMBF GenoMik and GenoMikPlus projects (Competence Network Göttingen), and the SysMO project COSMIC (PtJ-BIO/SysMO/P-D-01-06-13), www.sysmo.net.

References

Alsaker, K.V., Spitzer, T.R., Papoutsakis, E.T. (2004). Transcriptional analysis of *spo0A* overexpression in *Clostridium acetobutylicum* and its effect on the cell's response to butanol stress. J. Bacteriol. *186*, 1959–1971.

Alsaker, K.V., and Papoutsakis, E.T. (2005) Transcriptional program of early sporulation and stationary-phase events in *Clostridium acetobutylicum*. J. Bacteriol. *187*, 7103–7118.

Angert, E.R., Brooks, A.E., and Pace, N.R. (1996). Phylogenetic analysis of *Metabacterium polyspora*: clues to the evolutionary origin of daughter cell production in *Epulopiscium* species, the largest bacteria. J. Bacteriol. *178*, 1451–1456.

Arcuri, E.F., Wiedmann, M., and Boor, K.J. (2000). Phylogeny and functional conservation of σ^E in endospore-forming bacteria. Microbiology *146*, 1593–1603.

Bahl, H., Gottwald, M., Kuhn, A., Rale, V., Andersch, W., and Gottschalk, G. (1986). Nutritional factors affecting the ratio of solvents produced by *Clostridium acetobutylicum*. Appl. Environ. Microbiol. 52, 169–172.

Barth, H., and Aktories, K. (2005). Clostridial cytotoxins. In Handbook on Clostridia, P. Dürre, ed. (Boca Raton, FL: CRC), pp. 407–449.

Battistuzzi, F.U., Feijao, A., and Hedges, S.B. (2004). A genomic timescale of prokaryote evolution: insights into the origin of methanogenesis, phototrophy, and the colonization of land. BMC Evol. Biol. *4*, 44.

Böhringer, M. (2002). Molekularbiologische und enzymatische Untersuchungen zur Regulation des Gens der Acetacetat-Decarboxylase von *Clostridium acetobutylicum*. PhD thesis, University of Ulm, Germany.

Bowles, L.K., and Ellefson, W.L. (1985). Effects of butanol on *Clostridium acetobutylicum*. Appl. Environ. Microbiol. *50*, 1165–1170.

Burkholder, W.F., and Grossman, A.D. (2000). Regulation of the initiation of endospore formation in *Bacillus subtilis*. In Prokaryotic Development, Y.V. Brun and L.J. Shimkets, eds. (Washington, DC: American Society for Microbiology Press), pp. 151–166.

Chen, C.-K., and Blaschek, H.P. (1999). Effect of acetate on molecular and physiological aspects of *Clostridium beijerinckii* NCIMB 8052 solvent production and strain degeneration. Appl. Environ. Microbiol. *65*, 499–505.

Chen, J.-S. (1995). Alcohol dehydrogenase: multiplicity and relatedness in the solvent-producing clostridia. FEMS Micriobiol. Rev. *17*, 263–273.

Dürre, P., Bahl, H., and Gottschalk, G. (1988). Membrane processes and product formation in anaerobes. In Handbook on Anaerobic Fermentations, L.E. Erickson and D. Y.-C. Fung, eds. (New York: Marcel Dekker), pp. 187–206.

Dürre, P., Fischer, R.-J., Kuhn, A., Lorenz, K., Schreiber, W., Stürzenhofecker, B., Ullmann. S., Winzer, K., and Sauer, U. (1995). Solventogenic enzymes of *Clostridium acetobutylicum*: catalytic properties, genetic organization, and transcriptional regulation. FEMS Microbiol. Rev. *17*, 251–262.

Dürre, P. (2004) Solventogenesis by clostridia. In Strict and Facultative Anaerobes. Medical and Environmental Aspects, M.M. Nakano, P. Zuber, eds. (Wymondham, UK: Horizon Bioscience), pp. 329–342.

Dürre, P., and Hollergschwandner, C. (2004). Initiation of endospore formation in *Clostridium acetobutylicum*. Anaerobe *10*, 69–74.

Dürre, P. (2005a). Formation of solvents in clostridia. In Handbook on Clostridia, P. Dürre, ed. (Boca Raton, FL: CRC), pp. 671–693.

Dürre, P. (2005b). Sporulation in clostridia (genetics). In Handbook on Clostridia, P. Dürre, ed. (Boca Raton, FL: CRC), pp. 659–669.

Dürre, P. (2007a). Biobutanol: an attractive biofuel. Biotechnol. J. *2*, 1525–1534.

Dürre, P. (2007b). Clostridia. Encyclopedia of Life Sciences 2007, doi:10.1002/9780470015902.a0020370

Emeruwa, A.C., and Hawirko, R.Z. (1973). Poly-β-hydroxybutyrate metabolism during growth and sporulation of *Clostridium botulinum*. J. Bacteriol. *116*, 989–903.

Feustel, L., Nakotte, S., Dürre, P. (2004). Characterization and development of two reporter gene systems for *Clostridium acetobutylicum*. Appl. Environ. Microbiol. *70*, 798–803.

Fischer, R.J., Helms, J., and Dürre, P. (1993). Cloning, sequencing, and molecular analysis of the *sol* operon of *Clostridium acetobutylicum*, a chromosomal locus involved in solventogenesis. J. Bacteriol. *175*, 6959–6969.

Fontaine, L., Meynial-Salles, I., Girbal, L., Yang, X., Croux, C., and Soucaille, P. (2002). Molecular characterization and transcriptional analysis of *adhE2*, the gene encoding the NADH-dependent aldehyde/alcohol dehydrogenase responsible for butanol production in alcohologenic cultures of *Clostridium acetobutylicum* ATCC 824. J. Bacteriol. *184*, 821–830.

Fujita, M., and Losick, R. (2005). Evidence that entry into sporulation in *Bacillus subtilis* is governed by a gradual increase in the level and activity of the master regulator Spo0A. Genes Dev. *19*, 2236–2244.

Gerischer, U., and Dürre, P. (1992). mRNA analysis of the *adc* gene region of *Clostridium acetobutylicum* during the shift to solventogenesis. J. Bacteriol. *174*, 426–433.

Girbal, L., and Soucaille, P. (1998). Regulation of solvent production in *Clostridium acetobutylicum*. Trends Biotechnol. *16*, 11–16.

Haraldsen, J.D., and Sonenshein, A.L. (2003). Efficient sporulation in *Clostridium difficile* requires disruption of the σ^K gene. Mol. Microbiol. *48*, 811–821.

Harris, L.M., Blank, L., Desai, R.P., Welker, N.E., and Papoutsakis, E.T. (2001). Fermentation characterization and flux analysis of recombinant strains of *Clostridium acetobutylicum* with an inactivated *solR* gene. J. Ind. Microbiol. Biotechnol. *27*, 322–328.

Harris, L.M., Welker, N.E., and Papoutsakis, E.T. (2002). Northern, morphological, and fermentation analysis of *spo0A* inactivation and overexpression in *Clostridium acetobutylicum* ATCC 824. J. Bacteriol. *184*, 3586–3597.

Heap, J.T., Pennington, O.J., Cartman, S.T., Cater, G.P., and Minton, N.P. (2007). The ClosTron: a universal knock-out system for the genus *Clostridium*. J. Microbiol. Methods *70*, 452–464.

Hitchen, P.G., and Dell, A. (2006). Bacterial glycoproteomics. Microbiology *152*, 1575–1580.

Huang, I.-H., Waters, M., Grau, R.R., and Sarker, M.R. (2004). Disruption of the gene (*spo0A*) encoding sporulation transcription factor blocks endospore formation and enterotoxin production in enterotoxigenic *Clostridium perfringens* type A. FEMS Microbiol. Lett. *233*, 233–240.

Huang, I.-H., and Sarker, M.R. (2006). Complementation of a *Clostridium perfringens spo0A* mutant with wild-type *spo0A* from other *Clostridium* species. Appl. Environ. Microbiol. *72*, 6388–6393.

Keis, S., Shaheen, R., and Jones, D.T. (2001). Emended descriptions of *Clostridium acetobutylicum* and *Clostridium beijerinckii*, and descriptions of *Clostridium saccharoperbutylacetonicum* sp. nov. and *Clostridium saccharobutylicum* sp. nov. Int. J. Syst. Evol. Microbiol. *51*, 2095–2103.

Kitano, H. (2002). Systems biology: a brief overview. Science *295*, 1662–1664.

Kosaka, T., Nakayama, S., Nakaya, K., Yoshino, S., and Furukawa, K. (2007). Characterization of the *sol* operon in butanol-hyperproducing *Clostridium saccharoperbutylacetonicum* strain N1–4 and its degeneration mechanism. Biosci. Biotechnol. Biochem. *71*, 58–68.

Kroos, L. (2007). The *Bacillus* and *Myxococcus* developmental networks and their transcriptional regulation. Annu. Rev. Genet. *41*, 13–39.

Labbé, R.G. (2005). Sporulation (morphology) of clostridia. In Handbook on Clostridia, P. Dürre, ed. (Boca Raton, FL: CRC), pp. 647–658.

Lazebnik, Y. (2002). Can a biologist fix a radio? – Or, what I learned while studying apoptosis. Cancer Cell *2*, 179–182.

Lengeler, J.W., Drews, G., and Schlegel, H.G. (1999). Biology of the Prokaryotes (Stuttgart, Germany: Georg Thieme Verlag).

Losick, R., and Stragier, P. (1992). Crisscross regulation of cell-type-specific gene expression during development in *Bacillus subtilis*. Nature 355, 601–604.

Lund, B.M., and Peck, M.W. (2000). *Clostridium botulinum*. In The Microbiological Safety and Quality of Foods, B.M. Lund, A.C. Baird-Parker, and G.W.

Gould, eds. (Gaithersburg, MD: Aspen Publ. Inc.), pp. 1057–1109.

Lyristis, M., Boynton, Z.L., Petersen, D., Kan, Z., Bennett, G.N., and Rudolph, F.B. (2000). Cloning, sequencing, and characterization of the gene encoding flagellin, *flaC*, and the post-translational modification of flagellin, FlaC, from *Clostridium acetobutylicum* ATCC824. Anaerobe 6, 69–79.

McClane, B.A. (2005). Clostridial enterotoxins. In Handbook on Clostridia, P. Dürre, ed. (Boca Raton, FL: CRC), pp. 385–406.

Molle, V., Fujita, M., Jensen, S.T., Eichenberger, P., Gonzalez-Pastor, J.E., Liu, J.S., and Losick, R. (2003). The Spo0A regulon of *Bacillus subtilis*. Mol. Microbiol. 50, 1683–1701.

Moreira, A.R., Ulmer, D.C., and Linden, J.C. (1981). Butanol toxicity in the butylic fermentation. Biotechnol. Bioeng. Symp. *11*, 567–579.

Nair, R.V., Bennett, G.N., and Papoutsakis, E.T. (1994). Molecular characterization of an aldehyde/alcohol dehydrogenase gene from *Clostridium acetobutylicum* ATCC 824. J. Bacteriol. *176*, 871 – 885.

Nair, R.V., Green, E.M., Watson, D.E., Bennett, G.N., and Papoutsakis, E.T. (1999). Regulation of the *sol* locus genes for butanol and acetone formation in *Clostridium acetobutylicum* ATCC 824 by a putative transcriptional repressor. J. Bacteriol. *181*, 319–330.

Nakamura, S., Serikawa, T., Yamakawa, K., Nishida, S., Kozaki, S., and Sakaguchi, G. (1978). Sporulation and C2 toxin production by *Clostridium botulinum* type C strains producing no C1 toxin. Microbiol. Immunol. *22*, 591–596.

Nölling, J., Breton, G., Omelchenko, M.V., Makarova, K.S., Zeng, Q., Gibson, R., Lee, H.M., Dubois, J., Qiu, D., Hitti, J., GTC Sequencing Center Production, Finishing, and Bioinformatics Teams, Wolf, Y.I., Tatusov, R.L., Sabathe, F., Doucette-Stamm, L., Soucaille, P., Daly, M.J., Bennett, G.N., Koonin, E.V., and Smith, D.R. (2001). Genome sequence and comparative analysis of the solvent-producing bacterium *Clostridium acetobutylicum*. J. Bacteriol. *183*, 4823–4838.

Paredes, C.J., Alsaker, K.V., and Papoutsakis, E.T. (2005). A comparative genomic view of clostridial sporulation and physiology. Nature Rev. Microbiol. 3, 969–978.

Perego, M., and Hoch, J.A. (2002). Two-component systems, phosphorelays, and regulation of their activities by phosphatases. In *Bacillus subtilis* and Its Closest Relatives. From Genes to Cells, A.L. Sonenshein, J.A. Hoch, and R. Losick, eds. (Washington, DC: American Society for Microbiology), pp. 473–481.

Popoff, M.R., and Stiles, B.G. (2005). Clostridial toxins vs. other bacterial toxins. In Handbook on Clostridia, P. Dürre, ed. (Boca Raton, FL: CRC), pp. 323–383.

Ravagnani, A., Jennert, K.C.B., Steiner, E., Grünberg, R., Jefferies, J.R., Wilkinson, S.R., Young, D.I., Tidswell, E.C., Brown, D.P., Youngman, P., Morris, J.G., and Young, M. (2000). Spo0A directly controls the switch from acid to solvent production in solvent-forming clostridia. Mol. Microbiol. 37, 1172–1185.

Sauer, U., Treuner, A., Buchholz, M., Santangelo, J.D., and Dürre, P. (1994). Sporulation and primary sigma factor homologous genes in *Clostridium acetobutylicum*. J. Bacteriol. *176*, 6572–6582.

Sauer, U., and Dürre, P. (1995). Differential induction of genes related to solvent formation during the shift from acidogenesis to solventogenesis in continuous culture of *Clostridium acetobutylicum*. FEMS Microbiol. Lett. *125*, 115–120.

Sauer, U., Santangelo, J.D., Treuner, A., Buchholz, M., and Dürre, P. (1995) Sigma factor and sporulation genes in *Clostridium*. FEMS Microbiol. Rev. *17*, 331–340.

Santangelo, J.D., Kuhn, A., Treuner-Lange, A., and Dürre, P. (1998). Sporulation and time course expression of sigma-factor homologous genes in *Clostridium acetobutylicum*. FEMS Microbiol. Lett. *161*, 157–164.

Schaffer, S., Isci, N., Zickner, B., and Dürre, P. (2002). Changes in protein synthesis and identification of proteins specifically induced during solventogenesis in *Clostridium acetobutylicum*. Electrophoresis 23, 110–121.

Schmidt, M.A., Riley, L.W., and Benz, I. (2003). Sweet new world: glycoproteins in bacterial pathogens. Trends Microbiol. *11*, 554–561.

Siunov, A.V., Nikitin, D.V., Suzina, N.E., Dmitriev, V.V., Kuzmin, N.P., and Duda, V.I. (1999). Phylogenetic status of *Anaerobacter polyendosporus*, an anaerobic, polysporic bacterium. Int. J. Syst. Bacteriol. *49*, 1119–1124.

Sonenshein, A.L., Haraldsen, J.D., and Dupuy, B. (2005). RNA polymerase and alternative σ factors. In Handbook on Clostridia, P. Dürre, ed. (Boca Raton, FL: CRC), pp. 607–629.

Stephenson, K., and Lewis, R.J. (2005). Molecular insights into the initiation of sporulation in Gram-positive bacteria: new technologies for an old phenomenon. FEMS Microbiol. Rev. 29, 281–301.

Stragier, P. (2002). A gene odyssey: exploring the genomes of endospore-forming bacteria. In *Bacillus subtilis* and Its Closest Relatives. From Genes to Cells, A.L. Sonenshein, J.A. Hoch, and R. Losick, eds. (Washington, DC: American Society for Microbiology Press), pp. 519–525.

Thormann, K., and Dürre, P. (2001). Orf5/SolR: a transcriptional repressor of the *sol* operon of *Clostridium acetobutylicum*? J. Ind. Microbiol. Biotechnol. *27*, 307–313.

Thormann, K., Feustel, L., Lorenz, K., Nakotte, S., and Dürre, P. (2002). Control of butanol formation in *Clostridium acetobutylicum* by transcriptional activation. J. Bacteriol. *184*, 1966–1973.

Tomas, C.A., Beamish, J.A., and Papoutsakis, E.T. (2004). Transcriptional analysis of butanol stress and tolerance in *Clostridium acetobutylicum*. J. Bacteriol. *186*, 2006–2018.

Tummala, S.B., Tomas, C.A., and Papoutsakis, E.T. (2005). Genetic analysis of clostridia. In Handbook on Clostridia, P. Dürre, ed. (Boca Raton, FL: CRC), pp. 53–70.

Upreti, R.K., Kumar, M., and Shankar, V. (2003). Bacterial glycoproteins: functions, biosynthesis and applications. Proteomics 3, 363–379.

Vreeland, R.H., Rosenzweig, W.D., and Powers, D.W. (2000). Isolation of a 250 million-year-old halotolerant bacterium from a primary salt crystal. Nature *407*, 897–900.

Walter, K.A., Bennett, G.N., and Papoutsakis, E.T. (1992). Molecular characterization of two *Clostridium acetobutylicum* ATCC 824 butanol dehydrogenase isozyme genes. J. Bacteriol. *174*, 7149–7158.

Weizmann, C. (1915) Improvements in the bacterial fermentation of carbohydrates and in bacterial cultures for the same. Br. Patent No. 4845.

Welch, R.W., Rudolph, F.B., and Papoutsakis, E.T. (1989) Purification and characterization of the NADH-dependent butanol dehydrogenase from *Clostridium acetobutylicum* (ATCC 824). Arch. Biochem. Biophys. *273*, 309–318.

Winzer, K., Lorenz, K., Zickner, B., and Dürre, P. (2000). Differential regulation of two thiolase genes from *Clostridium acetobutylicum* DSM 792. J. Mol. Microbiol. Biotechnol. *2*, 531–541.

Wörner, K., Szurmant, H., Chiang, C., and Hoch, J.A. (2006). Phosphorylation and functional analysis of the sporulation initiation factor Spo0A from *Clostridium botulinum*. Mol. Microbiol. *59*, 1000–1012.

Wong, J., Sass, C., and Bennett, G.N. (1995). Sequence and arrangement of genes encoding sigma factors in *Clostridium acetobutylicum* ATCC 824. Gene *153*, 89–92.

Zhao, Y., and Melville, S.B. (1998). Identification and characterization of sporulation-dependent promoters upstream of the enterotoxin gene (*cpe*) of *Clostridium perfringens*. J. Bacteriol. 180, 136–142.

Zhao, Y.S., Tomas, C.A., Rudolph, F.B., Papoutsakis, E.T., and Bennett, G.N. (2005). Intracellular butyryl phosphate and acetyl phosphate concentrations in *Clostridium acetobutylicum* and their implications for solvent formation. Appl. Environ. Microbiol. *71*, 530–537.

Other Books of Interest

Index

E

F

G

H

I

L

M

N

O

P

R

S

T

U

V